Lipid Metabolism in the Healthy and Disease Heart

Developments in Molecular and Cellular Biochemistry

Series Editor: Naranjan S. Dhalla, Ph.D., FACC

1. V.A. Najjar (ed.): *Biological Effects of Glutamic Acid and Its Derivatives.* 1981 ISBN 90-6193-841-4
2. V.A. Najjar (ed.): *Immunologically Active Peptides.* 1981 ISBN 90-6193-842-2
3. V.A. Najjar (ed.): *Enzyme Induction and Modulation.* 1983 ISBN 0-89838-583-0
4. V.A. Najjar and L. Lorand (eds.): *Transglutaminase.* 1984 ISBN 0-89838-593-8
5. G.J. van der Vusse (ed.): *Lipid Metabolism in Normoxic and Ischemic Heart.* 1989 ISBN 0-7923-0479-9
6. J.F.C. Glatz & G.J. van der Vusse (eds.): *Cellular Fatty Acid-Binding Proteins.* 1990 ISBN 0-7923-0896-4
7. H.E. Morgan (ed.): *Molecular Mechanisms of Cellular Growth.* 1991 ISBN 0-7923-1183-3
8. G.J. van der Vusse & H. Stam (eds.): *Lipid Metabolism in the Healthy and Diseased Heart.* 1992 ISBN 0-7923-1850-1

KLUWER ACADEMIC PUBLISHERS – DORDRECHT / BOSTON / LONDON

Lipid Metabolism in the Healthy and Disease Heart

edited by

GER J. VAN DER VUSSE

Department of Physiology, Cardiovascular Research Institute Maastricht, Medical Faculty, University of Limburg, Maastricht, The Netherlands

and

HANS STAM

Department of Biochemistry, Faculty of Medicine and Health Sciences, Erasmus University Rotterdam, The Netherlands

Springer-Science+Business Media, B.V.

Library of Congress Cataloging-in-Publication Data

International Symposium on Lipid Metabolism in the Normoxic and
Ischaemic Heart (3rd : 1991 : Rotterdam, Netherlands)
Lipid metabolism in the healthy and diseased heart : proceedings
of the Third International Symposium on Lipid Metabolism in the
Normoxic and Ischemic Heart, September 9 & 10, 1991, Rotterdam, the
Netherlands / edited by G.J. van der Vusse, Hans Stam.
p. cm. -- (Developments in molecular and cellular
biochemistry ; 8)
"Reprinted from Molecular and cellular biochemistry."
ISBN 978-1-4613-6559-4 ISBN 978-1-4615-3514-0 (eBook)
DOI 10.1007/978-1-4615-3514-0
1. Coronary heart disease--Pathophysiology--Congresses.
2. Lipids--Metabolism--Congresses. 3. Blood lipids--Metabolism--Congresses. 4. Heart--Metabolism--Congresses. 5. Vascular
endothelium--Congresses. I. Vusse, G. J. van der. II. Stam, H.
III. Title. IV. Series: Developments in molecular and cellular
biochemistry ; v. 8.
[DNLM: 1. Endothelium, Vascular--metabolism--congresses.
2. Heart--physiopathology--congresses. 3. Ischemia--drug therapy.
4. Lipids--metabolism--congresses. 5. Myocardium--metabolism--congresses. W1 DE998D v.8]
RC685.C6I574 1992
616.1'2307--dc20
DNLM/DLC
for Library of Congress 92-23631
CIP

Printed on acid-free paper

Molecular and Cellular Biochemistry:

An International Journal for Chemical Biology in Health and Disease

CONTENTS

LIPID METABOLISM IN THE HEALTY AND DISEASED HEART

Molecular and Cellular Biochemistry **116**: 1, 1992.

Preface

During the past decades numerous efforts have been made to elucidate the details of cardiac lipid homeostasis under normal and pathophysiological conditions. Fatty acids are playing a key role in lipid metabolism in the heart. They serve as substrates to meet cardiac energy requirements either extracted from the extracellular space or released from endogenous triacylglycerol stores. Fatty acids, incorporated in phospholipids, also form important building blocks of cardiac membranes. The physical and chemical processes responsible for the maintenance of cardiac lipid homeostasis are characterized by a high degree of complexity. Moreover life-threatening alterations occur in hearts deprived of molecular oxygen, a situation leading to impaired oxidation of fatty acyl moieties, hydrolysis of phospholipids and other events challenging the balance of lipid synthesis and degradation. Concomitant accumulation of lipid intermediates, which are noxious at increased intracellular concentrations, has been reported to compromise cardiac function under less favorable circumstances.

With this knowledge in mind the Third International Symposium on Lipid Metabolism in the Normoxic and Ischemic Heart was held in Rotterdam, The Netherlands, during September 9–10, 1991 under the chairmanschip of Willem Hülsmann, Hans Stam, Jan Glatz and Ger J. van der Vusse. The topics of this meeting were focussed on:

1. Modulation of myocardial lipid metabolism.
2. Biological membranes: structure, function and turnover.
3. Pharmacological modification of myocardial fatty acid oxidation.
4. Myocardial vascular endothelium; contribution to myocardial lipid homeostasis.

Special attention was given to the interrelationship between carbohydrates and fatty acids as energy substrates for the heart under normoxic and (post)ischemic circumstances, the influence of diets, varying in their fatty acid composition, on cardiac function, and the significance of phospholipid topology, turnover, and methylation in general and the phosphatidylinositol pathway in particular on performance of the heart. The role of carnitine in cardiac function altered by lack of oxygen or by elevated levels of fatty acyl derivatives of carnitine and the modulatory effects of the endothelium on cardiac lipid homeostasis were also extensively discussed during the conference. This Focussed Issue of Molecular and Cellular Biochemistry is a collection of invited papers based on the lectures and a selection of posters presented during the meeting. It includes contributions of renowned investigators delineating recent results and discussing significant aspects of their data in an attempt to enlarge our insight in the complexity of cardiac lipid transport and metabolism, in the healthy and diseased myocardium.

Ger J. van der Vusse[1] and Hans Stam[2]

[1]Department of Physiology, Cardiovascular Research Institute Maastricht, University of Limburg, Maastricht, The Netherlands
[2]Department of Biochemistry, Faculty of Medicine and Health Sciences, Erasmus University Rotterdam, The Netherlands

Molecular and Cellular Biochemistry **116**: 3–9, 1992.

Paradoxical role of lipid metabolism in heart function and dysfunction

Naranjan S. Dhalla, Vijayan Elimban and Heinz Rupp
Division of Cardiovascular Sciences St. Boniface General Hospital Research Centre and Department of Physiology, Faculty of Medicine, University of Manitoba, Winnipeg, Manitoba, Canada R2H 2A6

Abstract

The heart utilizes fatty acids as a substrate in preference to glucose for the production of energy. The rate of fatty acid uptake and oxidation by heart muscle is controlled by the availability of exogenous fatty acids, the rate of acyl translocation across the mitochondrial membrane and the rate of acetyl-CoA oxidation by the citric acid cycle. Carnitine acyl-CoA tranferase appears to have an important function in coupling the fatty acid activation and acyl transfer to the oxidative phosphorylation. Activated fatty acids are also utilized for the synthesis of triglycerides and membrane phospholipids in the myocardium. The inhibition of long chain acyl-carnitine transferase I reduces the oxidation of fatty acids and promotes the synthesis of lipids in the myocardium. Accumulation of fatty acids and their metabolites such as long chain acyl-CoA and long chain acyl-carnitine has been associated with cardiac dysfunction and cell damage in both ischemic and diabetic hearts. Alterations in the composition of membrane phospholipids are also considered to change the activities of various membrane bound enzymes and subsequently heart function under different pathophysiological conditions. Chronic diabetes was found to be associated with increased plasma lipids, subcellular defects and cardiac dysfunction. Lowering the plasma lipids or reducing the oxidation of fatty acids by agents such as etomoxir, an inhibitor of palmitoylcarnitine transferase I was found to promote glucose utilization and remodel the subcellular membranous organelles in the heart. The crucial role of fatty acids in membrane phospholipids for the maintenance of structural integrity and production of energy for cardiac contractile activity as well as the toxic effects of fatty acids and their long chain acyl-derivatives support the concept of 'lipid paradox' in the myocardium. (Mol Cell Biochem **116:** 3–9, 1992)

Key words: fatty acid metabolism, membrane phospholipids, heart membranes, myocardial ischemia, diabetic heart

Introduction

It is well known that fatty acids are the preferred substrate for energy production in the heart and their oxidation under normal conditions accounts for about 60–70% of the oxidative metabolism. Several excellent reviews describing various processes involved in lipid metabolism in both healthy and diseased hearts are available in the literature [1–10]. Fats have important nutritional functions in addition to supplying a concentrated energy source; however, an excessive consumption of fats is considered to be bad for health in general and heart function in particular. The incidence of and death rate from coronary heart disease have been shown to bear a positive relationship with the intake of saturated fats as well as elevated levels of plasma lipids.

Address for offprints: N.S. Dhalla, Division of Cardiovascular Sciences, St. Boniface General Hospital Research Centre, 351 Tache Avenue, Winnipeg, Manitoba, Canada R2H 2A6

The association of the level of saturated fatty acids in the diet with atherosclerosis is mediated by increased circulating total cholesterol and low density lipoprotein-cholesterol levels. The observation that free fatty acid levels are high in plasma in acute myocardial infarction [11–14] has focussed the attention of very many investigators on the need for a better understanding of the lipid metabolism in both health and disease.

During the span of the past five decades, a great deal of work has been carried out concerning the role of lipid metabolism in energy production for the maintenance of heart function as well as with respect to the pathophysiology of heart disease. Some of the major areas of research in this regard are given in Table 1. It has now become clear that lipids are required for the energy needs and structural integrity of the heart; however, lipids are basically toxic substances and thus lowering the plasma as well as intracellular levels of free fatty acids and their intermediates can be seen beneficial for heart function. In fact there appears to be a 'lipid paradox' in the sense that low concentrations of free fatty acids are essential for the proper functioning of the heart whereas excessive amounts are deleterious. While free fatty acids have been reported to exert cardiodepressant and arrhythmogenic effects on the heart, these are essential components of membrane phospholipids, which are known to play a wide variety of roles in maintaining as well as regulating the heart function (Table 2). The purpose of this article is to highlight the importance of both neutral lipids and phospholipids in heart function as well as pathophysiology of heart dysfunction by illustrating some of the selected examples.

General concepts of lipid metabolism in the heart

Although heart is known to take up fatty acids from plasma in a concentration dependent manner [15], the rate of their utilization is influenced by heart work and coronary blood flow [16]. Furthermore, the rate of fatty acid utilization is also dependent upon the availability of alternative substrates, the status of oxidative respiration and plasma levels of certain hormones. While most of fatty acids in plasma are bound to albumin or other proteins, their extraction by the heart is considered to occur by a competition between some membrane proteins and albumin [17]. The preferential utilization of fatty acids by heart involves the inhibition of carbohydrate utilization at the levels of glucose transport, phosphofructokinase, hexokinase, glycogen phosphorylase and pyruvate dehydrogenase as well as stimulation of glycogen synthetase. Free fatty acids taken up by the heart are activated involving the formation of fatty acyl-CoA thioesters before further metabolism. These activated long chain fatty acids are transferred into the mitochondrial matrix by a carnitine-dependent process for β-oxidation and subsequent energy production [18]; the oxidation of activated short chain fatty acid may include conversion to long chain acyl-carnitine derivatives before transfer across the inner mitochondrial membrane. Activated fatty acids are also utilized for the synthesis of triglycerides and phospholipids in the cytosol. Thus free fatty acids are metabolized in the myocardium not only for the generation of energy required for heart function but are also rerouted for the synthesis of membrane phospholipids, deposition of fatty material and accumulation of some toxic derivatives.

Toxic effects of lipid metabolites

The arrhythmogenic, cardiac depressant and oxygen wasting effects of high concentrations of fatty acid are most probably due to the formation of increased level of long chain acyl-CoA or carnitine derivatives; however, a careful investigation in this regard needs to be carried out. In view of the detergent-like property of fatty acids and long chain acyl-derivatives, these lipid metabolites can be seen to induce myocardial cell injury. In this

Table 1. Major areas of investigation involving neutral lipids in heart function

1.	Dietary intake, storage and mobilization
2.	Transport and competition with other substrates
3.	Oxidation of fatty acids and energy production
4.	Incorporation in membrane phospholipids
5.	Fatty deposition and myocardial cell injury
6.	Accumulation of long chain acyl derivatives

Table 2. Involvement of phospholipids in heart function

1.	Membrane fluidity and permeability
2.	Storage of calcium in the membrane
3.	Anchoring of enzymes and proteins in the membrane
4.	Regulation of enzyme activity
5.	Precursor for prostaglandins
6.	Substrate for methyltransferases
7.	Signal transduction

Table 3. Effect of acetylcarnitine (25 μM) and palmitoylcarnitine (25 μM) on heart membrane ATPase activities

	Inhibition (%)	
	Acetylcarnitine	Palmitoylcarnitine
Sarcolemmal Na^+-K^+ ATPase	1 ± 0.7	74 ± 5.3
Sarcolemmal Ca^{2+}-stimulated ATPase	2 ± 1.5	56 ± 6.5
Sarcoplasmic reticular Ca^{2+}-stimulated ATPase	5 ± 2.6	45 ± 3.9

Each value is a mean ± S.E. of 6 experiments. Rat heart membranes were isolated and their activities were determined in the absence or presence of acyl-derivatives according to the methods described elsewhere [20].

regard, palmitoylcarnitine has been reported to inhibit sarcolemmal Na^+-K^+ ATPase and sarcoplasmic reticular Ca^{2+}-stimulated ATPase activities [19, 20]. The data shown in Table 3 indicate that the inhibitory effect of palmitoylcarnitine on sarcolemmal Na^+-K^+ ATPase was greater than that on sarcoplasmic reticular Ca^{2+}-stimulated ATPase or sarcolemmal Ca^{2+}-stimulated ATPase whereas a short chain acyl-derivative, acetylcarnitine, did not exert any effect on these membrane ATPase activities. It should be pointed out that carnitine deficiency in ischemic hearts has been associated with disturbance in fatty acid metabolism, accumulation of long chain acyl-derivatives and myocardial cell damage [21–23]. Likewise, alterations in membrane phospholipids by the activation of different types of phospholipase under various pathophysiological conditions can be seen to alter the lipid composition of heart membranes. It should be noted that phospholipids are not only essential for some enzyme activities but are also known to affect the activities of membrane bound enzymes [24–26]. For example, phospholipid N-methylation has been reported to alter the sarcolemmal Ca^{2+}-pump and Na^+-Ca^{2+} exchange as well as sarcoplasmic reticular Ca^{2+}-pump ATPase activities [27–29]. The results in Table 4 indicate that treatment of the sarcolemmal membrane with phosphatidylinositol-specific phospholipase C [30] released the 5′-nucleotidase from the membrane without affecting the Mg^{2+} ATPase or Na^+-K^+ ATPase activities. Thus it is evident that phospholipids not only affect the activities of certain enzymes but are also intimately involved in anchoring certain proteins and enzymes in the membrane. Furthermore, in view of the crucial role played by several transport enzymes in maintaining the ionic homeostasis in the cell, alterations in membrane-bound enzymes due to changes in the phospholipid composition of the membrane may lead to changes in myocardial metabolism and heart function.

Modification of lipid metabolism

Disturbances in lipid metabolism including accumulation of fatty acids and long chain acyl-derivatives of carnitine and coenzyme A have been reported to occur due to myocardial ischemia [31–36]. In view of the toxic effects of fatty acids and their metabolites on the myocardium, reduction in the dependence of the heart on lipid metabolism can be seen to exert beneficial effects. Such a modification of lipid metabolism can be achieved by lowering the plasma level of fatty acids by reducing the dietary intake as well as decreasing the fat mobilization by some pharmacological agents. Reduction in the intracellular concentrations of fatty acids and their metabolites can occur by interfering with the transport of fatty acids into the myocardium at the plasma mem-

Table 4. Effect of phosphatylinositol-specific phospholipase C on cardiac sarcolemmal enzymes

	Mg^{2+}-ATPase (μmol Pi/mg/h)	Na^+-K^+ ATPase (μmol Pi/mg/h)	5′-Nucleotidase (μmol adenosine/mg/min)
Untreated membranes	32.5 ± 2.0	8.7 ± 2.0	75.7 ± 8.4
Treated membranes	30.9 ± 2.4	8.3 ± 2.1	14.1 ± 3.4
Untreated supernatant	N.D.	N.D.	1.2 ± 0.8
Treated supernatant	N.D.	N.D.	57.4 ± 1.7

Each value is a mean ± S.E. of 4 experiments. N.D – not detectable. The methods employed in this study were the same as described elsewhere [30].

Table 5. Plasma lipids and cardiac subcellular mechanisms in rats treated with 12–15 mg/kg etomoxir for 12 weeks

	Control	Etomoxir-treated
Plasma triglycerides (μmol/L)	130 ± 10	83 ± 7*
Plasma free fatty acids (mmol/L)	0.54 ± 0.02	0.35 ± 0.02*
Heart to body wt ratio (mg/g)	2.18 ± 0.01	2.87 ± 0.15*
SR Ca^{2+}-stimulated ATPase (μmol Pi/mg/min)	121 ± 6.8	196 ± 5.2*
Myosin isozymes (%)		
V1	74 ± 2.8	89 ± 2.5*
V2	17 ± 1.3	7 ± 1.6*
V3	9 ± 0.6	4 ± 0.2*

Each value is a mean ± S.E. of 6 experiments. Etomoxir was given in drinking water and plasma and ventricular tissue were analyzed biochemically as described earlier [40]. * Significantly (P < 0.05) different from the control value.

brane level. Recent studies have also shown that carnitine palmitoyltransferase I inhibitors such as oxfenicine, 2-tetradecylglycidic acid, POCA (sodium 2-(5-)4-chlorophenyl)pentyl)-oxirane-2-carboxylate) exert their protective effects in the ischemic myocardium by lowering the intracellular concentrations of long chain acylcarnitine [37–39]. However, other investigators [36] have claimed that the protective effect of etomoxir is unrelated to changes in the levels of long chain acylcarnitine but is rather due to increased glucose use by the reperfused ischemic myocardium. A shift in myocardial substrate utilization from fatty acids to carbohydrates by chronic treatment of animals with etomoxir has also been reported to prevent redistribution of myosin isozymes and depression in sarcoplasmic reticular Ca^{2+}-stimulated ATPase in pressure-overloaded hypertrophied heart [40]. The data in Table 5 show an increase in myosin V_1 and sarcoplasmic reticular Ca^{2+}-stimulated ATPase activity whereas a decrease in both myosin V_2 and V_3 associated with myocardial hypertrophy and depressed plasma triglyceride and free fatty acid levels were seen upon chronic treatment of animals with etomoxir. Thus it appears that the protective effect of etomoxir in the ischemic heart may also be due to reduction in the fatty acid oxidation not only at the level of carnitine palmitoyltransferase I but may also be due to a decrease in the availability of plasma fatty acids for uptake in the myocardial cell. Nonetheless, the data in Table 5 support the view that reduction in the metabolism of fatty acids and associated increase in glucose

Table 6. Alterations in plasma lipids, heart function and subcellular mechanisms in diabetic rats

	Control	Diabetic
Plasma glucose (mg/dl)	156 ± 8	430 ± 21*
Plasma triglycerides (μmol/L)	130 ± 15	510 ± 34*
Plasma free fatty acids (mmol/L)	0.56 ± 0.03	0.81 ± 0.07*
LVSP (mm Hg)	150 ± 3	110 ± 4*
LVEDP (mm Hg)	3 ± 0.4	16 ± 2*
+ dP/dt (mm Hg/s)	5890 ± 140	3802 ± 125*
− dP/dt (mm Hg/s)	5525 ± 132	3418 ± 142*
SR Ca^{2+} uptake (nmol/mg/min)	74 ± 3.2	41 ± 3.0*
SR Ca^{2+}-stimulated ATPase (nmol Pi/mg/min)	125 ± 7	68 ± 4*
Myofibrillar Ca^{2+}-stimulated ATPase (μmol Pi/mg/min	0.92 ± 0.03	0.61 ± 0.04*
Myosin isoenzymes (%)		
V_1	76 ± 4.2	7 ± 1.1*
V_2	16 ± 1.4	16 ± 0.8
V_3	8 ± 0.5	77 ± 7.8*

Each value is a mean ± S.E. of 4 to 10 experiments. Rats were made diabetic by injecting 65 mg/kg (i.v.) streptozotocin for 8 weeks. After hemodynamic assessment, plasma and ventricular tissue were analyzed biochemically according to methods described elsewhere [15]. LVSP – left ventricular systolic pressure; LVEDP – left ventricular diastolic pressure; + dP/dt – rate of force development; – dP/dt – rate of relaxation; SR – sarcoplasmic reticular. * Significantly (P < 0.05) different from the control value.

oxidation may serve as a signal for the genetic apparatus concerned with the formation of subcellular organelles. Furthermore, reduction in the level of long chain acylcarnitine due to the inhibition of carnitine palmitoyltransferase I by agents such as etomoxir can also be seen to promote the formation of triglycerides and phospholipids from long chain acyl-CoA in the myocardium.

Lipid metabolism and diabetic heart dysfunction

Diabetes has been associated with high levels of plasma glucose and free fatty acids as well as increased glycogen and triglyceride contents in the myocardium [41, 42]. It is believed that high levels of plasma free fatty acids increase triglyceride synthesis and accumulation of glycogen in the diabetic heart. There is evidence that the β-oxidation of fatty acid is associated with increased levels of long chain acyl-CoA and acylcarnitine in the heart of diabetic animals [43, 44]. Normalization of fatty acid metabolism is associated with beneficial effects on the mechanical function of the diabetic heart [45–47]. Thus it appears that cardiac dysfunction in diabetes may be a consequence of hyperlipidemia where lack of insulin and high levels of plasma free fatty acids result in promoting oxidation of fatty acids and inhibiting glucose utilization. The importance of such an imbalance of substrate utilization in cardiac dysfunction is also apparent from the study in which dichloroacetate, an activator of pyruvate dehydrogenase, was found to promote glucose oxidation and reverse the depression of cardiac function in diabetes [48].

In view of the role of L-carnitine in the metabolism of fatty acids and the observed depression in the concentration of carnitine in the diabetic heart [49], treatment of diabetic animals with L-carnitine was found to lower the elevated levels of plasma lipids and improved cardiac performance [50]. It should be noted from Table 6 that cardiac dysfunction in chronic diabetes in rats is also associated with marked increases in plasma lipids as well as depression in both sarcoplasmic reticular Ca^{2+}-pump and myofibrillar ATPase activities in addition to a shift in the pattern of myosin isozymes. The defect in sarcoplasmic reticulum was prevented upon treating the diabetic animals with propionyl L-carnitine whereas the changes in myofibrils were not affected [51]. Treatment of diabetic animals with propionyl L-carnitine also partially or completely prevented the depression in sarcolemmal Na^+-Ca^{2+} exchange and Na^+-K^+ ATPase activities but had no effect on the depressed sarcolemmal Na^+-H^+ exchange and Ca^{2+}-pump activities in myocardium [52]. These studies indicate that some of the membrane defects and functional abnormalities in diabetic heart are the result of abnormal lipid metabolism as a consequence of carnitine deficiency.

Acknowledgements

The research work reported in this paper was supported by the Juvenile Diabetes Foundation International. Dr. Heinz Rupp was a Visiting Professor from the Institute of Physiology II, University of Tubingen, Germany.

References

1. Bing RJ: Cardiac metabolism. Physiol Rev 45: 171–213, 1965
2. Neely JR, Rovetto MJ, Oram JF: Myocardial utilization of carbohydrate and lipids. Prog Cardiovasc Disease 15: 289–329, 1972
3. Neely JR, Morgan HE: Relationship between carbohydrate and lipid metabolism and the energy balance of the heart muscle. Ann Rev Physiol 36: 413–459, 1974
4. Opie LH: Effects of regional ischemia on metabolism of glucose and fatty acids. Relative rates of aerobic and anaerobic energy production during myocardial infarction and comparison with effects of anoxia. Circ Res 38 (Suppl I): 152–186, 1976
5. Masoro EJ: Lipids and lipid metabolism. Ann Rev Physiol 39: 301–321, 1977
6. Opie LE: Role of carnitine in fatty acid metabolism of normal and ischemic myocardium. Am Heart J 97: 375–388, 1979
7. Randle PJ, Tubbs PK: Carbohydrate and fatty acid metabolism. In: RM Berne and N Sperelakis (eds.) The Cardiovascular System, Vol. I, The Heart. Am Physiol Soc, Bethesda, 1979, pp 805–844
8. Vary TC, Reibel DK, Neely JR: Control of energy metabolism of heart muscle. Ann Rev Physiol 43: 419–430, 1981
9. Liedtke JA: Alterations of carbohydrate and lipid metabolism in the acutely ischemic heart. Prog Cardiovasc Disease 23: 321–336, 1981
10. Van der Vusse GJ, Stam H: Lipid and carbohydrate metabolism in the ischemic heart. Basic Res Cardiol 82 (Suppl I): 149–154, 1987
11. Oliver MF, Kurien VA, Greenwood TW: Relation between serum free fatty acids and arrhythmias and death after acute myocardial infarction. Lancet 1: 710–715, 1968
12. Oliver MF: Metabolic response during impending myocardial infarction II. Clinical implications. Circulation 45: 491–500, 1972
13. Kjekshus JK, Mjos OD: Effect of free fatty acids on myocardial function and metabolism in the ischemic dog heart. J Clin Invest 51: 1767–1776, 1972
14. Opie LH, Tansey M, Kennelly BM: Proposed metabolic vicious circle in patients with large myocardial infarcts and high plasma free fatty acid concentrations. Lancet 2: 890–892, 1977
15. Neely JR, Bowman RH, Morgan HE: Effects of ventricular

pressure development and palmitate on glucose transport. Am J Physiol 216: 804–811, 1969

16. Spitzer JJ: Application of tracers in studying free fatty acid metabolism of various organs *in vivo*. Fed Proc 34: 2242–2245, 1975
17. Stein O, Stein Y: Lipid synthesis, intracellular transport and storage III. Electron microscopic radioautographic study of the rat heart perfused with tritiated oleic acid. J Cell Biol 36: 63–77, 1968
18. Fritz IB, Kaplan E, Yue KTN: Specificity of carnitine action on fatty acid oxidation by heart muscle. Am J Physiol 202: 117–121, 1962
19. Adams RJ, Cohen DW, Gupta S, Johnson D, Wallic ET, Wang T, Schwartz A: *In vitro* effects of palmitoyl carnitine on cardiac plasma membrane Na^+, K^+-ATPase and sarcoplasmic reticulum Ca^{2+}-ATPase and Ca^{2+}-transport. J Biol Chem 254: 12404–12410, 1983
20. Dhalla NS, Kolar F, Shah KR, Ferrari R: Effects of some L-carnitine derivatives on heart membrane ATPase. Cardiovasc Drugs 5: 25–30, 1991
21. Idell-Wenger JA, Grotoyohann LW, Neely JR: Coenzyme A and carnitine distribution in normal and ischemic hearts. J Biol Chem 253: 4310–4318, 1978
22. Feuvray D, Idell-Wenger JA, Neely JR: Effects of ischemia on rat myocardial function and metabolism in diabetes. Circ Res 44: 322–329, 1979
23. Shug AL, Thomsen JH, Folts JD, Bittar N, Klein MI, Koke JR, Huth PJ: Changes in tissue levels of carnitine and other metabolites during myocardial ischemia and anoxia. Arch Biochem Biophys 187: 25–33, 1978
24. Warren GB, Honslay MD, Metcalf JC, Birdsall NJM: Cholesterol is excluded from the phospholipid annulus surrounding an active calcium transport protein. Nature 255: 684–687, 1975
25. Caroni P, Zurini M, Clark A, Carafoli E: Further characterization and reconstitution of the purified Ca^{2+}-pumping ATPase of heart sarcolemma. J Biol Chem 258: 7305–7310, 1983
26. Zhao D, Dhalla NS: Purification and composition of Ca^{2+}/Mg^{2+} ATPase from rat heart plasma membrane. Mol Cell Biochem 107: 135–149, 1991
27. Panagia V, Okumura K, Makino N, Dhalla NS: Stimulation of Ca^{2+} pump in rat heart sarcolemma by phosphatidylethanolamine N-methylation. Biochim Biophys Acta 856: 383–387, 1986
28. Ganguly PK, Panagia V, Okumura K, Dhalla NS: Activation of Ca^{2+}-stimulated ATPase by phospholipid N-methylation in cardiac sarcoplasmic reticulum. Biochem Biophys Res Commun 130: 472–478, 1985
29. Panagia V, Makino N, Ganguly PK, Dhalla NS: Inhibition of Na^+-Ca^{2+} exchange in heart sarcolemmal vesicles by phosphatidylethanolamine N-methylation. Eur J Biochem 166: 597–603, 1987
30. Panagia V, Heyliger CE, Choy PC, Dhalla NS: Solubilization of rat heart sarcolemmal 5′-nucleotidase by phosphatidylinositol specific phospholipase C. Biochim Biophys Acta 640: 802–806, 1981
31. Chein KR, Han A, Sen A, Buja LM, Willerson JT: Accumulation of unesterified arachidonic acid in ischemic canine myocardium. Circ Res 48: 313–322, 1981
32. Corr PB, Gross RW, Sobel BE: Amphipathic metabolites and membrane dysfunction in ischemic myocardium. Circ Res 55: 135–154, 1984
33. Katz AM, Messineo FC: Lipid membrane interactions and the pathogenesis of ischemic damage in the myocardium. Circ Res 48: 1–16, 1981
34. Van der Vusse GJ, Prinen FW, Reneman RS: Uptake and tissue content of fatty acids in dog myocardium under normoxic and ischemic conditions. Circ Res 50: 538–546, 1982
35. Van Bilsen M, Van der Vusse GJ, Willemsen PHM, Coumans WA, Reneman RS: Fatty acid accumulation during ischemic reperfusion: Effects of pyruvate and POCA, a carnitine palmitoyltransferase I inhibitor. J Mol Cell Cardiol 23: 1437–1447, 1991
36. Lopaschuk GD, Wall SR, Olley PM, Davies NJ: Etomoxir, a carnitine palmitoyltransferase I inhibitor, protects hearts from fatty acid-induced ischemic injury independent of changes in long chain acylcarnitine. Circ Res 63: 1036–1043, 1988
37. Paulson DJ, Noonan JJ, Ward KM, Stanley H, Sherratt A, Shug AL: Effects of POCA on metabolism and function in the ischemic rat heart. Basic Res Cardiol 81: 180–187, 1986
38. Molaparast-Saless F, Liedtke AJ, Nellis SH: Effects of fatty acid blocking agents, oxfenicine and 4-bromocrotonic acid on performance in aerobic and ischemic myocardium. J Mol Cell Cardiol 19: 509–520, 1987
39. Hekimian G, Feuvray D: Reduction of ischemia-induced acylcarnitine accumulation by TDGA and its influence on lactate dehydrogenease release in diabetic rat hearts. Diabetes 35: 906–910, 1985
40. Rupp H, Elimban V, Dhalla NS: Modification of subcellular organelles in pressure-overloaded heart by etomoxir, a carnitine palmitoyltransferanse I inhibitor. FASEB J 6: 2349–2353, 1992
41. Denton RM, Randle PJ: Concentrations of glycerides and phospholipids in rat heart and gastronemius muscles. Biochem J 104: 423–434, 1967
42. Randle PJ, Garland PB, Hales CN, Newshoom EA, Denton RM, Pagson CI: Interactions of metabolism and the physiological role of insulin. Recent Prog Horm Res 22: 1–44, 1966
43. Lopaschuk GD, Tahiliani RVSV, Vadlamidi S, Katz S, McNeill JH: Cardiac sarcoplasmic reticulum function in insulin or carnitine treated rats. Am J Physiol 245: H969–H976, 1983
44. Pieper GM, Murray WJ, Salhany JM, Wu ST, Eliot RS: Salient effects of L-carnitine on adenine-nucleotide loss and coenzyme A acylation in the diabetic heart perfused with excess palmitic acid. A phosphorus-31 NMR and chemical extract study. Biochim Biophys Acta 804: 229–240, 1984
45. Rodrigues B, Goyal RK, McNeill JH: Effects of hydralazine on STZ-induced diabetic rats: Prevention of hyperlipidemia and improvement in cardiac function. J Pharmacol Exp Ther 237: 292–299, 1986
46. Tahiliani AG, McNeill JH: Prevention of diabetes-induced myocardial dysfunction in rats by methyl palmoxirate and triiodothyronine treatment. Can J Physiol Pharmacol 63: 925–931, 1985
47. Xiang H, Heyliger CE, McNeill JH: Effect of myoinositol and T_3 on myocardial lipids and cardiac function in streptozotocin-induced diabetic rats. Diabetes 376: 1542–1548, 1988
48. Nicholl TA, Lopaschuk GD, McNeill JH: Effects of free fatty acids and dichloracetate on isolated working diabetic rat heart. Am J Physiol 261: H1053–H1059, 1991
49. Vary TC, Neely JR: A mechanism for reduced myocardial carnitine levels in diabetic animals. Am J Physiol 243: H154–H158, 1982
50. Rodrigues B, Xiang H, McNeill JH: Effect of L-carnitine treatment on lipid metabolism and cardiac performance in chronically diabetic rats. Diabetes 37: 1358–1364, 1988

51. Ferrari R, Shah KR, Hata T, Beamish RE, Dhalla NS: Subcellular defects in diabetic myocardium: Influence of propionyl L-carnitine on Ca^{2+} transport. In: M Nagano and NS Dhalla (eds.) The Diabetic Heart. Raven Press, New York, 1991, pp 167–181

52. Dhalla NS, Dixon IMC, Shah KR, Ferrari R: Beneficial effects of L-carnitine and derivatives on heart membranes in experimental diabetes. In: R. Ferrari, S Dimauro and G Sherwood (eds.) L-Carnitine and its Role in Medicine: From Function to Therapy. Academic Press, London, 1992, pp 411–426

Molecular and Cellular Biochemistry **116**: 11–17, 1992.

Interrelationship between lactate and cardiac fatty acid metabolism

Ger J. van der Vusse and Monique J.M. de Groot
Department of Physiology, Cardiovascular Research Institute Maastricht, University of Limburg, Maastricht, The Netherlands

Abstract

This overview is presented, in the main, to summarize the following aspects of lactate and cardiac fatty acid metabolism:
1. The utilization of exogenous carbohydrates and fatty acids by the heart.
2. The competition between lactate and fatty acids in cardiac energy metabolism.
3. The effect of lactate on endogenous triacylglycerol homeostasis.
4. Lactate-induced impairment of functional recovery of the post-ischemic heart.
5. The effect of lactate on lipid metabolism in the ischemic and post-ischemic heart.
6. The consequences of hyperlactaemia for cardiac imaging.
(Mol Cell Biochem **116:** 11–17, 1992)

Key words: heart, ischemia, reperfusion, lipids, fatty acids, lactate

Introduction

To fulfill its energy requirements the heart extracts a variety of substrates from the extracellular compartment. Glucose, fatty acids, lactate, pyruvate, acetate, ketone bodies, and, to a minor extent, amino acids can serve as oxidizable substrates. Since the heart is capable of utilizing a number of different substrates it can be considered as an omnivorous organ [1]. The rate at which any of these energy-providing substances is consumed by the heart depends upon the concentration of the substrate in the vascular and interstitial compartment, the presence of alternative and competing substrates, the actual degree of mechanical activity of the heart, and the availability of circulating or locally released hormones which are capable of modulating cardiac metabolism [2, 3]. Early studies of Bing and colleagues [4, 5] indicated that under normal circumstances oxidation of fatty acids accounts for 70 to 100% of total oxygen consumption and, hence, energy production in the heart. Later studies taking into account that fatty acids are predominantly present in blood plasma, not in blood cells, showed that a contribution of lipid oxidation to cardiac ATP production is more likely on the order of 30 and 50% [6, 7].

Keul and colleagues [8, 9] found that in the human heart at rest the contribution of fatty acids to oxidative metabolism was 34% and that the remaining part of energy required was obtained from carbohydrates. Oxidation of glucose and lactate accounted for 31% and 29% of total oxygen consumption, respectively. The contribution of pyruvate, β-hydroxy butyrate and acetoacetate was limited. During physical exercise at a degree of 200 watt, resulting in an arterial lactate con-

Address for offprints: G.J. van der Vusse, Department of Physiology, University of Limburg, P.O. Box 616, 6200 MD Maastricht, The Netherlands

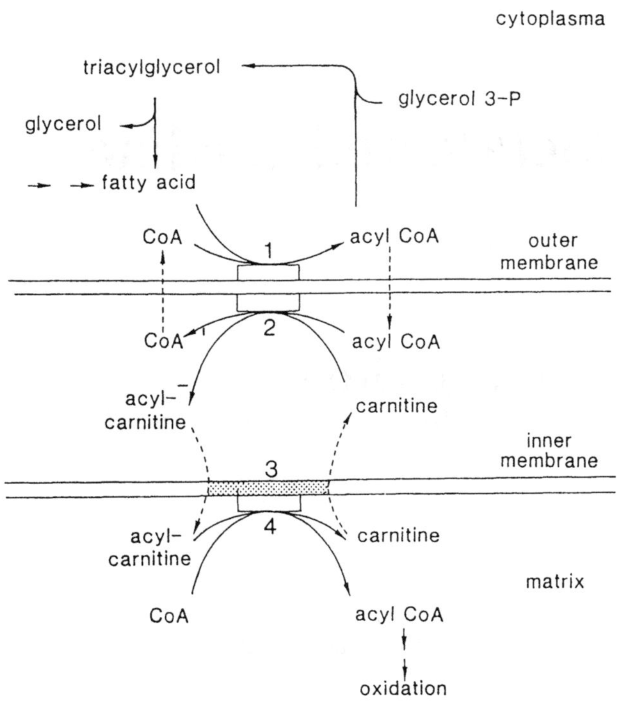

Fig. 1. Schematic representation of fatty acid activation, mitochondrial transport and oxidation in the heart. 1, 2, 3 and 4 refer to acyl-CoA synthetase, carnitine acyltransferase I, acylcarnitine translocase and carnitine acyltransferase II, respectively.

centration of 8 mmol/l, the relative contribution of circulating substrates to cardiac energy metabolism changed appreciably. The contribution of glucose and fatty acids fell to 16% and 21%, respectively. In contrast, lactate oxidation accounted for 62% of the total amount of oxygen consumed by the heart [8, 9]. The findings in human volunteers were supported by results obtained in awake dogs [10]. A significant role of lactate in cardiac energy metabolism can also be inferred from studies by Goodale *et al.* [11] and Hirche and Röhner [12]. Spitzer and Spitzer [13] reported that in dogs an early change in myocardial metabolism following severe hemorrhage is a significant decrease in fatty acid oxidation and an increased utilization of lactate for cardiac energy production.

Competition between lactate and fatty acids

The outcome of studies cited above suggests that lactate is capable of competing effectively with fatty acids for cardiac oxygen consumption. This particular aspect of cardiac metabolism has been studied in more detail by Spitzer [14] and Drake and coworkers [15]. The latter authors found that in anesthetized dogs the contribution of lactate to cardiac oxidative energy production increased to 87% when the arterial lactate concentration exceeded 4.5 mmol/l. The contribution of both glucose and fatty acid oxidation to ATP production in cardiac cells was minimal under these conditions [15]. Hence, Drake and coworkers [15] concluded that lactate instead of fatty acids is the preferred substrate of the heart provided that sufficient quantities of lactate are available from extracardiac sources. A level of 1 mmol lactate per liter blood in the body at rest is insufficient to meet the energy requirements of the heart. Obviously, when circulating lactate levels are increased 4 to 5-fold the amount extracted is sufficient to cover a substantial part of cardiac need for oxidizable substrates.

The molecular mechanism underlying the competition between lactate and fatty acids for cardiac oxygen consumption might be multifactorial. From findings obtained in dogs in multiple indicator dilution experiments, Rose and Goresky [16] proposed that enhanced lactate utilization by the heart leads to inhibition of acyl-CoA synthetase resulting in impaired oxidative degradation of fatty acids. This hypothesis has been refuted by experimental findings indicating that lactate induced impairment of fatty acid oxidation is accompanied by increased incorporation of labeled fatty acids in the endogenous triacylglycerol pool [17–22]. This observation strongly suggests that inhibition of fatty acid oxidation occurs beyond the step in which the fatty acyl-CoA complex is formed by acyl-CoA synthetase. Bielefeld and colleagues [17] have proposed that the activity of carnitine acyl-CoA transferase (CAT-I, see Fig. 1) located at the innerside of the mitochondrial outer membrane [23] becomes inhibited in the cardiac cell during increased utilization of lactate. It has been hypothetized that lactate stimulates cardiac acetyl-CoA carboxylase activity, which gives rise to elevated cytoplasmic levels of malonyl-CoA. The latter substance is thought to inhibit CAT-I activity. This notion is supported by the finding that acylcarnitine levels are increased in lactate perfused hearts [17]. As a result, transport of fatty acyl moieties across the mitochondrial inner membrane (Fig. 1) is impaired and part of fatty acyl-CoA is used for triacylglycerol formation. In addition to incorporation into the triacylglycerol pool, in dogs with high levels of circulating lactate part of fatty acids extracted by the heart was found to return to the extracellular space [16, 25]. Additional evidence for a crucial role of a carnitine related step in lactate-induced inhibition of fatty acid oxidation is the fact that the oxidation of octanoate, a fatty acid that does not require carnitine for entry into mitochondria, is insensitive to high levels of circulating lactate [17, 24]. Beside inhibition of CAT-I, the sarcoplasmic $NADH/NAD^+$ ratio most likely increases when substantial amounts of lactate are converted into pyruvate [26]. A concomitant increase of the mitochondrial redox state will lead to decreased activity of the β-oxidation pathway of fatty acyl moieties in the mitochondrial matrix.

Although evidence is accumulating that enhanced supply of lactate effectively inhibits cardiac fatty acid oxidation, less information is available on the effect of increased circulating fatty acid levels on the oxidation of lactate in the heart. When data obtained in studies performed by Drake and coworkers [15] are used to plot cardiac lactate oxidation as function of the actual arterial fatty acid concentration at elevated arterial lactate levels (8 to 10 mmol/l), a significant inverse relationship was found (Fig. 2). This observation suggests that the competition between fatty acid and lactate is reciprocal. In contrast, Liu and Spitzer [27] observed that lactate oxidation in beating myocytes, isolated from adult dog heart, was stimulated by the additon of fatty acids to the incubation medium. The significance of the latter finding remains to be elucidated.

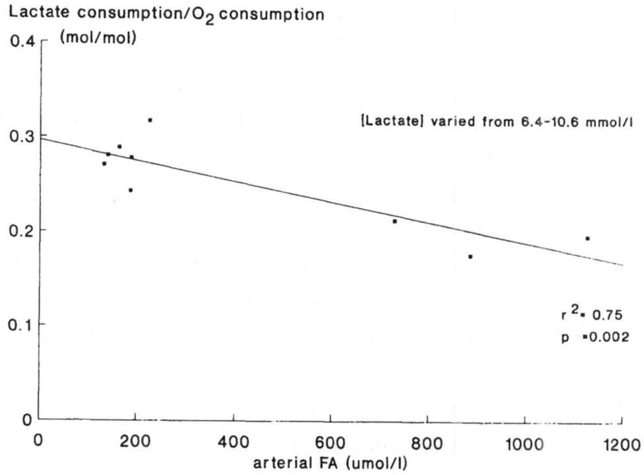

Fig. 2. The ratio cardiac lactate consumption/O_2 consumption as function of arterial fatty acid concentration in anesthetized dogs. A ratio of 0.33 indicates that all O_2 molecules extracted from the blood are used for lactate oxidation. Arterial lactate levels were artificially elevated by lactate infusion. Data are obtained from Table 1 in Drake *et al.* [15].

The effect of lactate on endogenous esterified lipid stores

From an energetic point of view triacylglycerols are the most important storage form of endogenous fatty acids in cardiac tissue. Cardiac phospholipids, containing about 5 times more fatty acyl moieties than the triacylglycerol pool, do not play a role of significance. As has been discussed above, elevated arterial lactate concentrations stimulate the incorporation of fatty acids extracted from extracellular sources into the cardiac triacylglycerol pool. However, it is unknown as to whether prolonged exposure of the heart to high circulating lactate levels results in lipidosis in cardiac cells. When incorporation of fatty acyl moieties in triacylglycerols continues, net accumulation of neutral lipids, i.e. lipidosis, will occur. Fat accumulation in cardiac cells can be prevented when hydrolysis of the endogenous triacylglycerol pool keeps pace with synthesis. In this respect recent findings of De Groot and colleagues [28] are worth mentioning. In isolated rat hearts, perfused in the presence of 11 mM glucose plus 5 mM lactate the endogenous triacylglycerol pool was well maintained over 120 min of normoxic perfusion. In contrast, control hearts perfused with glucose as sole substrate considerably depleted their triacylglycerol pool (Table 1). The release of glycerol into the cardiac venous system was also monitored. The loss of glycerol from glucose perfused hearts amounted to 0.8 (0.2–1.4) μmol/g wet weight per 120 min of antegrade perfusion (median values and 95% confidence limits, n = 13). Assuming a dry weight to wet weight ratio of 6.25, this finding indicates that all glycerol molecules released by complete hydrolysis of the endogenous triacylglycerol pool (Table 1) are washed out from the heart. The loss of glycerol from lactate-perfused hearts was found to be significantly higher, i.e. 2.1 (1.3–2.6) μmol per gram wet weight per 120 min of antegrade perfusion, despite unchanged tissue triacylglycerol levels. These data indicate that in lactate-perfused hearts hydrolysis of triacylglycerols is fully compensated by resynthesis of triacylglycerol molecules. The fatty acid moieties required for triacylglycerol formation are most likely derived from previously hydrolyzed triacylglycerols. In other words, exogenous lactate stimulates the turnover of the cardiac triacylglycerol pool. It should be kept in mind that in the present experimental set up no fatty acids were added to the perfusion medium. Hence, it is tempting to state that lactate-induced incorporation of exogenous radiolabeled fatty acids in cardiac triacylglycerols [17, 18, 20, 21] is also accompanied by increased hydrolysis of triacylglycerols to prevent cardiac lipidosis. Since lactate decreased the half-time values of radioactivity incorporated in the endogenous triacylglycerol pool from 91 to 51 min in dog hearts perfused in situ with ^{131}I-heptadecanoic acid the conclusion can be drawn that also in the *in vivo* heart the turnover rate of the neutral lipid pool is enhanced by lactate [21].

Both the hydrolytic and synthetic pathway have to be considered to clarify the mechanism of lactate-induced acceleration of cardiac triacylglycerol turnover [26, 29]. The action of lysosomal lipases is thought to be responsible for cardiac endogenous triacylglycerol hydrolysis [30], although contribution of neutral lipases cannot be excluded [30–34]. Regulation of the activity of cardiac triacylglycerol hydrolases is achieved by both hormonal and metabolic agents. Among others, intracellular fatty acids may act as feed-back inhibitors of the hydrolytic activity [30, 31]. Since increased glycerol release is associated with enhanced tissue fatty acid levels in lactate perfused hearts [28] it is highly unlikely that endogenous fatty acids inhibit intracellular lipases *in vivo*. It should be noted that the level of fatty acids needed to block lipase activity *in vitro* [31, 32] is appreciably higher than the content measured in lactate perfused hearts [26, 28].

Increased triacylglycerol hydrolysis is most likely closely related to the rate of triacylglycerol synthesis. Enhanced cytoplasmic levels of glycerol-3-phosphate, a substance required for triacylglycerol formation, may stimulate by mass action the production of neutral fat. Since lactate perfused hearts show an elevated glycerol-3-phosphate content (Table 1), this mechanism should be considered in addition to enhanced levels of acyl-CoA as outlined above. Elevation of the cytoplasmic level of glycerol-3-phosphate is most likely caused by an enhanced NADH/NAD$^+$ ratio which will shift the equilibrium between dihydroxy aceton phosphate and glycerol-3-phosphate to the latter. It is hypothetized that the newly formed pool of triacylglycerol in lactate perfused hearts are highly susceptible to lipolytic enzymes and, hence, degradation of triacylglycerols may occur at an increased rate [26].

Since triacylglycerol cycling is considered to be a 'futile' process lactate-induced triacylglycerol turnover may pose a burden on cardiac energy metabolism. As mentioned above, lactate perfused hearts release about 17 nmol glycerol per gram wet weight per min. As synthesis of 1 mol triacylglycerol requires the splitting of 7 mol ATP, the extra ATP consumption in lactate perfused hearts is on the order of 120 nmol of ATP per gram wet weight per min. The calculated oxygen consumption in antegradely perfused rat hearts is about 7 μmol O_2/gram wet weight per min [26], which corresponds with 42 μmol/gram wet weight per min ATP produced by rephosphorylation of ADP. From this calculation it can be inferred that lactate induced triacylglycerol-fatty acid cycling is inconsequential for cardiac energy metabolism as only 0.3% of ATP produced is used in this 'futile' process.

The effect of lactate on function and lipid metabolism in the ischemic and reperfused heart

High circulating lactate levels (5 mM) were very well tolerated by the isolated rat heart [28]. When compared with glucose perfused hearts, mechanical function was

Table 1. Myocardial content of triacylglycerols, fatty acids and glycerol-3-phosphate of rat hearts immediately after a 10 min stabilization period (t = 0) and after 10 min stabilization followed by 120 min of antegrade perfusion (t = 120) (ref. 28)

	Glucose (11 mM)		Glucose (11 mM) plus lactate (5 mM)	
time (min)	0	120	0	120
Triacylglycerol	15.2	3.3*	21.6	19.0
	(13.7–25.7)	(2.6–6.6)	(15.1–31.9)	(11.5–34.4)
Fatty acids	72	70	115	217*
	(40–241)	(61–114)	(64–174)	(168–267)
Glycerol-3-phosphate	0.12	0.35*	0.61	0.94*
	(0.06–0.30)	(0.20–0.79)	(0.40–0.90)	(0.67–1.41)

Tissue content of triacylglycerol is expressed as μmol fatty acid equivalents per gram dry weight of tissue, fatty acids as nmol per gram dry weight and glycerol-3-phosphate as μmol per gram dry weight. Asterisk indicates significantly different from values in corresponding hearts freeze-clamped at 0 min ($p < 0.05$). Data refer to median values and 95% confidence limits. The number of experiments varies from 7 to 9 for triacylglycerols and fatty acids. The number of experiments varies from 10 to 13 for glycerol-3-phosphate.

better preserved in lactate hearts during long term perfusion (up to 2 hr). Complete cessation of flow (ischemia) resulted in a rapid decline of mechanical activity of the oxygen-deprived heart.

Restoration of flow in both glucose and glucose plus lactate (5 mM) perfused hearts resulted in adequate hemodynamic performance when the duration of preceding ischemia did not exceed 15 min (Table 2). However, significant differences in rate-pressure product were observed between glucose and lactate hearts when the ischemic insult lasted for 30 min. Mechanical performance was virtually absent in lactate perfused hearts, while glucose perfused hearts showed adequate recovery. Reperfusion following 45 min of ischemia also depressed functional recovery of glucose perfused hearts (Table 2). At present, it is uncertain whether lactate induced impairment of post-ischemic functional outcome is related to damage inflicted upon the heart prior to readmission of flow or during the immediate period of reperfusion. Enhanced susceptibility of the isolated heart to ischemia by elevated levels of extracellular lactate was also observed by Bünger and colleagues [35].

Since both fatty acid and lactate degradation are entirely oxygen dependent in the ischemic heart oxidation of both types of substrates is impaired. In contrast, lactate production by anaerobic conversion of glucose derived from either endogenous glycogen or from extracellular sources leads to significant accumulation of lactate in the flow-deprived heart. It has been hypothetized by Neely and Grotyohann [36] that anaerobically produced lactate negatively influences the tolerance of cardiac structures towards ischemia. It is unknown as to whether high tissue levels of lactate influence ischemia-induced changes in lipid metabolism in the affected heart. Recent experiments performed by De Groot [37] indicate that in hearts perfused with 5 mM lactate prior to no-flow ischemia tissue fatty acid levels started to increase after 30 min of flow-deprivation. It should be noted that 15 min after the onset of ischemia tissue fatty acids significantly declined in these hearts. In contrast, hearts perfused with glucose as sole substrate prior to ischemia did not show the transient decrease in tissue fatty acid content at 15 min of no-flow ischemia. It remains to be established whether these subtle changes in lipid homeostasis contribute to the degree of damage inflicted upon the heart during the ischemic episode.

Table 2. The effect of exogenous lactate on cardiac rate-pressure product in the reperfused heart following 15, 30 and 45 min of no-flow ischemia

Substrates:	Glucose (11 mM)	Glucose (11 mM) plus lactate (5 mM)
pre-I	3583 (3334–3714)	3546 (3325–3808)
15 I/R	3194 (3028–3382)	3399 (2984–3739)[a]
30 I/R	2779 (0–3462)[a]	176 (0–1962)[a,b]
45 I/R	2099 (0–2297)[a]	0 (0–1053)[a,b]

Rate-pressure product (heart rate times developed left ventricular pressure) is expressed as kPa/min. Data refer to median values and 95% confidence limits. Number of experiments varies from 5 to 8. Pre-I refers to 15 min prior to ischemia. 15 I/R, 30 I/R and 45 I/R refer to values measured at the end of the reperfusion phase (35 min) preceded by 15, 30 and 45 min of ischemia, respectively.
[a]) indicates significantly different from pre-ischemic value;
[b]) indicates significantly different from corresponding glucose hearts ($p < 0.05$).

Detailed analysis of ischemic and reperfused cardiac tissue and of the post-ischemic perfusate indicates that lactate perfused hearts show alterations in their lipid homeostasis markedly different from those in glucose perfused hearts during reperfusion [38].

Tissue fatty acid levels returned to control pre-ischemic values in glucose perfused hearts after reinstallation of flow. In contrast, in lactate hearts the tissue content of fatty acids continued to increase. The change in fatty acid level was not associated by a measurable decline in the tissue triacylglycerol and phospholipid pool. This observation can be explained by the fact that the quantity of accumulated fatty acids corresponds with on the order of 2% of the total amount of esterified fatty acids. Obviously, the current analytical techniques are not sensitive enough to detect such small changes in the esterified lipid pool. Enhanced tissue fatty acid levels can be caused by impaired oxidation of these substances and/or an imbalance between hydrolysis and resynthesis of esterified lipids.

Mickle and coworkers [39] found that exogenous fatty acids were preferentially utilized in the post-ischemic dog heart despite high blood levels of lactate. Later studies performed by Renstrom and colleagues [40] in swine hearts in situ are supporting the notion that fatty acids are the preferred oxidative substrates during the initial reperfusion phase. Lactate and pyruvate oxidation failed to recover to pre-ischemic levels. These data are not in favor of impaired fatty acid oxidation as cause of accumulation of fatty acids in reperfused cardiac tissue. It is worth mentioning that in the study of De Groot and Van der Vusse [38] elevated levels of fatty acids in tissue were found to be accompanied by enhanced release of glycerol in the lactate perfused post-ischemic heart, suggesting lactate induced triacylglycerol-fatty acid cycling.

Combining the findings in *ex vivo* rat heart preparations and the results obtained in in situ dog and swine hearts, the conclusion can be drawn that in the reperfused heart lactate influences the metabolic fate of fatty acids derived from either extracellular or intracellular sources in a different manner. Exogenous fatty acids are subject to oxidation while fatty acids released from endogenous stores are reincorporated in the neutral fat pool with a concomitant increase of cellular fatty acid levels.

Although discrete alterations in cardiac lipid homeostasis have been identified in lactate perfused hearts during the post-ischemic phase, it remains to be established whether these changes are causally related to the severe depression of mechanical function after reinstallation of flow when lactate levels in the perfusion medium are high.

Plasma lactate and the interpretation of metabolic imaging techniques of cardiac lipid homeostasis

The use of radiolabeled fatty acids has been advocated to assess metabolic behavior of the heart in the clinical setting with non-invasive techniques such as positron emission tomography (PET)and single photon emission computer tomography (SPECT). The first technique requires ^{11}C-labeled fatty acids, while radioactive iodine labeled fatty acid analogues are commonly used in SPECT studies. Applying the PET technique in anesthetized dogs, Wijns and coworkers [41] established that cardiac time activity curves of ^{11}C-palmitate were significantly changed when the arterial lactate concentration was artificially raised. They concluded that lactate reduces the oxidation of radiolabeled palmitate in cardiac cells and enhances the incorporation of label in endogenous lipid stores. Scintigraphic studies performed by Visser and colleagues [18, 21] with iodine-labeled fatty acid analogues in dogs and human volunteers provided the same kind of information.

It should be kept in mind that a variety of physiological or pathophysiological conditions are found to be associated with enhanced levels of lactate in the blood [42]. Infusion of lactate containing buffers in the experimental and clinical setting (e.g. lactate-Ringer solutions during cardiopulmonary bypass) and moderate to severe physical exercise performed by cardiac patients [43] easily lead to elevated blood lactate concentrations. In addition, circulatory shock, hypoxemia and endotoxin administration are also associated with hyperlactaemia. Under these circumstances cardiac fatty acid oxidation is most likely inhibited in favor of oxidative degradation of lactate. Hence, interpretation of clinical data obtained with PET and SPECT techniques should be done with care when increased arterial lactate levels in the patients under investigation are likely to occur.

Acknowledgements

The authors are greatly indebted to Mrs. Karlijn Dickison and Mrs. Emmy van Roosmalen for help in preparing the manuscript and to Peter Willemsen and Will Coumans for their skillful help in assessing the biochemical substances. This study was supported (in part) by the Netherlands Heart Foundation, grant no. 90.072.

References

1. Taegtmeyer H: Carbohydrate interconversions and energy production. Circulation 72: IV-1 – IV-8, 1985
2. Neely JR, Rovetto MJ, Oram JF: Myocardial utilization of carbohydrate and lipids. Progr Cardiovasc Dis 15: 289–329, 1972
3. Neely JR, Rovetto MJ, Morgan HE: Relationship between carbohydrate and lipid metabolism and the energy balance of heart muscle. Annu Rev Physiol 36: 413–459, 1974
4. Bing RJ, Siegel A, Ungar J, Gilbert M: Metabolism of the human heart II. Studies on fat, ketone and aminoacid metabolism. Amer J Med 16: 504–515, 1954
5. Blain J, Schaeffer H, Siegel A, Bing RJ: Studies of myocardial metabolism. Amer J Med 20: 820–833, 1956
6. Ballard F, Danforth W, Nagele S, Bing RJ: Myocardial metabolism of fatty acids. J Clin Invest 39: 717–723, 1960
7. Gordon RS, Cherkes A: Unesterified fatty acids in human blood plasma. J Clin Invest 35: 206–212, 1956
8. Keul J, Doll E, Steim H, Homburger H, Kern H, Reindell H: Uber den Stoffwechsel des Menschlichen Herzens. I. Die Substrat versorgung des gesundes menschlichen Herzens in Ruhe, während und nach körperlichen Arbeit. Pflügers Arch. 282: 1–27, 1965
9. Keul J, Doll E, Steim H, Fleer U, Reindell H: Uber den Stoffwechsel des Menslichen Herzens. III. Der oxydative Stoffwechsel des menschlichen Herzens unter verschiedenen Arbeitsbedingungen. Pflügers Arch 282: 43–53, 1965
10. Lochner W, Nasseri M: Uber den venösen Sauerstoffdruck, die Einstellung der Coronardurchblutung und den Kohlenhydrat-

stoffwechsel des Herzens bei Muskelarbeit. Pflügers Arch 269: 407–416, 1959
11. Goodale WT, Olson R, Hackel DB: Myocardial glucose, lactate and pyruvate metabolism of normal and failing hearts studies by coronary sinus catheterization in man. Fed Proc 9: 49, 1950
12. Hirche H, Röhner G: Anderungen der Substrat aufnahme des Herzmuskels bei induzierten Anderungen der arteriellen Substrat Konzentration. Pflügers Arch 278: 408–422, 1963
13. Spitzer JJ, Spitzer JA: Myocardial metabolism in dogs during hemorrhagic shock. Amer J Physiol 222: 101–105, 1972
14. Spitzer JJ: Effect of lactate infusion on canine myocardial free fatty acid metabolism *in vivo*. Am J Physiol 226: 213–217, 1974
15. Drake AJ, Haines JR, Noble MIM: Preferential uptake of lactate by the normal myocardium in dogs. Cardiovasc Res 14: 65–72, 1980
16. Rose CP, Goresky CA: Constraints on the uptake of labeled palmitate by the heart. The barriers at the capillary and sarcolemmal surfaces and the control of intracellular sequestration. Circ Res 41: 534–545, 1977
17. Bielefeld DR, Vary TC, Neely JR: Inhibition of carnitine palmitoyl CoA transferase activity and fatty acid oxidation by lactate and oxfenicine in cardiac muscle. J Mol Cell Cardiol 17: 619–625, 1985
18. Duwel CMB, Visser FC, Van Eenige MJ, Westra G, Roos JP: The influence of lactate and dipyridamole on myocardial fatty acid metabolism traced with the radioiodinated fatty acid 17-I-123-iodoheptadecanoic acid in man. Nucl-Med 29: 28–34, 1990
19. Duwel CMB, Visser FC, Van Eenige MJ, Den Hollander W, Roos JP: The fate of 17-I-123 iodoheptadecanoic acid during lactate loading; its oxidation is strongly inhibited in favor of its esterification. A radiochemical study in the normal canine heart. Nucl-Med 29: 24–27, 1990
20. Visser FC, Duwel CMB, Van Eenige MJ, Roos JP, Knoop FF, Van der Vusse GJ: Biochemistry of radioiodinated non-esterified fatty acids. Mol Cell Biochem 88: 185–190, 1989
21. Visser FC, Van Eenige MJ, Duwel CMB, Roos JP: Radioiodinated free fatty acids: can we measure myocardial metabolism? Europ J Nucl Med 12: S20–S23, 1986
22. Reske SN, Schön S, Schmitt W, Machulla HJ, Knopp R, Winkler C: Effect of myocardial perfusion and metabolic interventions on cardiac kinetics of phenylpentadecanoic acid (IPPA)-I-123. Europ J Nucl Med 12 (suppl.): 27–31, 1986
23. Murthy MSR, Pande SV: MalonylCoA binding site and the overt carnitine palmitoyltransferase activity reside on the opposite sides of the outer mitochondrial membrane. Proc Natl Acad Sci USA 84: 378–382, 1987
24. Forsey RGP, Reid K, Brosnan JT: Competition between fatty acids and carbohydrate or ketone bodies as metabolic fuels for the isolated perfused heart. Canad J Physiol Pharmacol 65: 401–406, 1987
25. Duwel CMB, Visser FC, Van Eenige MJ, Roos JP: Variables of myocardial back-diffusion, determined with 17-iodo-131 heptadecanoic acid in the normal dog heart. Mol Cell Biochem 88: 191–194, 1989
26. Van der Vusse GJ, De Groot MJM: The effect of exogenous lactate and pyruvate on triacylglycerol turnover in the normoxic heart. In: S Sideman, R Beyar, AG Kléber (eds.) Cardiac electrophysiology, circulation and transport. Kluwer Acad Publ Boston, pp 351–361, 1991
27. Liu M.-S, Spitzer JJ: Oxidation of palmitate and lactate by beating myocytes isolated from adult dog heart. J Mol Cell Cardiol 10: 415–426, 1978
28. De Groot MJM, Willemsen PHM, Coumans WA, Van Bilsen M, Van der Vusse GJ: Lactate-induced stimulation of myocardial triacylglycerol turnover. Biochim Biophys Acta 1006: 111–115, 1989
29. Schoonderwoerd K, Broekhoven-Schokker S, Hülsmann WC, Stam H: Enhanced lipolysis of myocardial triglycerides during low-flow ischemia and anoxia in the isolated rat heart. Basic Res Cardiol 84: 165–173, 1989
30. Hülsmann WC, Stam H, Breeman WAP: Acid and neutral lipases involved in endogenous lipolysis in small intestine and heart. Biochem Biophys Res Commun, 162: 440–448, 1981
31. Severson DL, Hurley B: Regulation of rat heart triacylglycerol ester hydrolases by free fatty acids, fatty acyl CoA and fatty acylcarnitine. J Mol Cell Cardiol 14: 467–474, 1982
32. Stam H, Hülsmann WC: Regulation of lipases involved in the supply of substrate fatty acids for the heart. Europ Heart J 6: 158–167, 1985
33. Goldberg DI, Khoo JC: Activation of myocardial neutral triglyceride lipase and neutral cholesterol esterase by cAMP-dependent protein kinase. J Biol Chem 260: 5879–5882, 1985
34. Schoonderwoerd K, Broekhoven-Schokker S, Hülsmann WC, Stam H: Stimulation of myocardial neutral triglyceride lipase activity by adenosine-3′, 5′-monophosphate: involvement of glycogenolysis. Basic Res Cardiol 82(suppl. I), 29–35, 1987
35. Bünger R, Mallet RT, Hartman DA: Pyruvate-enhanced phosphorylation potential and inotropism in normoxic and post-ischemic isolated working heart. Eur J Biochem 180: 221–233, 1989
36. Neely JR, Grotyohann LW: Role of glycolytic products in damage to ischemic myocardium. Dissociation of adenosine triphosphate levels and recovery of function of reperfused ischemic hearts. Circ Res 55: 816–824, 1984
37. De Groot MJM: The effect of lactate on the normoxic, ischemic and reperfused heart. Thesis. University of Limburg, Maastricht, The Netherlands, 1992
38. De Groot MJM, Van der Vusse GJ: The effect of lactate on triacylglycerol (TG) metabolism in ischemic and reperfused hearts. J Mol Cell Cardiol 22 (suppl. III): S. 114, 1990
39. Mickle DAG, Del Nido PJ, Wilson GJ, Harding RD, Romaschin AD: Exogenous substrate preference of the post-ischaemic myocardium. Cardiovasc Res 20: 256–263, 1986
40. Renstrom B, Nellis SH, Liedtke AJ: Metabolic oxidation of pyruvate and lactate during early myocardial reperfusion. Circ Res 66: 282–288, 1990
41. Wijns W, Schwaiger M, Huang S.-C, Buxton DC, Hansen H, Selin C, Keen R, Phelps ME, Schelbert HR: Effects of inhibition of fatty acid oxidation on myocardial kinetics of ^{11}C-labeled palmitate. Circ Res 65: 1787–1797, 1989
42. Moret PR: Opening remarks. In: PR Moret, J Weber, JC Haissly, H Denolin (eds.) Lactate physiologic, methodologic and pathologic approach. Springer Verlag Berlin, pp 195–196, 1980
43. Messin R: Peripheral lactic acid production in heart disease. In: PR Moret, J Weber, JC Haissly, H Denolin (eds.) Lactate physiologic, methodologic and pathologic approach. Springer Verlag, Berlin, pp 195–196, 1980

Molecular and Cellular Biochemistry **116**: 19–25, 1992.

Dietary modulation of lipid metabolism and mechanical performance of the heart

John S. Charnock, Peter L. McLennan and Mahinda Y Abeywardena
CSIRO, Division of Human Nutrition, Glenthorne Laboratories, O'Halloran Hill, South Australia 5158

Abstract

Sudden Cardiac Death resulting from sustained ventricular fibrillation or malignant cardiac arrhythmia has been linked to the type of dietary fat intake in several economically well developed countries where high levels of saturated fatty acids are common. Experimental studies with the small non-human primate marmoset monkey have clearly demonstrated the health benefit of substituting polyunsaturated fatty acids (PUFA's) for dietary saturated fatty acids. Heart rate and blood pressure are lowered, while the left ventricular ejection fraction and the electrical threshold for the induction of ventricular fibrillation are both increased after prolonged feeding of PUFA enriched diets. All these changes in heart function reduce the risk of developing malignant cardiac arrhythmias.

The fatty acid composition of cardiac membrane phospholipids is profoundly altered by these changes in dietary lipid intake. In particular the proportions of arachidonic acid (AA), eicosapentaenoic acid (EPA) and docosahexaenoic acid (DHA) are altered in such a way that the production of myocardial eicosanoids is affected. Although the changes in proportion of these long-chain PUFA's in cardiac phosphatidyl ethanolamine and phosphatidyl inositol are not identical, the shift in balance between these substrates or inhibitors of cyclo-oxygenase activity leads to relatively greater production of prostacyclin (PGI_2) than thromboxane (TXA_2).

The effect of the omega-3 PUFA's of fish oil is proportionally greater than that of linoleic acid (LA; 18:2, ω6) rich sunflower seed oil, particularly during ischaemia, and probably reflects the different nutritionally induced changes in cardiac membrane fatty acid composition by these different types of dietary PUFA's. (Mol Cell Biochem **116:** 19–25, 1992).

Key words: dietary fats, heart function, cardiac-membrane composition, myocardial eicosanoids

Introduction

Sudden Cardiac Death remains a major medical problem despite some recent declines in mortality and morbidity from Ischaemic Heart Disease [1]. A notable feature of these acute deaths is that they are often described as both 'sudden and unexpected' and are thought to arise from sustained ventricular fibrillation or malignant cardiac arrhythmia [2, 3]. Despite extensive trials of many different anti-arrhythmic agents, they have generally failed to demonstrate any significant reduction in mortality associated with cardiac arrhythmia [4]. In fact several recent studies have demonstrated that the drug treatments employed to reduce arrhythmia have in fact led to enhanced mortality [5, 6].

Conversely there is considerable experimental evidence from carefully controlled nutritional studies in laboratory animals which demonstrate that relatively

Address for offprints: J.S. Charnock, CSIRO, Division of Human Nutrition, Glenthorne Laboratories, O'Halloran Hill, South Australia 5158

simple modifications in dietary lipid intake both change the fatty acid composition of cardiac membranes [7, 8] and alter the vulnerability of the heart to develop arrhythmias when blood flow to the myocardium is restricted [9, 10]. Saturated fatty acids from either animal or plant sources are found to be detrimental whereas polyunsaturated fatty acids (PUFA's) are generally beneficial [9, 10]. This is particularly evident when the PUFA's are of the long chain omega-3 (ω3) series commonly derived from cold water fish [10, 11].

These experimental studies offer strong support for the epidemiological evidence of Kromhout *et al.* [12] and Burr *et al.* [13], both of whom have shown marked reductions in mortality from heart disease after increased consumption of fatty fish or fish oil supplements. That such relatively simple dietary measures may be so successful is particularly encouraging as reliable diagnostic predictors of Sudden Cardiac Death such as the angiographic determination of left ventricular ejection fraction (LVEF), or the electrical threshold for the induction of sustained ventricular fibrillation (VFT) by programmed stimulation [14, 15] require sophisticated medical technology which is not readily applicable to wide scale screening procedures.

However, such diagnostic techniques can be applied to animal studies where nutritional modifications can also be well controlled. In addition, such animal studies provide access to cardiac tissues and their cell membranes and hence a level of investigative biochemistry not otherwise possible. It thus becomes possible to examine the underlying mechanisms responsible for the nutritionally induced changes in cardiac function [16]. In this paper we will review our recent findings of the effects of habitual change in dietary lipid intake on the cardiac function of the marmoset monkey *Callithrix jacchus,* as well as relate the changes to altered cardiac membrane phospholipid fatty acid composition and myocardial eicosanoid production which we feel are central to the alteration in the vulnerability of the heart to develop arrhythmia when under ischaemic stress.

Methods

Because of the increased risk of 'heart attack' in man after middle age [17], and the established role of age in our previous studies of cardiac arrhythmia in rats subjected to occlusion of a major coronary artery [9, 10], the experimental results to be described here were obtained from mature marmosets. All animals were at least three years old at the time of study, and had received their various lipid supplemented diets for a minimum of 24 months prior to investigation.

The preparation, storage and fatty acid composition of all diets, along with the details of the feeding regime employed have been described in full in several recent publications from this laboratory [8, 18, 19]. Similarly the application of gated blood pool angiography to the heart of this small non-human primate in order to measure several parameters of left ventricular function has also been described in previous publications [20, 21]. Cardiac arrhythmias were induced in anaesthetised adult marmosets by programmed electrical stimulation using a protocol of burst pacing plus two premature stimuli of variable current applied to the left ventricle [22].

A total lipid extract of thawed cardiac muscle was prepared by treating the tissue with 2:1 chloroform : methanol at room temperature for 60 min. Phospholipids were separated from other lipid moieties by routine thin layer chromatography; and fatty acid methyl esters (FAME's) were prepared by heating with methanolic boron trifluoride as described previously [7, 8, 23, 24]. The FAME's were analysed by gas chromatography and the individual fatty acids identified by comparison of retention times with those of authentic standards [8, 24]. The stable metabolites of myocardial prostacyclin (PGI_2) and thromboxane A_2, that is 6-keto $PGI_{1\alpha}$ and

Table 1. Fatty acid composition of experimental diets

Fatty acid		SF	SSO	FO
Myristic	14:0	2.4	1.0	2.8
Palmitic	16:0	21.2	12.0	19.5
Palmitoleic	16:1ω7	1.6	0.8	4.2
Stearic	18:0	24.9	10.3	7.0
Oleic	18:1ω9	28.2	21.9	22.0
Linoleic	18:2ω6	10.9	52.4	10.8
Linolenic	18:3ω3	1.4	1.1	1.4
Eicosapentaenoic	20:5ω3	–	–	4.7
Docosapentaenoic	22:5ω3	–	–	1.1
Docosahexaenoic	22:6ω3	–	–	15.6
P/S		0.3	2.3	1.1
$\sum$ω6/ω3		7.8	47.6	0.5

The lipid supplemented diets SF, SSO and FO are described in the text. Each diet contained 94% primate meal supplemented with 6% by weight sheep fat (SF), sunflower seed oil (SSO) or fish oil (FO). The P/S ratio is determined from the sum of the total polyunsaturated fatty acids (PUFA's) divided by the sum of the total saturated fatty acids present. Values given are the relative % total extractable fatty acids.

TXB_2 were determined by radio immunoassay procedures described in full in previous publications from this laboratory [23, 24].

Results

Marmoset diets

The lipid supplemented diets were fabricated to be isocaloric and provide 28% of their total energy as 'fat'. In this regard they are very close to the most desirable level of total fat recommended by the American Heart Association. In general the experimental diets contained comparable amounts of either animal fat (SF), sunflower seed oil (SSO) or fish oil (FO), thus yielding diets rich in either saturated fatty acids, polyunsaturated fatty acids of the ω6 family or long-chain polyunsaturated fatty acids of the ω3 family respectively. The complete fatty acid composition of these experimental diets has been described in detail in several recent publications [8, 25]. However, a summary of their major fatty acid components is given in Table 1. It is important to recognise that not only do these diets have different P/S ratios, but that their $\sum$ω6/ω3 PUFA distribution ranges from 47.6 for the SSO diet, to 7.8 for the SF diet but is only 0.5 for the FO diet.

Heart rate and blood pressure

A previous study had shown that introduction of saturated animal fats into the diets of young marmosets had led to a significant increase in their heart rates after six months of feeding the lipid supplemented diet. In that study the increase in heart rate was only accompanied by a small increase in systolic or diastolic blood pressure [21]. In the present study with older animals fed different lipid supplemented diets for 24 months, the increase in heart rate and blood pressure in the SF group was much more marked. This means that substitution of either ω6 or ω3 PUFA's for saturated animal fats leads to significant benefits in these important aspects of cardiovascular function (Table 2).

Table 2. Effect of dietary lipids on cardiac function in adult marmosets

Cardiac function	Diet supplements		
	SF	SSO	FO
Heart rate (beats min^{-1})	259	250	226
Systolic BP (mm Hg)	94	84	84
Diastolic BP (mm Hg)	68	58	60
LVEF (% filled volume)	46	57	55
PFR (ml $min^{-1}g^{-1}$ heart wgt $\times 10^2$)	140	200	173
EDV (ml g^{-1} heart wgt $\times 10^3$)	215	240	250

Dietary supplements (SF, SSO & FO) are described in the text and in Table 1.
LVEF is left ventricular ejection fraction; PFR is peak filling rate and EDV is end diastolic volume. All values are the average obtained from six animals per group.

Ventricular function

A gated blood-pool radionuclide angiography procedure was developed in order to measure left ventricular ejection fraction (LVEF), peak filling rate (PFR) and end diastolic volume (EDV) in surviving marmosets [21]. After 24 months of feeding of the experimental diets it is quite clear that PUFA enriched diets lead to significant improvement in all three parameters of cardiac performance (Table 2). In fact it can be seen that there is more than a 20% improvement in the LVEF of PUFA fed animals compared to age matched marmosets which had consumed a saturated fat (SF) enriched diet. Similar improvements in the EDV and PFR of the PUFA fed groups demonstrates considerable benefit in the mechanical performance of the hearts of these animals under normoxic conditions.

Ventricular fibrillation threshold

Programmed electrical stimulation was employed to determine the stability of the heart, that is the ventricular fibrillation threshold (VFT) of adult marmosets after 30 months of feeding the experimental diets. Ventricular fibrillation was induced equally in all dietary groups, but Table 3 shows that the addition of PUFA's into the diet greatly improved the stability of the heart, ie. the VFT in both PUFA fed groups was markedly higher than that in the SF fed group. In addition, the episodes of VF in the PUFA fed groups were predominantly selflimiting with a marked reduction in the number of episodes of VF that were sustained and required defibrillation. Strikingly, after prolonged feeding of the FO supplement, the incidence of sustained VF in ischaemic hearts was NIL compared to either 43% in the SF or 11% in the SSO supplemented group. These results demonstrate the extent of benefit achieved by introducing PUFA's and particularly those of the ω3 family into the diet of this non-human primate.

Cardiac muscle fatty acid composition

The fatty acid composition of cardiac muscle membranes after dietary manipulation is given in Table 4. Generally the saturated fatty acid components (palmitic and stearic acids) of cardiac phospholipids are not greatly altered by dietary change. On the other hand a lipid supplement of saturated fat (SF) raises the level of the major monounsaturated component – oleic acid (OA; 18:1, ω9) in cardiac phospholipids, but depresses the level of linoleic acid (LA; 18:2, ω6). This latter essential fatty acid (EFA) is even further depressed by substituting fish oil (FO) for either the saturated fat (SF) or the polyunsaturated sunflower seed oil (SSO) enriched diet. Arachidonic acid (AA; 20:4, ω6) is a major component of membranes where it acts as a precursor for Series-2 prostaglandin synthesis. Feeding polyunsaturated fatty acid (PUFA) supplements – whether they be of the ω6 family in sunflower seed oil (SSO) or the ω3 family in fish oil (FO) significantly lowers the level of AA in cardiac muscle membranes. However, only feeding FO leads to significant accumulation of eicosapentaenoic acid (EPA; 20:5, ω3) and docosahexaenoic acid (DHA; 22:6, ω3) although some DHA is found after feeding the SF diet. As both EPA and DHA are either substrates for the enzymic synthesis of Series-3 prostaglandins or inhibitors of cyclo-oxygenase, it is clear that such dietary induced changes in membrane composition can alter the availability of substrates for Series 2 and Series 3 prostaglandin production.

This is achieved by significant changes in the ω6/ω3 PUFA ratio of membranes which can occur with little alteration in their total saturated fatty acid: unsaturated fatty acid ratio. This can be seen even more sharply by examining the fatty acid profiles of some individual cardiac membrane phospholipids like phosphatidyl ethanolamine (PE) or phosphatidyl inositol (PI) which have different PUFA profiles.

Table 3. Effect of dietary lipids on ventricular fibrillation in normoxic and ischaemic hearts of the adult marmoset

Ventricular fibrillation		Diet supplements		
		SF	SSO	FO
Threshold	Normoxic	8.3	15.1	15.0
	Ischaemic	6.8	13.3	13.4
Incidence	Normoxic	43	13	8
	Ischaemic	45	13	NIL

Dietary supplements (SF, SSO & FO) are described in the text and in Table 1.
Threshold value is given in milli amps. Incidence is the proportion (%) of animals demonstrating sustained VF.

Cardiac muscle phosphatidyl ethanolamine

Phosphatidyl ethanolamine (PE) is a major component of cardiac muscle membranes accounting for about one third of the total membrane lipids of the marmoset [7]. It is a rich source of the metabolites of both the essential fatty acids, linoleic (18:2, ω6) and α-linolenic acid (18:3, ω3) as it contains appreciable amounts of AA and DHA respectively (Table 5).

There are significant changes in the PUFA profile of PE after feeding different lipid supplemented diets.

Table 4. Effect of dietary lipids on the fatty acid composition of cardiac muscle membranes of the adult marmoset

Fatty acids	Diet supplements		
	SF	SSO	FO
16:0	14.0	13.4	16.6
18:0	20.1	22.9	19.9
18:1ω9	12.6	9.3	8.4
18:2ω6	21.9	33.4	9.3
20:3ω6	1.2	0.6	–
20:4ω6	14.1	11.3	8.8
20:5ω3	–	–	7.9
22:4ω6	0.6	0.8	–
24:0	0.9	1.5	–
22:5ω3	2.7	1.0	1.3
22:6ω3	5.0	2.3	20.9
∑ Sat FA	35.0	37.8	36.5
∑ Unsat FA	58.1	58.7	56.6
∑ PUFA	45.5	49.4	48.2
∑ Sat FA/Unsat FA	0.60	0.62	0.64
∑ PUFA/Sat FA	1.30	1.31	1.43
∑ ω6PUFA	37.8	46.1	18.1
∑ ω3PUFA	7.7	3.3	30.1
∑ ω6/ω3PUFA	4.9	14.0	0.7

Dietary supplements (SF, SSO & FO) are described in the text and in Table 1.
Values given are relative % total phospholipid fatty acids.

This becomes most apparent by contrasting the results obtained after either PUFA supplement, with those found after feeding the SF diet. After feeding the SSO diet rich in ω6 PUFA, there is a significant increase in AA and a small decrease in DHA; whereas after feeding the FO diet containing long chain ω3 PUFA's there is a significant decrease in AA as well as a significant increase in DHA. In addition, FO feeding results in appreciable EPA in this PE fraction. Consequently the ∑ω6/ω3 fatty acid ratio of marmoset PE is much greater after the SSO diet than after the FO diet (6.4 vs 0.7 respectively). As will be seen later (Table 6) these changes in AA, EPA and DHA are associated with marked changes in the production of myocardial eicosanoids.

Cardiac muscle phosphatidyl inositol

Although phosphatidyl inositol (PI) only makes up about 3% of the total phospholipids of the cardiac membranes of the marmoset [7] it is thought to play a central role in trans-membrane signalling [26, 27]. Like PE, the polyunsaturated fatty acid profile of PI is also altered by introduction of different PUFA's into the diet. However, unlike the changes seen in PE, the ω6 PUFA rich SSO diet does not increase the AA content of PI, whereas the omega-3 PUFA rich FO diet again decreases the level of AA.

Table 5. Effect of dietary lipids on cardiac muscle phospholipids of the adult marmoset

Phospholipid	Major PUFA's	Diet supplements		
		SF	SSO	FO
Phosphatidyl Ethanolamine	18:2ω6	15.7	20.0	12.0
	20:4ω6	16.9	20.3	11.1
	20:5ω3	–	–	6.1
	22:5ω3	2.8	1.8	–
	22:6ω3	5.6	4.5	26.3
	∑ω6/ω3PUFA	3.9	6.4	0.7
Phosphotidyl Inositol	18:2ω6	7.5	20.5	5.6
	20:4ω6	22.3	18.4	16.4
	20:5ω3	–	–	8.6
	22:5ω3	1.2	0.1	1.1
	22:6ω3	1.1	0.1	9.0
	∑ω6/ω3PUFA	12.9	19.4	1.2

Dietary supplements (SF, SSO & FO) are described in the text and in Table 1.
Values given are relative % total phospholipid fatty acids.

On the other hand, while there is virtually no detectable EPA or DHA in marmoset cardiac PI after feeding the SSO diet, both EPA and DHA (in approximately equal proportions) are found after feeding the FO diet. Consequently the ∑ω6/ω3 ratio of marmoset PI is nearly 200 × greater after the SSO diet than after the FO diet (Table 5). Perhaps then, even more than in PE, this shift in ω6:ω3 ratio in PI can hava a profound effect upon the availability of membrane substrates for myocardial eicosanoid production. If PI is particularly involved in transmembrane signalling [26], then changes in the fatty acid composition of this phospholipid may have effects disproportionate to the level of PI in cardiac membranes.

Myocardial eicosanoid production

Several recent experiments in our laboratory have shown that eicosanoid production in the marmoset monkey, both *in vitro* and *in vivo,* is similar to that of man [19, 23, 24].

The effect of different dietary lipid supplements on myocardial prostacyclin (PGI_2) and thromboxane (TXA_2) production *in vitro* is given in Table 6. By comparison with the levels found after feeding the SF diet, both ω6 and ω3 PUFA dietary supplements significantly reduce PGI_2 production. However, the reduction found after feeding the FO diet is proportionally much greater than that following the SSO diet.

Similarly, after feeding both PUFA containing diets, the levels of myocardial TXA_2 production are also much less than that observed after feeding the SF diet. Again the level after the FO diet is less than that after the SSO diet. Consequently in comparison to the effect of a saturated fat enriched diet, the ratio of myocardial PGI_2

Table 6. Effect of dietary lipids on myocardial eicosanoid production in the adult marmoset

Basal production eicosanoids	Diet supplements		
	SF	SSO	FO
PGI_2	167	114	88
TXA_2	69	24	12
PGI_2/TXA_2	2.4	4.8	7.3

Dietary supplements (SF, SSO & FO) are described in the text and in Table 1.
Values are $pg \cdot mg^{-1}$ dry heart wgt. Adapted from Abeywardena *et al.* 1991; Ref [24] this paper.

vs TXA_2 is greatly altered by feeding PUFA dietary supplements. The alteration is clearly towards an increase in the 'balance' of PGI_2 over TXA_2, with the greatest effect being seen after feeding the fish oil supplement.

It should also be noted that in contrast to some other studies in experimental animals [28] these changes in eicosanoid 'balance' remain within the physiological rather than the pharmacological range of these biologically active compounds.

Conclusions

It is evident that the substitution of either ω6 or ω3 PUFA's for saturated fatty acids into the diets of adult marmoset monkeys, leads to significant beneficial effects upon many aspects of their cardiac performance. In particular both the mechanical performance and the electrical stability of their hearts are greatly improved, with the effect of ω3 PUFA's being relatively greater than that of ω6 PUFA's. This is particularly noticeable under conditions of ischaemia when the incidence of sustained ventricular fibrillation in the FO supplemented group was NIL compared to 13% in the SSO group and 45% after prolonged feeding of the SF diet. It should also be remembered that in these experiments the total amount of ω3 PUFA's in the FO diet was only about half that of the ω6 PUFA in the SSO diet.

These important beneficial changes in the physiology of the marmoset heart are accompanied by major changes in the PUFA profiles of their cardiac membrane phospholipids. Although these changes in fatty acid composition are not identical in different phospholipids, it is apparent that the greatest alterations are in those polyunsaturated fatty acid components which are concerned with the tissues capacity to generate Series-2 and Series-3 eicosanoids, namely AA, EPA and DHA. This is born out by the direct measurement of myocardial ability to generate PGI_2 and TXA_2 *in vitro*, where it is quite clear that in comparison to the effects of a saturated fat enriched diet (SF), both PUFA supplemented diets lead to an improvement in the balance between the production of PGI_2 over TXA_2. However, as the reduction in TXA_2 production after an ω3 PUFA rich diet is greater than that after an ω6 PUFA supplement, the final 'balance' between PGI_2 and TXA_2 production is greatest following the feeding of a fish oil dietary supplement.

As the work of Parratt and his colleagues in dogs has clearly demonstrated that PGI_2 has anti-arrhythmic effects while TXA_2 promotes arrhythmia [28], it is not unreasonable to suggest that these shifts in eicosanoid balance are of great benefit to the heart of the marmoset in reducing its susceptibility to develop sustained ventricular fibrillation and cardiac arrhythmia when under ischaemic stress.

It seems clear then, that the dietary induced changes in the fatty acid composition of the membrane of the marmoset heart play an important role in reducing the risk of developing cardiac arrhythmia. As this primate has been shown to be a good animal model for the study of human cardiovascular disease [30, 31], it may well be that similar dietary induced changes in man may be responsible for the reduction in mortality found in post-myocardial infarction patients advised to include-fatty fish or fish oil supplements into their diets [13, 29].

References

1. Lown B: Sudden Cardiac Death: The major challenge confronting contemporary cardiology. Am J Cardiol 43: 313–328, 1979
2. Keefe DL, Schwartz J, Somberg JC: The substrate and the trigger: The role of myocardial vulnerability in sudden cardiac death. Am Heart J 113: 218–225, 1987
3. Myerberg RJ, Kessler KM, Bassett AL, Castellanos A: A biological approach to Sudden Cardiac Death: Structure, Function and Cause. Am J Cardiol 63: 1512–1516, 1989
4. Anderson JL: Sudden Cardiac Death, Ventricular Arrhythmias and Antiarrhythmic Therapy. Aust NZJ Med 16: 409–412, 1986
5. Echt DS, Liebson PR, Michell BR, Peters RW: *et al.* and the CAST investigators. Mortality and morbidity in patients receiving Encainide, Flecainide or Placebo: The Cardiac Arrhythmia Suppression Trial. New Engl J Med 324: 781–788, 1991
6. Goldstein RE, Boccuzzi SJ, Cruess D *et al.*: Diltiazein increases late-onset congestive heart failure in post-infarction patients with early reduction in ejection fraction. Circulation 83: 52–60, 1991
7. Charnock JS, McIntosh GH, Abeywardena MY, Russell GR: Changes in fatty acid composition of the cardiac phospholipids of the cotton-eared marmoset after feeding different lipid supplements. Ann Nutr Metab 29: 83–95, 1985
8. Charnock JS, Abeywardena MY, McLennan PL: Tissue specific differences in the fatty acid composition of the marmoset monkey. Comp Biochem Physiol 92A: 299–304, 1989
9. McLennan PL, Abeywardena MY, Charnock JS: The influence of age and dietary fat in an animal model of Sudden Cardiac Death. Aust NZ J Med 19: 1–5, 1989
10. McLennan PL, Abeywardena MY, Charnock JS: Reversal of the arrhythmogenic effects of long-term saturated fatty acid intake by dietary n-3 and n-6 polyunsaturated fatty acids. Am J Clin Nutr 51: 53–59, 1990
11. McLennan PL, Abeywardena MY, Charnock JS: Dietary fish oil prevents ventricular fibrillation following occlusion and reperfusion. Am Heart J 116: 709–711, 1988

12. Kromhout D, Bosschieter EB, Coulander C ce L: The inverse relation between fish consumption and 20-year mortality from coronary heart disease. New Engl J Med 312: 1205–1209, 1985
13. Burr ML, Fehily AM, Gilbert JF, Rogers S *et al.*: Effects of changes in fat, fish and fibre intakes on death and myocardial reinfarction. Diet and Reinfarction Trial (DART). Lancet ii, 757–761, 1989
14. Pitt B: Sudden Cardiac Death: Role of left ventricular dysfunction. Ann N.Y. Acad Sci 382: 218–222, 1983
15. Lesch M, Kehoe RF: Predicability of Sudden Cardiac Death. New Engl J Med 310: 255–257, 1984
16. Charnock JS, Abeywardena MY, McLennan PL: Dietary lipids, membrane composition and cardiac function. Colloque INSERM 195: 135–144, 1989
17. Hetzel BS, Charnock JS: Aging and the heart. In: DM Prinsley and HH Sandstead (eds.) Nutrition and Aging. Alan R Liss Inc New York 1990, pp 21–37
18. McLennan PL, Abeywardena MY, Charnock JS, McMurchie EJ: Dietary lipid modulation of myocardial β-adrenergic mechanisms, Ca^{2+}-dependent automaticity and arrhythmogenesis in the marmoset. J Cardiovasc Pharmacol 10: 293–300, 1987
19. Abeywardena MY, Fischer S, Schweer H, Charnock JS: *In vivo* formation of metabolites of prostaglandin I_2 and I_3 in the marmoset monkey (Callithrix jacchus) following dietary supplementation with tuna fish oil. Biophys Acta 1003: 161–166, 1989
20. McIntosh GH, Barnden LR, Buttfield IH, Charnock JS: Gated blood pool studies of cardiac function in the rat and marmoset. J Nucl Med 24: 728–731, 1983
21. Charnock JS, McLennan PL, McIntosh GH, Barnden LR, Buttfield IH: Radionuclide angiographic study of the influence of dietary lipid supplements on cardiac function in the marmoset. Cardiovasc Res 21: 369–376, 1987
22. McLennan PL, Bridle TM, Abeywardena MY, Charnock JS: Dietary lipid modification of ventricular fibrillation threshold in the marmoset monkey. J Mol Cell Cardiol 22: p viii (Abstract #23) 1990
23. Abeywardena MY, McLennan PL, Charnock JS: Effects of variable lineolate intake on aortic PGI_2-like activity and fatty acid composition: Differences between the rat and marmoset monkey. Prostaglandins Leukotrienes and Med 25: 209–221, 1986
24. Abeywardena MY, McLennan PL, Charnock JS: Differential effects of dietary fish oil on myocardial prostaglandin I_2 and thromboxane A_2 production. Am J Physiol 260: H379–385, 1991
25. Charnock JS, Abeywardena MY, Poletti VM, McLennan PL: Differences in fatty acid composition of various tissues of the marmoset monkey after different lipid supplemented diets. Comp Biochem Physiol 101A: 387–393, 1992
26. Hokin LE: Receptors and phosphoinositide-generated second messengers. Ann Rev Biochem 54: 205–235, 1985
27. Chap H, Simon MF, Plantavid M, Mauco G, Flauvel and Douste-Blazy L: Platelet Phosphoinositides. Colloque INSERM 152: 21–32, 1987
28. Parratt JR, Wainwright CL, Coker SJ, Zeitlin IJ: Eicosanoids and cardiac arrhythmia. Biomed Biochim Acta 47: 813–818, 1988
29. Burr ML, Fehily AM: Fatty fish and heart disease: A randomized controlled trial. In Health Effects of ω3 Polyunsaturated Fatty Acids in Seafoods. World Rev Nutr Diet 66: 306–312, 1991
30. Dreizen S, Levy BM, Bernick S: Diet-induced atherosclerosis in the marmoset. Proc Soc Exptl Biol Med 143: 1218–1223, 1973
31. Abbey M, Clifton PM, McMurchie EJ, McIntosh GH, Nestel PJ: The effect of a high fat/cholesterol diet with or without eicosapentaenoic acid on plasma lipids, lipoproteins and lipid transfer activity in the marmoset. Atheroscleros 81: 163–174, 1990

Molecular and Cellular Biochemistry **116**: 27–31, 1992.

The substrate specificity of phosphoinositide-phospholipase C in rat heart sarcolemma

Johanna T.A. Meij and Vincenzo Panagia
Division of Cardiovascular Sciences, St. Boniface General Hospital Reserach Centre and Department of Anatomy, Faculty of Medicine, University of Manitoba, Winnipeg, Canada

Abstract

In rat cardiac sarcolemmal membranes a phosphoinositide-specific phospholipase C (PLC) was found to be present. The enzyme hydrolysed exogenous [^{3}H-]phosphatidylinositol 4,5-biphosphate ([^{3}H-]PtdIns(4,5)P_2) in an optimized assay mixture containing 15 μg SL protein, 100 mM NaCl, 1 mM free Ca^{2+}, 14 mM Na-cholate and 20 μM [^{3}H-]PtdIns-(4,5)P_2 (400–500 dpm/μl) in 30 mM HEPES-Tris buffer (pH 7.0). The average specific activity was 9.14 ± 0.55 nmol·mg^{-1}·2.5 min^{-1}. The addition of Mg^{2+} to the assay mixture did not change PLC activity but increased the relative amounts of dephosphorylated inositol products. In the absence of Na^+ and at a low Ca^{2+} concentration (0.3 μM), Mg^{2+} also enhanced the intraSL levels of PtdIns4P and PtdIns, and, moreover, inhibited PLC activity (IC_{50}~0.07 mM). PtdIns4P seemd to be a good substrate for the rat SL PLC (23.07 ± 1.57 nmol·mg^{-1}·2.5 min^{-1}) whereas PtdIns was hydrolysed at a very low rate (0.36 ± 0.08 nmol·mg^{-1}·2.5 min^{-1}). Unlike PtdIns(4,5)P_2, PLC-dependent PtdIns4P and PtdIns hydrolysis was not inhibited by Ca^{2+} concentrations over 1 mM. The possibility of distinct isozymes being responsible for the different hydrolytic activities is discussed. (Mol Cell Biochem **116**: 27–31, 1992).

Key words: phospholipase C, phosphoinositides, phosphatidylinositol 4,5-bisphosphate, phosphatidylinositol 4-phosphate, rat heart, sarcolemma

Abbreviations: DAG – *sn*-1,2-diacylglycerol; EGTA – ethyleneglycol-O,O'-bis(aminoethyl)-N,N,N',N',-tetraacetic acid; Ins(1,4,5)P_3 – inositol 1,4,5-trisphosphate; Ins$P$ – inositol monophosphate (unidentified isomer); InsP_2 – inositol bisphosphate (unidentified isomer); InsP_3 – inositol trisphosphate (unidentified isomer); InsP_x – any inositol phosphate; PLC – phospholipase C; PtdIns – phosphatidylinositol; PtdIns(4,5)P_2 – phosphatidylinositol 4,5-bisphosphate; PtdIns4P – phosphatidylinositol 4-monophosphate; SL – sarcolemma

Introduction

In many tissues a membrane-bound phospholipase C (EC 3.1.4.10) specific for phosphoinositides [phosphatidylinositol (PtdIns), phosphatidylinositol 4-monophosphate (PtdIns4P), phosphatidylinositol 4,5-bisphosphate (PtdIns(4,5)P_2)] is involved in the trans-membrane conduction of signals from cell surface receptors [1]. Receptor-stimulated hydrolysis of PtdIns (4,5)P_2 leads to the production of two second messengers, inositol 1,4,5-trisphosphate (Ins(1,4,5)P_3) and *sn*-1,2-diacylglycerol (DAG) [1]. In the heart, Ins

Address for offprints: V. Panagia, Division of Cardiovascular Sciences, St. Boniface G.H. Research Centre, R3020 – 351 Tache Avenue, Winnipeg (MB), Canada R2H 2A6

$(1,4,5)P_3$ is able to mobilize Ca^{2+} from the sarcoplasmic reticulum, and DAG activates protein kinase C [2].

In contrast to the multitude of studies reporting receptor-dependent activation of phosphoinositide-phospholipase C (PLC) in intact cardiomyocytes [2, 3], little information is available on the regulatory mechanism and the catalytic properties of this enzyme. Studies in different heart preparations suggest the association of PLC with a regulatory GTP-binding protein, which links the enzyme to a stimulated receptor [4–6]. Some biochemical properties of a PtdIns-hydrolysing PLC activity have been described in rat heart homogenate [7], whereas cytoplasm- as well as sarcolemma-associated PLC activities utilizing PtdIns4*P* and PtdIns $(4,5)P_2$ as substrate have been investigated in guinea-pig heart [8].

The present study examines some characteristics of the phosphoinositide-PLC activity which is located at the surface membrane of rat cardiomyocytes and therefore is the one most likely involved in receptor signal transduction. According to current views, the receptor-coupled PLC hydrolyses only, or preferentially, PtdIns-$(4,5)P_2$ [1]. With purified PLC enzymes, however, it has been shown that under certain conditions PtdIns and PtdIns4*P* may also serve as substrates [9]. The focus of this study is, therefore, on establishing whether the phosphoinositide-PLC activity of rat heart sarcolemma (SL) is specific for PtdIns$(4,5)P_2$ or is also able to hydrolyse other phosphoinositides under the same assay conditions.

Materials and methods

Materials

Phosphatidylinositol 4,5-bisphosphate, [inositol-2-^{3}H (N)]- (136.9 GBq/mmol), phosphatidylinositol 4-phosphate, [inositol-2-^{3}H(N)]- (118.4 GBq/mmol), and phosphatidylinositol, L-α-[myo-inositol-2-^{3}H(N)]- (192.4 GBq/mmol) were obtained from DuPont/New England Nuclear (Mississauga Ont., Canada). Phosphatidylinositol 4,5-bisphosphate, phosphatidylinositol 4-phosphate and phosphatidylinositol were purchased from Sigma Chem. Comp. (St. Louis MO, U.S.A.). Dowex 1X8 (formate form, 100–200 mesh) was from BioRad Labs. (Mississauga Ont., Canada). CytoScint™ is a product of ICN Biomedicals Inc. (Mississauga Ont., Canada). All other chemicals were of analytical grade.

Preparation of cardiac sarcolemmal membranes

Male Sprague-Dawley rats weighing 300–400 g were used in all experiments. After decapitation of the animals, the hearts were quickly excised, the atria were removed and the ventricular tissue was used to isolate membranes by the method of Pitts [10]. This membrane fraction is highly enriched in SL [11], as is also indicated by the presence of several receptors [12–14], including the α_1-adrenoreceptor [14] which is coupled to phosphoinositide-PLC [2]. The final pellet was resuspended in 250 mM sucrose/10 mM histidine (pH 7.4) and rapidly frozen in liquid N_2. Protein measurements were done according to Lowry *et al.* [15] using bovine serum albumin as standard.

Phosphoinositide phospholipase C assay

Substrate was prepared for each experiment by mixing an aliquot of [^{3}H-]PtdIns$(4,5)P_2$, [^{3}H-]PtdIns4*P* or [^{3}H-]PtdIns with an aliquot from the stock solution (in chloroform) of the respective cold substrate. The mixture was evaporated to dryness under a stream of N_2 and redissolved in 10% Na-cholate (232 mM) to a substrate concentration of 331.5 μM. The substrate solution was kept under N_2 gas overnight at 0–4° C and was, shortly before addition, diluted to 160 μM substrate/112 mM Na-cholate. A sample was taken to determine the specific activity.

Unless stated otherwise, the reaction mixture contained 30 mM HEPES-Tris (pH 7.0), 100 mM NaCl, 2 mM EGTA, 3.13 mM $CaCl_2$ ([free Ca^{2+}] = 1 mM), 15 μg SL protein, 14 mM Na-cholate and 20 μM [^{3}H-]-PtdIns$(4,5)P_2$, [^{3}H-]PtdIns4*P* or [^{3}H-]PtdIns (400–500 dpm/μl) in a final volume of 40 μl. The samples were incubated at 37° C for 2.5 min and the reaction was terminated by the addition of 144 μl ice-cold $CHCl_3$/CH_3OH/HCl (1:2:0.2, v/v) to each sample. Blanks were performed under identical conditions except that SL membranes were added after stopping the reaction. Phases were separated by adding 48μl of 2 M KCl and 48μl of $CHCl_3$ [16]. After mixing for 30 sec and 5 min centrifugation in a microcentrifuge (Eppendorf), the upper phase was aspired and applied to a 300μl column of Dowex 1X8 (formate form, 100–200 mesh). After the columns were rinsed with water, the inositol phosphates were eluted according to Berridge *et al.* [17] with 1 ml of each gradient step, and quantitated by liquid scintillation counting in 10 volumes of CytoScint™. In some experiments the [^{3}H-]phosphoinositides in the lower phase were analyzed as described before [18].

Free calcium concentrations were determined with the computer program Skin (copyright 1984, John Axline Vr).

Results

Characterization of rat cardiac sarcolemmal PtdIns(4,5)P_2-phospholipase C

To develop a standard procedure for the measurement of phosphoinositide-PLC in rat cardiac SL membranes the reaction was characterized for its protein-, time-, detergent-, pH- and cation-dependence. At pH 7.0 and pCa 3 the protein dependence was linear up to 25 μg. Using 15 μg SL protein per assay, the time-course of InsP_3 production (Fig. 1) showed linearity up to 5 min incubation, followed by a plateau which persisted up to 30 min. InsP_2 and Ins$P$ formation increased gradually up to 20 min, and maximally consisted of 33% of the total inositol phosphates (Fig. 1). Inclusion of 15 mM LiCl (an inhibitor of Ins monophosphatase [17]) in the incubation mixture did not improve the recovery of inositol phosphates. The detergent (Na-cholate) concentration was tested over a range of 0 to 28 mM. In the absence of cholate no appreciable PtdIns(4,5)P_2 hydrolysis occurred, whereas the optimum was found to be 14 mM (0.6%, w/v), which is close to the critical micellar concentration (CMC) of cholate [19]. Other detergents were not tested. pH-Dependence experiments showed an optimum at pH 7.0. The production of inositol phosphates was linear with the concentration of substrate (PtdIns(4,5)P_2) over the range tested (0.1 to 200 μM). 20 μM was chosen as substrate concentration for the standard assay.

Several cations were examined for their effect on the PtdIns(4,5)P_2 hydrolysis. Na^+ (0–200 mM) stimulated the hydrolysis of PtdIns(4,5)P_2 dose-dependently, to a maximum of 4-fold of the corresponding value in the absence of Na^+, at concentrations ≥ 100 mM. The optimum Ca^{2+} concentration appeared to be 1 mM (pCa 3). Addition of Mg^{2+} to the standard reaction mixture did not alter PLC activity. However, Mg^{2+} changed the percent distribution of the [^{3}H-]inositol phosphates, typically (1.5 mM Mg^{2+}) from 83.0%, 14.7% and 2.3% to 38.5%, 46.6% and 14.9% for InsP_3, InsP_2, and Ins$P$,

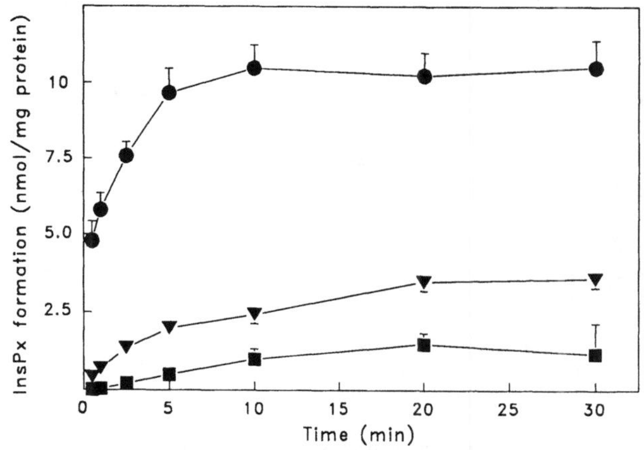

Fig. 1. Time-course of the formation of different inositol phosphates from phosphatidylinositol 4,5-bisphosphate in rat cardiac sarcolemmal membranes. SL membranes were incubated for the indicated times with 20 μM [^{3}H-]PtdIns(4,5)P_2, and inositol phosphates (InsP_x) were separated as described under Materials and methods. Points represent the means, bars the SEM of triplicate experiments in 12 different SL preparations. -●- InsP_3, -▼- InsP_2, -■- Ins$P$.

respectively. A dose-dependent inhibition (0 to 5 mM Mg^{2+}, IC_{50}~0.07 mM) of PtdIns(4,5)P_2 hydrolysis by Mg^{2+} became apparent in the absence of Na^+ and at a low Ca^{2+} concentration (0.3 μM). That was accompanied by alterations of the SL [^{3}H-]phosphoinositide levels which, typically (0 Na^+, 0.3 μM Ca^{2+} and 1.5 mM Mg^{2+}) shifted from 85.8%, 8.4% and 5.8% to 64.2%, 27.6% and 8.2% for PtdIns(4,5)P_2, PtdIns4P, and PtdIns, respectively; the percentage of dephosphorylated inositol products increased under these conditions. Therefore, Mg^{2+} was not included in the standard phosphoinositide-PLC assay.

Substrate specificity of the rat cardiac sarcolemmal phospholipase C

To establish the specificity of the phosphoinositide-PLC activity, PtdIns4P and PtdIns were tested as substrates under the standard assay conditions. As it can be seen in Fig. 2, at any time-point of incubation the hydrolysis of 20 μM PtdIns4P (23.07 ± 1.57 nmol·mg^{-1}·2.5 min^{-1}) was about double that of 20 μM PtdIns(4,5)P_2 (9.14 ± 0.55 nmol·mg^{-1}.2.5 min^{-1}), whereas PtdIns was hydrolysed at a very low rate (0.36 ± 0.08 nmol·mg^{-1}.2.5 min^{-1}). As expected, the main products from PtdIns4P and PtdIns hydrolysis were InsP_2, and Ins$P$, respectively (data not shown).

The effect of Ca^{2+} on the hydrolysis of PtdIns4P and

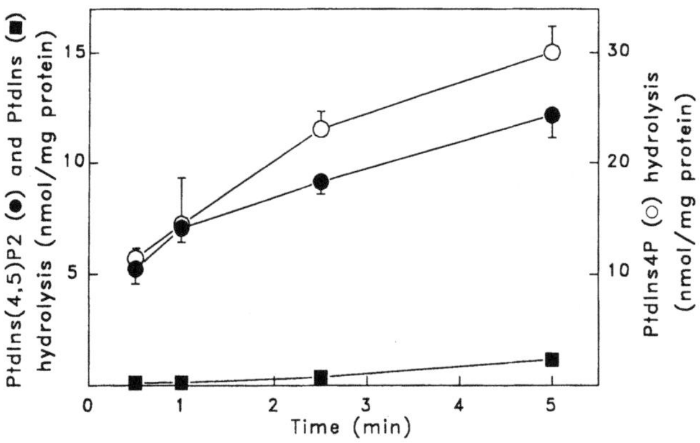

Fig. 2. Time-course of the hydrolysis of different phosphoinositides in rat cardiac sarcolemmal membranes. SL membranes were incubated for the indicated times with 20 μM [^{3}H-]PtdIns(4,5)P_2, [^{3}H-]PtdIns4*P* or [^{3}H-]PtdIns, respectively, as described under Materials and methods. Points represent the means, bars the SEM of triplicate experiments in 3–12 different SL preparations.

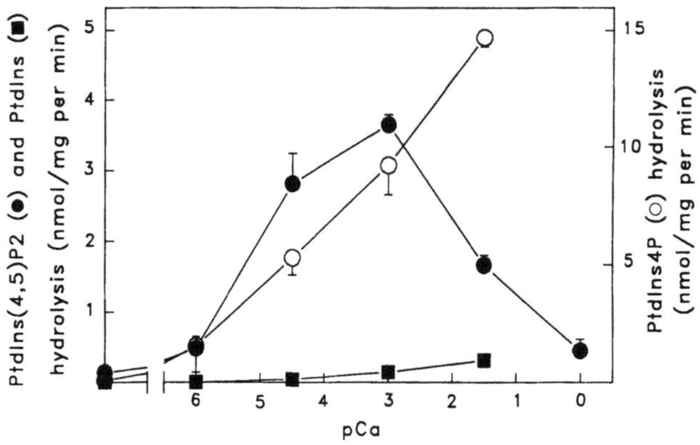

Fig. 3. Effect of the free Ca^{2+} concentration on the hydrolysis of different phosphoinositides in rat cardiac sarcolemmal membranes. SL membranes were incubated for 2.5 min with 20 μM [^{3}H-]PtdIns(4,5)P_2, [^{3}H-]PtdIns4*P* or [^{3}H-]PtdIns, respectively, as described under Materials and methods. Points represent the means, bars the SEM of triplicate experiments in 3 different SL preparations.

PtdIns was different from that on PtdIns(4,5)P_2 hydrolysis (Fig. 3). Whereas Ca^{2+} concentrations over 1 mM were inhibitory for PtdIns(4,5)P_2 hydrolysis, PtdIns4*P* and PtdIns hydrolysis was further stimulated.

Discussion

The present study describes some properties of the phosphoinositide-PLC activity associated with rat heart SL. The results related to pH, Ca^{2+}, and Na^+ were similar to those reported for the guinea-pig SL enzyme [8]. At variance, however, in the absence of Na^+ and at low Ca^{2+} concentrations the rat SL PLC was inhibited by Mg^{2+}. In particular, Mg^{2+} increased the percentage of dephosphorylated inositol products and enhanced the intraSL levels of PtdIns4*P* and PtdIns, which were found to be a good and a poor substrate for SL PLC, respectively (Fig. 2). PtdIns(4,5)P_2 dephosphorylation to PtdIns4*P* (giving InsP_2 upon hydrolysis) and to PtdIns (reducing the amount of substrate available to PLC) could explain both Mg^{2+}-induced effects, namely, PLC inhibition and the altered formation of inositol phosphates. On the other hand, in different cell types Mg^{2+} stimulated not only the dephosphorylation of membranal PtdIns(4,5)P_2 [16, 20] but also the Ins(1,4,5)P_3 5-phosphatase activity, which is largely confined to the plasma membrane [21] and increases InsP_2 and Ins*P* formation [20–22]. Therefore, our results can also be explained by a combination of PtdIns(4,5)P_2 and InsP_3 dephosphorylation.

In both rat and guinea-pig heart, SL PLC utilizes PtdIns(4,5)P_2 and PtdIns4*P* as substrate. In rat, however, PtdIns4*P* seemed to be a better substrate than PtdIns(4,5)P_2 when tested under the same assay conditions. In addition, the Ca^{2+}-dependence of PtdIns4*P* hydrolysis was different from that of PtdIns(4,5)P_2 hydrolysis (Fig. 3), a distinction not observed in guinea-pig [8]. Accordingly, a subtle diversity may exist among species with respect to the hydrolysis of intraSL phosphoinositides and thus it may be advisable to define the characteristics of the PLC activity in each species. Also, the observed difference in Ca^{2+}-dependence of the PtdIns4*P* and PtdIns(4,5)P_2 hydrolysis could be interpreted as an indication for the occurrence either of separate isoenzymes or of distinctly regulated sites on the same PLC enzyme molecule. Our recent findings that, unlike PtdIns4*P*, PtdIns(4,5)P_2 hydrolysis was significantly depressed in post-infarcted failing hearts [23] support this view.

Compared to the polyphosphoinositides, PtdIns seemed to be a substrate of negligible importance for both rat as well as guinea-pig SL PLC [8]. The previously observed PtdIns-hydrolysing activity in rat heart had a sharp pH-optimum at 5.5 and was associated with the soluble fraction [7]. Four soluble isoforms were identified in bovine heart cytosol and found capable of hydrolysing all three phosphoinositides [24]. In this regard, four PLC isozymes have been classified in mammalian tissues on the basis of their deduced amino acid sequences and immunological cross-reactivity. Different amounts of PLC-β_1, -γ_1, and -δ_1 forms were found in rat heart supernatants [25], but it is still unknown whether or not the PLC form(s) expressed in SL membranes and in the cytosol are the same, and which is the role of the cytosolic enzymes in signal transduction. Future studies should disclose the relevance of phosphoinositide-PLC that are not localized at the SL membrane or are not PtdIns(4,5)P_2-specific.

Acknowledgements

This research was supported by a grant to V. Panagia from the Medical Research Council of Canada. J.T.A. Meij was a Fellow of the Medical Research Council of Canada.

References

1. Meldrum E, Parker PJ, Carozzi A: The PtdIns-PLC superfamily and signal transduction. Biochim Biophys Acta 1092: 49-71, 1991
2. Meij JTA, Panagia V: Catecholamines and heart disease: status of phospholipid signalling pathways. In: PK Ganguly (ed.) Catecholamines and heart disease. CRC Press, Boca Raton, 1991, pp 245–266
3. Brown JH, Jones LG: Phosphoinositide metabolism in the heart. In: JW Putney (ed.) Phosphoinositides and receptor mechanisms. Alan R. Liss Inc., New York, 1986, pp 245–270
4. Masters SB, Martin MW, Harden TK, Brown JH: Pertussis toxin does not inhibit muscarinic-receptor-mediated phosphoinositide hydrolysis of calcium mobilisation. Biochem J 227: 933–937, 1985
5. Renard D, Poggioli J: Mediation by GTPγS and Ca^{2+} of inositol trisphosphate generation in rat heart membranes. J Mol Cell Cardiol 22: 13–22, 1990
6. Meij JTA, Bezstarosti K, Panagia V, Lamers JMJ: Phorbol ester and the actions of phosphatidylinositol 4,5-bisphosphate specific phospholipase C and protein kinase C in microsomes prepared from cultured cardiomyocytes. Mol Cell Biochem 105: 37–47, 1991
7. Schwertz DW, Halverson J, Palmer JW, Feinberg H: Characterization of phospholipase C-mediated phosphatidylinositol degradation in rat heart ventricle. Arch Biochem Biophys 253: 388–398, 1987
8. Edes I, Kranias EG: Characterization of cytoplasmic and membrane-associated phosphatidylinositol 4,5-bisphosphate phospholipase C activities in guinea pig ventricles. Basic Res Cardiol 85: 78–87, 1990
9. Rhee SG, Suh P-H, Ruy S-H, Lee SY: Studies of inositol phospholipid-specific phospholipase C. Science 244: 564–550, 1989
10. Pitts BJR: Stoichiometry of sodium-calcium exchange in cardiac sarcolemmal vesicles. J Biol Chem 245: 6232–6235, 1979
11. Panagia V, Taira Y, Ganguly PK, Tung S, Dhalla NS: Alterations in phospholipid N-methylation of cardiac subcellular membranes due to experimentally-induced diabetes in rats. J Clin Invest 86: 777–784, 1990
12. Dixon IMC, Lee S-L, Dhalla NS: Nitrendipine-binding in congestive heart failure due to myocardial infarction. Circ Res 66: 782–788, 1990
13. Zhao D, Dhalla NS: [^{35}S]ATPγS binding sites in the purified heart sarcolemma membrane. Am J Physiol 258: C185–C188, 1990
14. Vetter R, Dai J, Mesaeli N, Panagia V, Dhalla NS: Role of sulfhydryl groups in phospholipid methylation reactions of cardiac sarcolemma. Mol Cell Biochem 103: 85–96, 1991
15. Lowry OH, Rosenbrough NJ, Fare AL, Randall RJ: Protein measurement with the folin phenol reagent. J Biol Chem 193: 325–334, 1951
16. Jackowski S, Rettenmeier CW, Sherr CJ, Rock CO: A guanine nucleotide-dependent phosphatidylinositol 4,5-diphosphate phospholipase C in cells transformed by the v-*fms* and v-*fes* oncogenes. J Biol Chem 261: 4978–4985, 1986
17. Berridge MJ, Downes CP, Hanley MR: Lithium amplifies agonist-dependent phosphatidylinositol responses in brain and salivary glands. Biochem J 206: 587–595, 1985
18. Meij JTA, Lamers JMJ: Phorbolester inhibits α_1-adrenoceptor mediated phosphoinositide breakdown in cultured neonatal rat ventricular myocytes. J Mol Cell Cardiol 21: 661–668, 1989
19. Neugebauer JM: Detergents: an overview. Methods in Enzymol 182: 239–253, 1990
20. Chien MM, Cambier JC: Divalent cation regulation of phosphoinositide metabolism. Naturally occurring B lymphoblasts contain a Mg^{2+}-regulated phosphatidylinositol-specific phospholipase C. J Biol Chem 265: 9201–9207, 1990
21. Downes CP, Hawkins PT, Stephens L: Principles of inositol lipid-mediated signalling. Identification of the stimulated reaction in intact cells, its substrate supply and the metabolism of inositol phosphates. In: RH Michell, AH Drummond, CP Downes (eds.) Inositol lipids in cell signalling. Academic Press, San Diego, 1989, pp 3–38
22. Downes CP, Mussat MC, Michell RH: The inositol trisphosphate phosphomonoesterase of the human erythrocyte membrane. Biochem J 203: 169–177, 1982
23. Meij JTA, Afzal N, Panagia V, Dhalla NS: Changes in phospholipase C activity in congestive heart failure. J Mol Cell Cardiol 23 (Suppl. III), S.67 (Abstr.), 1991
24. Low MG, Weglicki WB: Resolution of myocardial phospholipase C into several forms with distinct properties. Biochem J 215: 325–334, 1983
25. Rhee SG, Kim H, Suh P-H, Choi WC: Multiple forms of phosphoinositide-specific phospholipase C and different modes of activation. Biochem Soc Trans 19: 337–341, 1991

Molecular and Cellular Biochemistry **116**: 33–37, 1992.

Long term incubation of cardiac myocytes with oleic acid and very-low density lipoprotein reduces heparin-releasable lipoprotein lipase activity

Brian Rodrigues, Michael R. Spooner and David L. Severson
Department of Pharmacology & Therapeutics, Faculty of Medicine, The University of Calgary, Calgary, Alberta, Canada T2N 4N1

Abstract

An exogenous [^{3}H]triolein emulsion was hydrolyzed by intact cardiac myocytes with functional LPL located on the cell surface. This surface-bound LPL could be released into the medium when cardiac myocytes were incubated with heparin. Incubation of cardiac myocytes with VLDL, or the products of TG breakdown, oleic acid or 2-monoolein, did not increase LPL activity in the medium. However, incubation of cardiac myocytes with either VLDL or oleic acid for > 60 min did reduce heparin-releasable LPL activity. In the heart, this inhibitory effect of FFA could regulate the translocation of LPL from its site of synthesis in the cardiac myocyte to its functional site at the capillary endothelium. (Mol Cell Biochem **116:** 33–37, 1992).

Key words: lipoprotein lipase, cardiac myocytes, free fatty acids

Abbreviations: LPL – lipoprotein lipase, TG – triacylglycerol, FFA – free fatty acids, VLDL – very-low density lipoprotein

Introduction

Lipoprotein lipase (LPL) is located at the luminal surface of the vascular endothelium and catalyzes the hydrolysis of the triacylglycerol (TG) component of circulating lipoproteins to yield free fatty acids (FFA) and 2-monoacylglycerols [1]; these lipolytic products are then available for tissue utilization. Endothelial cells cannot synthesize LPL [2] so in the adult heart, LPL is synthesized and processed by glycosylation in cardiac myocytes [2] and then translocated to its functional site at the capillary endothelium [3]. Heparin can displace functional LPL from heparan sulfate proteoglycan binding sites on the surface of vascular endothelial cells in whole perfused hearts [4]. The addition of heparin to incubations of isolated cardiac myocytes from adult rat hearts also resulted in the rapid release of LPL into the medium [5–7], presumably by displacement of the enzyme from heparan sulfate proteoglycan binding sites on the surface of myocardial cells. This heparin-releasable cell surface pool of LPL would be the precursor of the functional endothelium bound enzyme. Recent evidence has suggested that at least a portion of the heparan sulfate proteoglycan molecules that bind LPL are

Address for offprints: D. Severson, Faculty of Medicine, The University of Calgary, 3330 Hospital Drive, N.W., Calgary, Alberta T2N 4N1, Canada covalently linked to the cell surface of cardiac myocytes by a glycosyl-phosphatidylinositol membrane anchor [6]. The rapid heparin-induced release of LPL into the medium of cardiac myocytes was followed by a second, slower phase of secretion that was dependent on protein synthesis and microtubular function [5, 7], and thus likely represents the intracellular processes that replace the LPL that had been displaced from surface binding sites.

LPL is subject to regulation by a variety of metabolic factors. For example, FFA derived from the degradation of TG-rich lipoprotein such as chylomicrons and very low density lipoproteins (VLDL) displaced LPL bound to cultured pig aortic endothelial cells [8–10]; it was proposed that this mechanism would prevent the excessive delivery of lipolytic products to tissue (parenchymal) cells by limiting the subsequent degradation of TG-rich lipoproteins. The observation that elevations in circulating FFA correlated with increases in plasma LPL activity in humans [11] is consistent with this proposal. The objective of our investigation was to determine if incubation of cardiac myocytes with VLDL or lipolytic products (oleic acid or 2-monoolein) would displace LPL into the medium or reduce heparin-releasable LPL activity in some manner, so that the translocation of LPL to its functional site at the vascular endothelium in the whole heart would be compromised, thus providing an additional mechanism to regulate the degradation of TG-rich lipoproteins under conditions of enhanced lipolysis.

Experimental procedures

Preparation and incubation of cardiac myocytes. Calcium-tolerant cardiac myocytes were isolated from rat hearts (ventricles) by the procedure of Kryski *et al.* [12] except that the collagenase concentration in solution B was reduced to 100 U/ml. The yield of myocytes and viability (% of rod shaped cells excluding Trypan blue, ranging from 75–85%) was determined microscopically. Cardiac myocytes were finally suspended in Joklik-minimal essential medium supplemented with 1.2 mM $MgSO_4$, 1 mM DL-carnitine, 1.5 mM $CaCl_2$, and 1% (w/v; 0.15 mM) fatty acid-free bovine serum albumin to a cell density of 0.4×10^6 cells/ml. Cardiac myocytes were incubated at 37° C under an atmosphere of 95% O_2-5% CO_2 with the additions (heparin, VLDL, oleic acid, or 2-monoolein) as indicated in the text; 1 ml aliquots were removed at various times and centrifuged for 10 sec at 15,000 g in an Eppendorf microcentrifuge [5]. The supernatant (medium) was decanted and stored at − 80° C for subsequent measurements of LPL activity [5].

LPL assay. LPL activity in the medium of cardiac myocytes was determined by measuring the hydrolysis of a sonicated [^{3}H]triolein substrate emulsion and is routinely expressed as nmol oleate released/h/10^6 cells [13]. The standard assay conditions were: 0.6 mM glycerol tri[9,10-^{3}H]oleate (1 mCi/mmol), 25 mM Pipes (pH 7.5), 0.05% (w/v) albumin, 50 mM $MgCl_2$, 2% (v/v) heat-inactivated chicken serum [13] to provide the specific LPL activator apolipoprotein CII, and 100 μl of medium in a total volume of 400 μl. The release of radiolabelled oleate was measured by liquid-liquid partitioning after a 30 min incubation at 30° C [13]. All LPL assays were performed in duplicate under conditions where the reaction was linear with respect to time and the volume of medium that was assayed.

Hydrolysis of exogenous [^{3}H]triolein by intact cardiac myocytes. An appropriate quantity of [^{3}H]triolein (glycerol tri [9,10-^{3}H]oleate) solution in toluene was dried under nitrogen. Joklik medium as detailed above was added to the tube and the triolein was dispersed by sonication (75 W; Braun Sonic 1510) for 4×30 sec at 4° C, followed by the addition of heat-inactivated chicken serum. Myocytes were preincubated for 10 min in the absence and in the presence of 5 U/ml heparin, centrifuged and the respective cell pellets were resuspended in fresh Joklik medium. An equal volume of these myocyte cell suspensions (0.4×10^6 cells/ml) was added to the triolein substrate suspension to give a final concentration of 1.5 mM [^{3}H]triolein (approx. 16,000 dpm/nmol) and 4% (v/v) chicken serum, and then incubated for various times at room temperature (22° C) under an atmosphere of 95% O_2 and 5% CO_2. An incubation temperature of 22° C was chosen so that the rate at which LPL was replaced at the cell surface after preincubation with heparin would be reduced [7]. At times ranging from 5–70 min, 1 ml of the myocyte suspension was removed and centrifuged at 15,000 g for 15 sec. The presence of hydrolysed [^{3}H]oleate in the medium was determined by adding 3 ml of a fatty acid extraction solution [13] to 400 μl of supernatant. After vortex mixing and centrifugation to allow for phase separation, the content of radioactive sodium [^{3}H]oleate in a sample (0.5 ml) of the upper phase was determined by liquid scintillation counting [13].

Materials. Collagenase (Worthington Type II) was obtained from Technicon Canada Ltd., Richmond, B.C. and heparin (hepalean; 1000 U/ml was purchased from Organon Teknika, Toronto, Ont. Joklik-minimal essential medium was purchased from Gibco Canada, Burlington, Ont. Very-low density lipoprotein (VLDL; 1 mg protein/ml), 2-monoolein, and oleid acid were obtained from Sigma Chemical Co., St. Louis, MO. A stock solution of 2-monoolein was prepared in 0.01% Triton X-100. An oleic acid:albumin solution was prepared by first making the potassium salt of oleic acid in ethanol, and then adding appropriate amounts to Joklik medium containing 1% (w/v) defatted albumin to obtain a 6:1 molar ratio of oleic acid (0.9 mM) to albumin (0.15 mM).

Results and discussion

It has been presumed that the rapid, heparin-induced release of LPL into the incubation medium of cardiac myocytes [5–7] was due to displacement of the enzyme from heparan sulfate proteoglycan binding sites on the surface of myocardial cells that were similar to sites described for endothelial cells [14] and adipocytes [15], based on similar glycosaminoglycan specificities [5]. The demonstration that a substantial amount of LPL activity was released into the medium of cardiac myocytes after incubation with a phosphatidylinositol-specific phospholipase C [6] provided further evidence that LPL was located at the cell surface of myocytes. Direct evidence for the presence of active LPL on the surface of cardiac myocytes was provided by the observation that exogenous [^{3}H]triolein added to intact myocytes was hydrolyzed in a time-dependent manner (Fig. 1). The hydrolysis of exogenous triolein was completely dependent upon the presence of serum in the incubation medium (results not shown), indicating that the degradation was catalyzed by LPL. Preincubation of myocytes with heparin prior to re-suspension in fresh medium containing exogenous triolein resulted in a markedly reduced rate of hydrolysis (Fig. 1), indicating that the hydrolysis of triolein was catalyzed by a heparin-releasable LPL located on the cell surface as opposed to degradation by intracellular LPL after internalization of the exogenous substrate. A heparin-releasable LPL on the cell surface of cultured mesenchymal cells from neonatal hearts has also been observed to degrade VLDL added to the culture medium [16, 17]. These results with cardiac myocytes (Fig. 1) and cultured cardiac mesenchymal cells [16, 17] are in marked contrast to the suggestion of Pradines-Figueres *et al.* [18] that LPL in cardiac tissue is present in an intracellular 'cryptic' (inactive) state that is aggregated within storage vesicles.

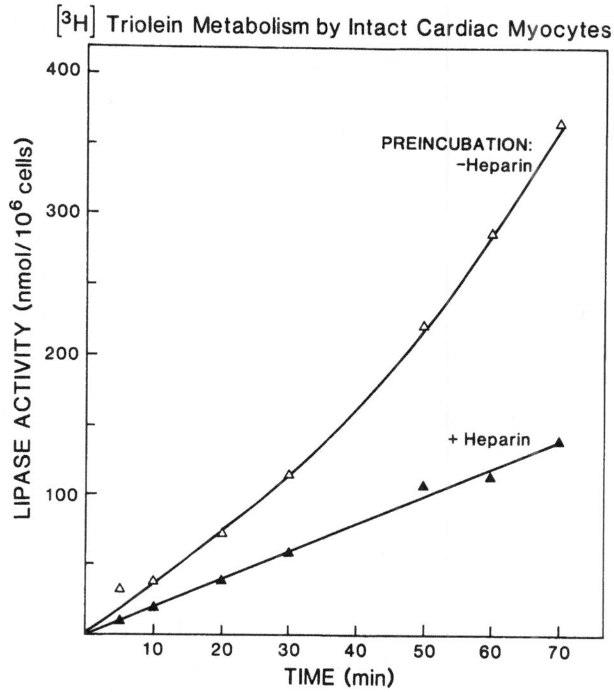

Fig. 1. Cardiac myocytes were preincubated for 5 min in the absence (△) and in the presence (▲) of 5 U/ml heparin, centrifuged and resuspended into fresh medium containing 1.5 mM [^{3}H[triolein and 4% (v/v) chicken serum. At the indicated times of incubation at 22° C, aliquots were removed and the content of [^{3}H]oleate released was determined (lipase activity). Similar results were obtained in three additional experiments.

Cardiac myocytes were next incubated with VLDL in order to determine if lipoprotein degradation by functional surface-bound LPL was associated with displacement of active LPL into the medium, as observed with LPL bound to cultured endothelial cells [8–10]. A brief (5 min) incubation with heparin resulted in a substantial release of LPL activity into the medium of cardiac myocytes (Table 1). In contrast, the addition of 0.02 mg/ml VLDL resulted in no appreciable increase in medium LPL activity. By comparison, the same concentration of VLDL resulted in the release of LPL activity into the medium of cultured endothelial cells that was 80% of heparin-releasable activity [8]. The direct addition of the products of TG hydrolysis, FFA (0.9 mM oleic acid) and 2-monoacylglycerol (0.1 mM monolein), also did not increase medium LPL acivity (Table 1), once again in contrast to results obtained with cultured endothelial cells where oleic acid resulted in the release of LPL (measured as either catalytic activity or as radioiodinated protein) into the culture medium that was 70–80% of the heparin-releasable LPL [8, 9]. This discrepancy in the ability of VLDL or its lipolytic product, FFA, to displace LPL bound to the cell surface could reflect differences in the nature of the heparan sulfate proteoglycan binding sites for LPL on the surface of endothelial cells and cardiac myocytes. In particular, it should be noted that experiments with endothelial cells first required the binding of an exogenous, purified milk LPL preparation to cultured endothelial cells from pig aorta [8–10], since endothelial cells do not synthesize LPL [2] and thus have no endogenous LPL on their cell surface. It is possible, therefore, that exogenous milk LPL may bind to different cell surface sites on a cultured cell than endogenous LPL produced by a parenchymal cell like the cardiac myocytes. Perfusion of hearts with oleic acid has also been shown not to release significant quantities of LPL activity into the perfusate (B. Rodrigues and D.L. Severson, unpublished observations).

Since VLDL or lipolytic products had no effect on LPL release into the medium of cardiac myocytes during a short-term incubation of 5 min (Table 1), the effect of a longer term incubation on heparin-releasable activity was studied next (Fig. 2). Cardiac myocytes were first preincubated with heparin to displace surface-bound LPL into the medium (Fig. 2, A and C). After centrifugation and resuspension into fresh Joklik medium, the heparin-preincubated myocytes were incubated for up to 120 min with a second addition of heparin (Fig. 2, B and D). Under these conditions, the slow heparin-induced rate of LPL release into the medium is a function of the intracellular processes that replace the LPL that was displaced from the cell surface by the first

Table 1. Effect of VLDL, oleic acid or 2-monoolein on LPL activity in the medium of cardiac myocytes. Myocytes were incubated at 37° C for 5 min with the indicated additions; after centrifugation, LPL activity in the supernatant (medium) was determined. Results are the mean ± SEM for the number of experiments indicated in parentheses

Additions to incubation	LPL activity in medium ($nmol/h/10^6$ cells)	
None	73 ± 11	(8)
Heparin (5 U/ml)	543 ± 43	(8)
VLDL (0.02 mg/ml)	116 ± 14	(3)
Oleic acid (0.9 mM)	56	(2)
2-Monoolein (0.1 mM)	109 ± 22	(3)

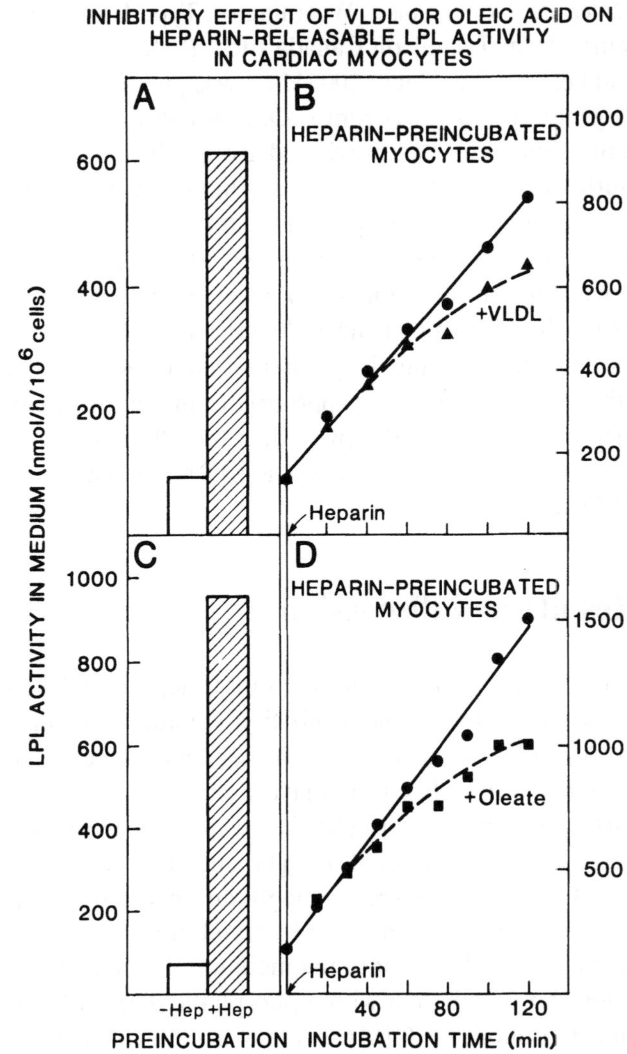

Fig. 2. Cardiac myocytes were preincubated for 10 min in the absence (open bars) or in the presence of 5 U/ml heparin (hatched bars); after centrifugation, LPL activity was determined in the incubation medium (panels A and C). Heparin-preincubated myocytes were resuspended into fresh medium containing 5 U/ml heparin with either no further additions (●, panels B and D), 0.036 mg/ml VLDL (▲, panel B), or 0.9 mM oleate (6:1 molar ratio to albumin; ■, panel D). At the indicated times of incubation, LPL activity in the medium was determined after centrifugation. Similar results were obtained in three additional experiments where heparin-preincubated myocytes were incubated with 0.6 mM oleate and 0.1 mM 2-monoolein.

preincubation with heparin [7]. The presence of VLDL (panel B) or oleic acid (panel D) decreased heparin-releasable LPL activity after an incubation time of about 60 min. Similar results were obtained when cardiac myocytes were incubated with both 0.6 mM oleate (4:1 molar ratio to albumin) and 0.1 mM monoolein. Friedman *et al.* [19] have also observed that a 3–5 h incubation of cultured cardiac mesenchymal cells with VLDL decreased LPL activity. Consequently, the delivery of FFA to cardiac myocytes over a long term (> 60 min) may inhibit either the synthesis, processing or secretion of LPL and thus reduce translocation of the enzyme to its functional site at the capillary endothelium of the whole heart. Such a mechanism would be complementary to short term mechanisms of regulation that operate at the endothelium, such as the reduction in LPL activity by FFA due to product inhibition and the diminished activation of LPL by its serum cofactor, apolipoprotein CII [20].

Acknowledgements

This work was supported by a grant from the Medical Research Council of Canada. BR is the recipient of Fellowship awards from the Heart and Stroke Foundation of Canada and the Alberta Heritage Foundation for Medical Research.

References

1. Eckel RH: Lipoprotein lipase. A multifunctional enzyme relevant to common metabolic diseases. New Engl J Med 320: 1060–1068, 1989
2. Camp L, Reina M, Llobera M, Vilaro S, Olivecrona T: Lipoprotein lipase: cellular origin and functional distribution. Am J Physiol 258: C673–C681, 1990
3. Blanchette-Mackie EJ, Masuno H, Dwyer NK, Olivecrona T, Scow RO: Lipoprotein lipase in myocytes and capillary endothelium of heart: immunocytochemical study. Am J Physiol 256: E818–E828, 1989
4. Ben-Zeev O, Schwalb H, Schotz MC: Heparin-releasable and nonreleasable lipoprotein lipase in the perfused rat heart. J Biol Chem 256: 10550–10554, 1981
5. Severson DL, Lee M, Carroll R: Secretion of lipoprotein lipase from myocardial cells isolated from adult rat hearts. Mol Cell Biochem 79: 17–24, 1988
6. Braun JEA, Severson DL: Diabetes reduces heparin- and phospholipase C-releasable lipoprotein lipase from cardiomyocytes. Am J Physiol 260: E477–E485, 1991
7. Severson DL, Carroll R: Effect of taxol on the heparin-induced secretion of lipoprotein lipase from cardiac myocytes. Mol Cell Biochem 88: 17–22, 1989
8. Saxena U, Witte LD, Goldberg IJ: Release of endothelial cell lipoprotein lipase by plasma lipoproteins and free fatty acids. J Biol Chem 264: 4349–4355, 1989
9. Saxena U, Goldberg IJ: Interaction of lipoprotein lipase with glycosaminoglycans and apolipoprotein CII: effects of free fatty acids. Biochim Biophys Acta 1043: 161–168, 1990
10. Saxena U, Witte LD, Goldberg IJ: Tumor necrosis factor induced release of endothelial cell lipoprotein lipase. Arteriosclerosis 10: 470–476, 1990
11. Peterson J, Bihain BE, Bengtsson-Olivecrona G, Deckelbaum RJ, Carpentier YA, Olivecrona T: Fatty acid control of lipoprotein lipase: a link between energy metabolism and lipid transport. Proc Natl Acad Sci USA 87: 909–913, 1990
12. Kryski A, Kenno KA, Severson DL: Stimulation of lipolysis in rat heart myocytes by isoproterenol. Am J Physiol 248: H208–H216, 1985
13. Ramirez I, Kryski AJ, Ben-Zeev O, Schotz MC, Severson DL: Characterization of triacylglycerol hydrolase activities in isolated myocardial cells from rat heart. Biochem J 232: 229–236, 1985
14. Shimada K, Gill PJ, Silbert JE, Douglas WHJ, Fanburg BL: Involvement of cell heparan sulfate in the binding of lipoprotein lipase to cultured bovine endothelial cells. J Clin Invest 68: 995–1002, 1981
15. Cisar LA, Hoogerwerf AJ, Cupp M, Rapport CA, Bensadoun A: Secretion and degradation of lipoprotein lipase in cultured adipocytes. J Biol Chem 264: 1767–1774, 1989
16. Chajek T, Stein O, Stein Y: Lipoprotein lipase of cultured mesenchymal rat heart cells. Hydrolysis of labelled very low density lipoprotein triacylglycerol by membrane-supported enzyme. Biochim Biophys Acta 528: 466–474, 1978
17. Friedman G, Chajek-Saul T, Stein O, Noe L, Etienne J, Stein Y: Beta-adrenergic stimulation enhanced translocation, processing and synthesis of lipoprotein lipase in rat heart cells. Biochim Biophys Acta 877: 112–120, 1986
18. Pradines-Figueres, Vannier C, Ailhaud G: Lipoprotein lipase stored in adipocytes and muscle cells is a cryptic enzyme. J Lipid Res 31: 1467–1476, 1990
19. Friedman G, Stein O, Stein Y: Lipoprotein lipase of cultured mesenchymal rat heart cells. Modulation of enzyme activity by VLDL added to the culture medium. Biochim Biophys Acta 573: 521–534, 1979
20. Bengtsson G, Olivecrona T: Lipoprotein lipase. Mechanism of product inhibition. Eur J Biochem 106: 557–562, 1980

Molecular and Cellular Biochemistry **116**: 39–45, 1992.

Carnitine palmitoyltransferase in the heart is controlled by a different mechanism than the hepatic enzyme

George A. Cook and Michael D. Lappi[1]
Department of Pharmacology, College of Medicine, The University of Tennessee, Memphis – The Health Science Center, 874 Union Avenue, Memphis, TN 38163, U.S.A.; [1] Department of Physiology, The Ohio State University, Columbus, Ohio, U.S.A.

Abstract

Diminished sensitivity of hepatic carnitine palmitoyltransferase to inhibition by malonyl-CoA in the fasting and diabetic states is a well-recognized aspect of the regulatory mechanism for *hepatic* fatty acid oxidation. Inhibition of myocardial carnitine palmitoyltransferase by malonyl-CoA may play an important role in regulation of fatty acid oxidation in the heart, but there has been a discrepancy in data relating to changes in malonyl-CoA sensitivity of the myocardial carnitine palmitoyltransferase during fasting. Analysis of malonyl-CoA inhibition of myocardial carnitine palmitoyltransferase in fasting and fed states under a variety of conditions has indicated that under no condition could any difference be found in malonyl-CoA sensitivity that was attributable to fasting. Proteolysis of the outer carnitine palmitoyltransferase led to artifactual changes in sensitivity due to the appearance of partial inhibition. We have concluded that the sensitivity of myocardial carnitine palmitoyltransferase to malonyl-CoA does not change during fasting. Changes in fatty acid oxidation in the heart are probably due to changes in malonyl-CoA concentrations or to other inhibitors. (Mol Cell Biochem **116:** 39–45, 1992)

Key words: carnitine palmitoyltransferase, malonyl-CoA, heart, liver, fatty acid oxidation, regulation

Carnitine palmitoyltransferase (palmitoyl-CoA:L-carnitine O-palmitoyltransferase, E.C. 2.3.1.21) catalyzes the transfer of fatty acyl groups from acyl-CoA to carnitine preceding the transport of acylcarnitine across the mitochondrial inner membrane for oxidation of the acyl group. This enzyme is a rate-controlling enzyme for the hepatic fatty acid oxidation pathway in the fed state in the presence of malonyl-CoA [1–2]. An important aspect of the regulation of the hepatic enzyme is the ten-fold change in the K_i value for malonyl-CoA that occurs during fasting [3] and in the diabetic state [4] which makes the enzyme insensitive to inhibition by malonyl-CoA so that the reaction it catalyzes is no longer the rate-limiting step in fatty acid oxidation [2]. Additional work from our laboratory and others has indicated that for the myocardial carnitine palmitoyltransferase the K_i value for malonyl-CoA does not change during fasting [3, 5], suggesting an important difference in regulation of hepatic and myocardial carnitine palmitoyltransferase. Recently, however, data have been published which suggest that the malonyl-CoA K_i value of carnitine palmitoyltransferase from the heart is diminished during fasting [6]. In this paper we have re-evaluated the effects of fasting on myocardial carnitine palmitoyltransferase as well as its sensitivity to inhibition by malonyl-CoA when assayed under a varie-

Address for offprints: G. A. Cook, Department of Pharmacology, The University of Tennessee, Memphis, 874 Union Avenue, Memphis, Tennessee 38163, U.S.A.

ty of conditions. These results indicate that the activity of the myocardial enzyme does not change during fasting, nor is its sensitivity to inhibition by malonyl-CoA altered. We also present an explanation for previous results that had suggested a change.

Materials and methods

Male Spague-Dawley rats (150–250 g), obtained from Harlan Industries, Inc. (Indianapolis, IN, U.S.A.), were fed Purina Rat Chow (Ralston Purina Co., Richmond, IN, U.S.A.) *ad libitum* or were fasted for 48 hr. All animals were allowed free access to water. Rats were decapitated before organs were removed for preparation of mitochondria by the method of Johnson and Lardy [7]. This procedure was modified for isolation of heart mitochondria as indicated previously [3]. Some mitochondria were incubated at 37° C in isolation medium with Nagarse at 5 μg/ml. After 10 min of incubation, the proteolytic activity was stopped by addition of 200 μl of 20% bovine serum albumin per ml of incubation volume, followed by addition of 40 ml of ice cold isolation medium, and mitochondria were subsequently sedimented by centrifugation at 8500 xg for 10 min then resuspended in the isolation medium. Protein determination for all experiments was by a biuret method [8].

For most experiments involving heart mitochondria, a spectrophotometric assay for carnitine palmitoyltransferase activity was used that was similar to that used by Fiol *et al.* [6]. These assays contained 50 mM imidazole, either 150 mM KCl or 300 mM sucrose for osmotic balance, 0.2 mM dithionitrobenzoic acid (DTNB), 0 to 2 mg/ml albumin, 0.86 mM EDTA, 10–100 μM decanoyl-CoA, 50 μM malonyl-CoA (in some assays), 3 mM carnitine, and 0.1 mg/ml heart mitochondria. Assays were conducted at 37° C at pH 7.0 in a total volume of 2 ml, and were analyzed spectrophotometrically at 412 nm using an extinction coefficient of 13.6 mM^{-1} cm^{-1}. Acyl-CoA hydrolase activity was excluded from the measurement by subtraction. In some experiments (liver mitochondria), the method of Bremer [9] was used for the assay of carnitine palmitoyltransferase, as modified and reported previously [4]. Each of these assays contained, in a total volume of 1 ml: 82 mM sucrose, 70 mM KCl, 70 mM imidazole, 1 μg antimycin A, 0.5 mM L-carnitine (0.4 mCi of L-[methyl-^{3}H] carnitine), 40 μM palmitoyl-CoA, and 2 mg bovine serum albumin. Malonyl-CoA was present in some of these assays.

Palmitoyl-CoA, imidazole, L-carnitine hydrochloride, hepes, antimycin A, glutathione, DTNB, EDTA, essentially fatty acid-free bovine serum albumin, malonyl-CoA, and Nagarse (subtilisin BPN', P 4789), were purchased from Sigma Chemical Company (St. Louis, MO, U.S.A.). L-[methyl-^{3}H] Carnitine hydrochloride was obtained from Amersham Corporation (Arlington Heights, IL, U.S.A.).

Results and discussion

Effects of ionic strength and albumin

The study of kinetic characteristics of the outer carnitine palmitoyltransferase of heart mitochondria was begun using decanoyl-CoA as substrate since it has been shown that this substrate does not form micelles at concentrations below 100 μM and it does not cause swelling of isolated mitochondria [6]. We initially examined the effects of albumin and ionic strength (KCl at 150 mM) as well as increasing concentrations of decanoyl-CoA on the activity of carnitine palmitoyltransferase. Figure 1 indicates that, in the fed state, ionic strength slightly decreases activity while albumin increases activity. The effect of albumin may be due to protection from inhibition at high substrate concentration since the effect was greater at higher concentrations of decanoyl-CoA. These results suggest that decanoyl-CoA is not as likely to bind to albumin as palmitoyl-CoA does since extensive binding of the substrate would have decreased activity. We did not see any indication of sigmoid kinetics with decanoyl-CoA; the K_m for decanoyl-CoA was 40 ± 10 μM (n = 5). Percentage inhibition of activity by malonyl-CoA was not altered by KCl or albumin.

In the fasting state (Fig. 2), the effects of ionic strength and albumin on carnitine palmitoyltransferase activity were identical to those in the fed state. Activity of the enzyme was identical in the fed and fasting states and sensitivity to inhibition by malonyl-CoA was not altered by fasting under any of the assay conditions tested (Figs 1 and 2).

Effects of glutathione redox state

Because of the possibility that glutathione could affect

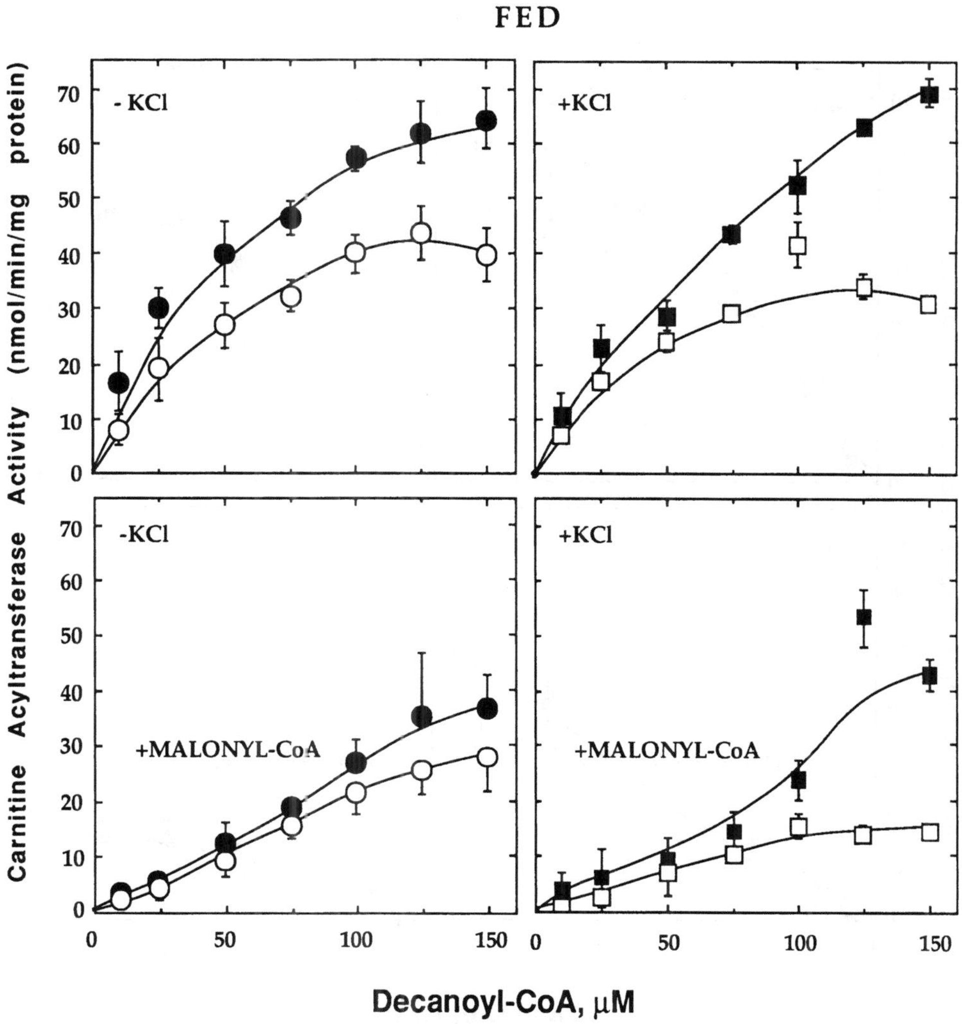

Fig. 1. Effects of ionic strength and albumin on the outer carnitine palmitoyltransferase of fed rat heart mitochondria. Carnitine palmitoyltransferase was assayed under spectrophotometric conditions (Materials and methods) using decanoyl-CoA as substrate with either 150 mM KCl (+ KCl) or 300 mM sucrose (− KCl) to preserve osmolarity. Malonyl-CoA was present in some assays at a concentration of 50 μM. Open symbols represent assays conducted in the absence of albumin; filled symbols represent assays in which 2 mg/ml albumin was present. Points are means ± SEM (n = 4).

the activity of carnitine palmitoyltransferase or its sensitivity to inhibition by malonyl-CoA, we assayed the enzyme in the absence of glutathione and in the presence of reduced (GSH) or oxidized (GSSG) glutathione. These assays were conducted using palmitoyl-CoA in the radiochemical assay since experiments with decanoyl-CoA had revealed no particular problems associated with the use of palmitoyl-CoA and since the radiochemical assays could be done more rapidly with greater sensitivity. As shown in Fig. 3, there were again no differences in carnitine palmitoyltransferase activity or its sensitivity to inhibition by malonyl-CoA produced by fasting animals for 48 hr. Although there were no significant differences in assays conducted with oxidized or reduced glutathione compared with controls, assays containing GSH in the absence of malonyl-CoA tended to produce greater activity while assays containing GSH in the presence of malonyl-CoA tended to produce less activity, so GSH caused greater sensitivity to inhibition, but this was similar in both the fed and fasting states (Fig. 3).

Conditions producing partial inhibition of carnitine palmitoyltransferase

A range of malonyl-CoA concentrations from 0 to 100 μM were used to evaluate further the inhibition of

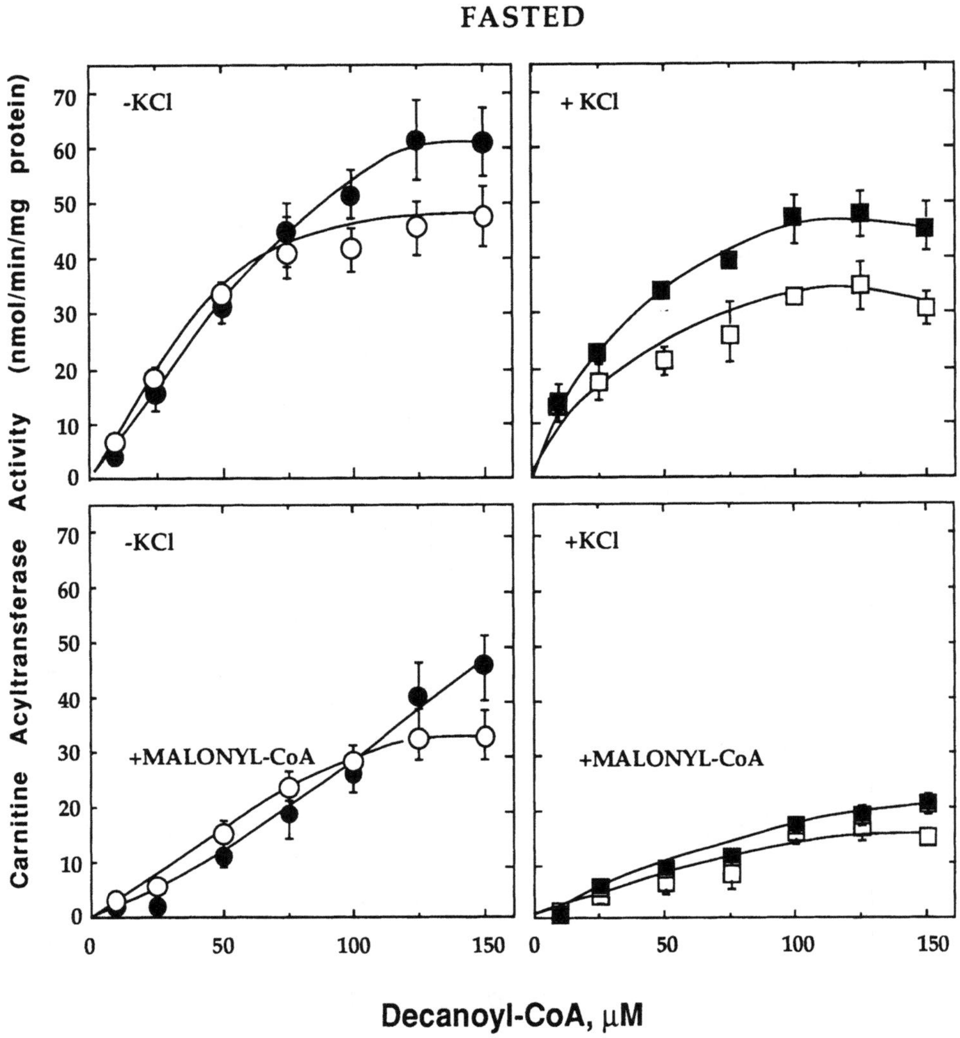

Fig. 2. Effects of ionic strength and albumin on the outer carnitine palmitoyltransferase in heart mitochondria from fasting rats. The enzyme was assayed as in Fig. 1. Symbols are exactly as in Fig. 1.

the myocardial carnitine palmitoyltransferase in the fed and fasting states. The data presented in Fig. 4a seem at first to suggest that this enzyme is more sensitive to inhibition by malonyl-CoA in the fasting state. However, careful examination of this figure suggests that increasing malonyl-CoA concentrations above 100 μM would not produce complete inhibition of activity as it does with the enzyme from liver (maximum inhibition = 100 ± 5%, n = 6). Interestingly, these plots of malonyl-CoA inhibition in heart mitochondria are similar to studies previously published using heart mitochondria (see Fig. 4 of reference 6) in which it was suggested that sensitivity to malonyl-CoA inhibition is greater in the fed state. It should be noted that in those studies the greatest inhibition observed did not exceed 40%. When our data from Fig. 4a were first converted to % inhibition vs. [malonyl-CoA] and then replotted in a double-reciprocal fashion (similar to the conversion of a Michaelis-Menten plot to a Lineweaver-Burk plot). The I_{50} derived from the intersection of lines at the x-axis in the fed and fasting states were identical (17 μM malonyl-CoA) indicating identical malonyl-CoA sensitivity in these states. The maximum inhibition at the theoretical infinite concentration of malonyl-CoA at the intersection of lines with the y-axis indicated that there was partial inhibition in both these state (I_{max} in the fed state = 42% and I_{max} in the starved stated = 71%). When we applied this same method of evaluation to previous results (Fig. 4 of reference 6) we found I_{50} values of 4.3 μM in the fed state and 2.8 μM in the starved state with I_{max} values of 44% in the fed state and 29% in the fasting state, so these data actually suggest a

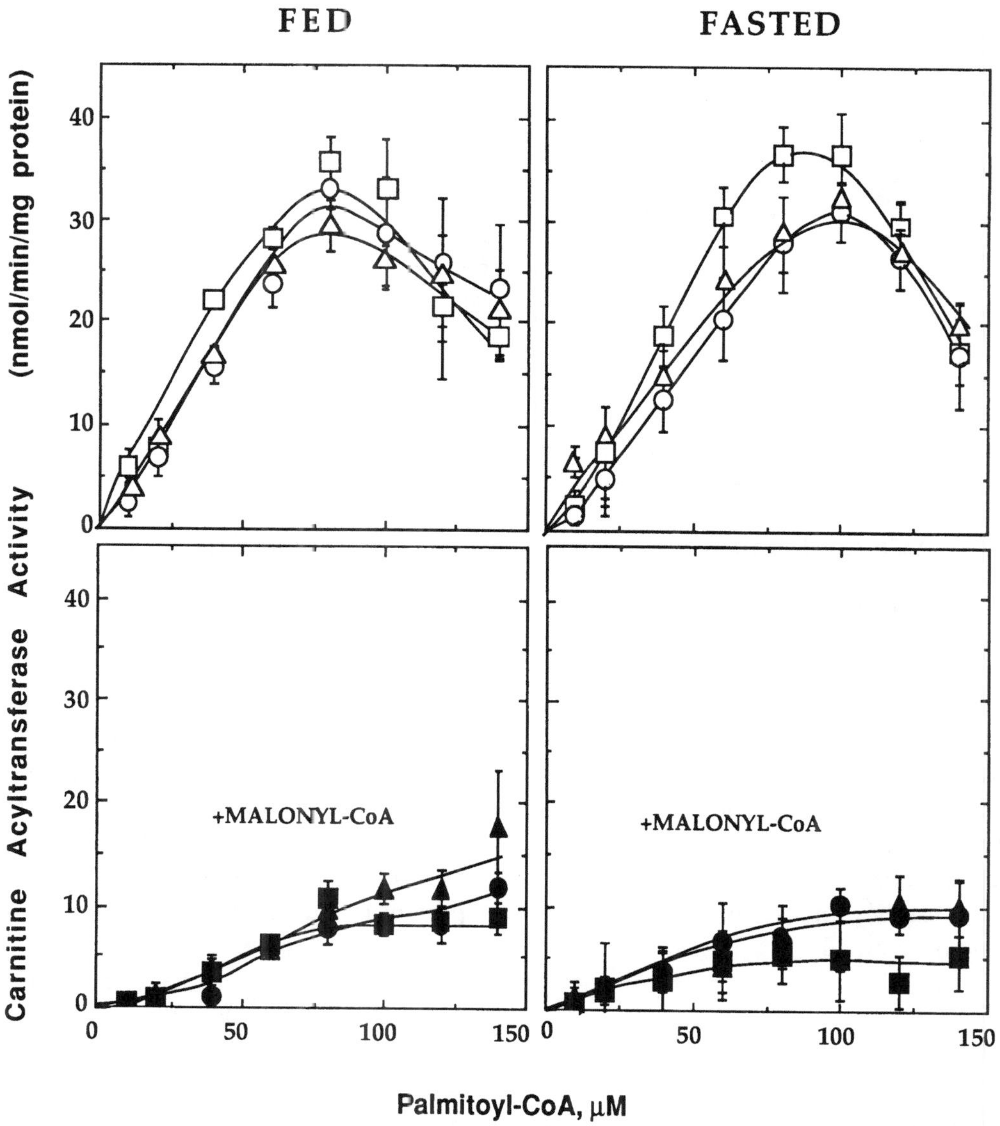

Fig. 3. Effects of reduced and oxidized glutathione on the outer carnitine palmitoyltransferase in heart mitochondria from fed and fasting rats and inhibition of the enzyme by malonyl-CoA. Carnitine palmitoyltransferase was assayed using the radiochemical assay described in Materials and methods with palmitoyl-CoA as substrate. Square symbols represent assays containing 5 mM reduced glutathione (GSH). Circles represent assays containing 5 mM oxidized glutathione (GSSG). Triangles represent assays in which no glutathione was added. Filled symbols represent the presence of 50 μM malonyl-CoA. Point are means ± SEM (n = 4).

greater sensitivity in the fasting state though the difference is exceedingly small. The much lower values for I_{50} obtained by these investigators reflects the lower concentrations of acyl-CoA used as the competitive substrate.

Murthy and Pande have previously reported a loss of the ability of malonyl-CoA to inhibit the myocardial carnitine palmitoyltransferase when heart mitochondria are isolated using proteases [10]. We examined the phenomenon of partial inhibition further using liver mitochondria that had been exposed to Nagarse and liver mitochondria exposed to vigorous homogenization after isolation. In the Nagarse-treated liver mitochondria, the I_{max} was decreased from 100% to 40% without a change in I_{50} values, and in the mechanically disrupted mitochondria, the I_{max} was decreased to 38% while I_{50} values were increased from 15 μM to 30 μM. Some inner carnitine palmitoyltransferase activity could have been exposed in the latter experiment. These results suggest that great care should be taken in selecting conditions for isolation of mitochondria and for evaluation of results. Previous data suggesting changes in sensitivity to malonyl-CoA of heart mitochondria in fasting [6] were obtained using a method for

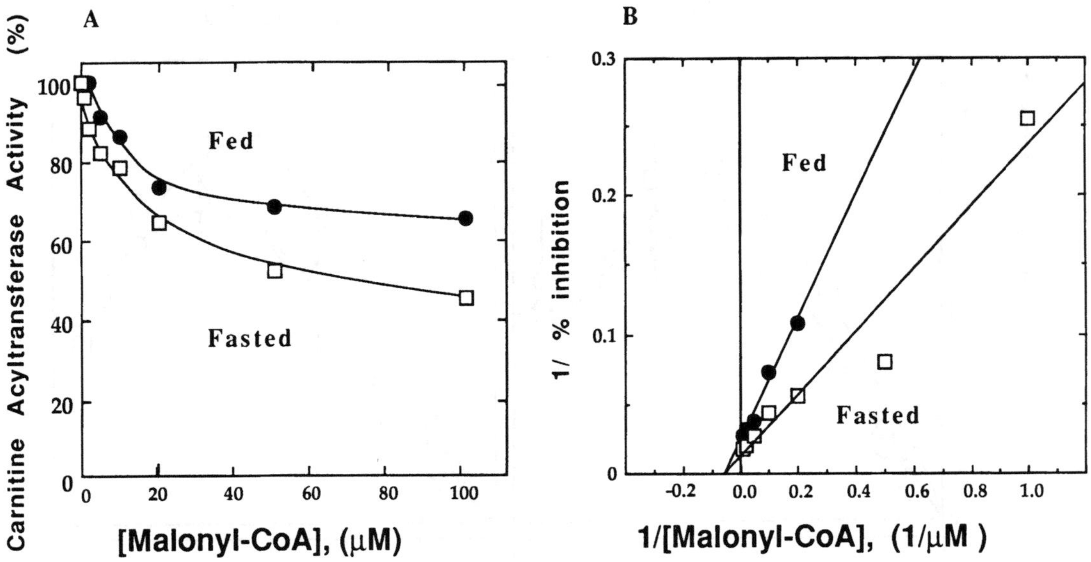

Fig. 4. Inhibition by malonyl-CoA of the outer carnitine palmitoyltransferase from hearts of fed and fasting rats. (A) Linear plot of activity versus inhibitor concentration. (B) Double-reciprocal replot of data from A after conversion to % inhibition vs. [malonyl-CoA] showing differences in partial inhibition (see text).

mitochondria isolation that included digestion by collagenase [11]. Our studies suggest that contamination of collagenase with other less specific proteases could produce partial inhibition and changes in sensitivity of carnitine palmitoyltransferases to inhibition by malonyl-CoA, and that the use of proteases in isolation of mitochondria should be avoided.

Conclusions

These studies confirm previous work indicating that fasting does not have the same effects on the myocardial outer carnitine palmitoyltransferase that it does on the liver enzyme [3]. While the liver enzyme increases somewhat in activity and changes its sensitivity to inhibition by malonyl-CoA during fasting, neither of these changes occur with this enzyme in the heart, no matter what assay conditions are used for its evaluation. Apparently, previous results indicating a change in malonyl-CoA sensitivity in the heart [6] were artifactually affected by the production of partial inhibition during the mitochondrial isolation procedure since the re-evaluation of those data by alternate means (double reciprocal plots) indicated no decrease in malonyl-CoA sensitivity in the fasting state. The fact that one can artificially produce damage to liver mitochondria that will result in partial inhibition and/or loss of malonyl-CoA sensitivity seems to confirm this conclusion.

These studies indicate that the regulatory processes that are active in the fed/fasting transition in the liver are not present in the heart. Regulation of fatty acid oxidation in the heart must therefore rely on different mechanisms. It is expected that the mitochondrial outer carnitine palmitoyltransferase of the heart plays an important role in regulation of fatty acid oxidation in the heart just as the hepatic enzyme does in the liver since the myocardial enzyme has a much greater affinity for malonyl-CoA than does the liver enzyme and because the malonyl-CoA concentration in the heart is known to be changed in concert with altered rates of fatty acid oxidation [12].

Acknowledgements

This work was supported by a Grant-in-Aid (#88–1003) from the American Heart Association with contributions derived in part from the AHA, Tennessee Affiliate. Additional support was received through grant #HL-40929 from the National Institutes of Health.

References

1. McGarry JD, Foster DW: Regulation of hepatic fatty acid oxidation and ketone body production. Annu Rev Biochem 49: 395–420, 1980
2. Cook GA, Otto DA, Cornell NW: Differential inhibition of

ketogenesis by malonyl-CoA in mitochondria from fed and starved rats. Biochem J 192: 955–958, 1980

3. Cook GA: Differences in the sensitivity of carnitine palmitoyltransferase to inhibition by malonyl-CoA are due to differences in K_i values. J Biol Chem 259: 12030–12033, 1984
4. Cook GA, Gamble MS: Regulation of carnitine palmitoyltransferase by insulin results in decreased activity and decreased apparent K_i values for malonyl-CoA. J Biol Chem 262: 2050–2055, 1987
5. Paulson DJ, Ward KM, Shug AL: Malonyl-CoA inhibition of carnitine palmitoyltransferase in rat heart mitochondria. FEBS Lett 176: 381–384, 1984
6. Fiol CJ, Kerner J, Bieber LL: Effect of malonyl-CoA on the kinetics and substrate cooperativity of membrane-bound carnitine palmitoyltransferase of rat heart mitochondria. Biochim Biophys Acta 916: 482–492, 1987
7. Johnson D, Lardy H: Isolation of liver or kidney mitochondria. Methods Enzymol 10: 94–96, 1967
8. Layne E: Spectrophotometric and turbidimetric methods for measuring proteins, III. Biuret method. Methods Enzymol 3: 450–451, 1957
9. Bremer J: The effect of fasting on the activity of liver carnitine palmitoyltransferase and its inhibition by malonyl-CoA. Biochim Biophys Acta 665: 628–631, 1981
10. Murthy MSR, Pande SV: Malonyl-CoA binding site and the overt carnitine palmitoyltransferase activity reside on the opposite sides of the outer mitochondrial membrane. Proc Natl Acad Sci U.S.A. 84: 378–382, 1987
11. Toth PP, Ferguson-Miller SM, Suelter CH: Isolation of highly coupled heart mitochondria in high yield using a bacterial collagenase. Methods Enzymol 125: 16–27, 1986
12. Bielefeld DR, Vary TC, Neely JR: Inhibition of carnitine palmitoyl-CoA transferase activity and fatty acid oxidation by lactate and oxfenicine in cardiac muscle. J Mol Cell Cardiol 17: 619–625, 1985

Molecular and Cellular Biochemistry **116**: 47–52, 1992

Myocardial cell vulnerability to exogenous phospholipase attack

Terje K. Steigen, Truls Myrmel, Kirsti Forsdahl and Terje S. Larsen
Department of Medical Physiology, Institute of Medical Biology, University of Tromsø, 9000 Tromsø, Norway

Abstract

Myocardial cell vulnerability to phospholipase C (PC-PLC) attack was investigated in three different preparations of rat myocardial cells: triacylglycerol (TG)-loaded, hypothermic/rewarmed and energy depleted myocytes. The attack by PC-PLC was evaluated as PC-PLC induced glycerol output due to the combined action of phospholipase C and intracellular lipases. PC-PLC induced glycerol output was significantly higher ($p < 0.05$) in all three myocyte preparations, compared to their respective controls. Cell morphology (% rod shaped myocytes) of TG-loaded or hypothermic/rewarmed myocytes was not different from their controls, whereas energy depleted myocytes almost exclusively were rounded up, due to hypercontraction of the myofilaments. Hypothermic/rewarmed and energy depleted myocytes showed a significantly higher release of lactate dehydrogenase (LDH), compared to their controls although the difference was much more pronounced in the latter. Finally, the cellular contents of ATP were maintained both in TG-loaded and hypothermic rewarmed myocytes, while energy depleted myocytes contained only about 25% of the normal ATP level. These results demonstrate that attack from exogenously added phospholipases can occur, not only in seriously damaged cardiac myocytes, but in myocytes with a more subtle damage as well. (Mol Cell Biochem **116:** 47–52, 1992)

Key words: rat cardiomyocytes, phospholipase C, hypothermia, triacylglycerol, adenosine triphosphate, lactate dehydrogenase, glycerol

Introduction

Two apparent contradictory theories have been put forward regarding the role of phospholipases during irreversible myocardial cell damage. One of these argues that phospholipase activation may be an initial step in the chain of events leading to cell damage [1, 2], whereas the other claims that phospholipase activation is merely a consequence of the dramatic structural changes occurring in ischemic tissue, including rupture of the cell membrane [3].

In a previous study [4] we have found that an exogenously added bacterial phospholipase C (PC-PLC) markedly accelerates glycerol production in energy depleted (rounded up) rat myocardial cells. Unpublished results from our own laboratory (Larsen and Forsdahl) show that this process involves PC-PLC catalyzed degradation of phospholipid, followed by hydrolysis of the resulting diacylglycerol moiety to fatty acids and glycerol. On the other hand, PC-PLC does not influence glycerol production in normal, rod shaped myocytes.

The purpose of the present study was to examine the accessibility of PC-PLC, not only to energy depleted myocytes, but also to triacylglycerol (TG) loaded and

Address for offprints: T K. Steigen, Department of Medical Physiology, Institute of Medical Biology, University of Tromsø, 9000 Tromsø, Norway hypothermic/rewarmed myocytes. TG-loaded myocytes are characterized by elevated intracellular concentration of free fatty acids 17, Larsen and Forsdahl, unpublished, which could distort the lipid bilayer configuration of the sarcolemma [5]. Likewise, hypothermia/rewarming is reported to induce phase separations of the membrane lipids [6, 7]. Our hypothesis was that, although the structural changes induced by fat loading or hypothermia/rewarming are much less dramatic than those observed in energy depleted myocytes, they would still favor phospholipase attack.

Materials and methods

Isolation of myocytes

Adult male Sprague-Dawley rats (250–350 g), fed ad libitum, were used in the study. Isolation of myocytes followed the procedure described by Larsen *et al.* [8], involving retrograde perfusion of the hearts with calcium free, Joklik modified minimum essential medium (MEM, Gibco) containing 200 mU/ml collagenase (Worthington type II), followed by mechanical dispersion of the ventricles in the same medium. Liberated cells were washed repeatedly to remove non-myocytes and damaged myocytes and finally suspended in Krebs-Henseleit bicarbonate buffer (pH 7.4, 37° C) supplemented with 1 mM DL-carnitine, 1% bovine serum albumin. The yield from two hearts ranged from 20–26 million cells which contained in average 87% rod shaped cells.

Incubation and sampling procedures

a) Triacylglycerol loaded myocytes
Triacylglycerol (TG) loaded myocytes were prepared as described above, except that 0.6 mmol/l albumin bound oleate (fatty acid:albumin = 4:1) was included in the cell isolation medium. This resulted in a marked elevation in the intracellular TG content of freshly isolated myocytes, from 58.6 ± 9.4 (controls) to 125.3 ± 12.7 nmol/10^6 cells (TG-loaded). Cells prepared without addition of oleate served as controls.

Immediately after isolation 3.5 ml aliquots of cell suspension (controls and TG-loaded myocytes) containing approximately 0.5 · 10^6 cells/ml were transferred to 25 ml Erlenmeyer flasks and incubated for 60 min on a linear shaking waterbath (37° C) in the absence (determination of basal glycerol output) or presence of 10 mU/ml PC-PLC (final concentration). The flasks were kept under metal tents with a 95% air-5% CO_2 atmosphere which was continuously renewed by blowing the gas mixture through the tents.

At the end of the incubation period 2 ml of cell suspension were transferred to 5 ml plastic tubes which were rapidly cooled on ice (0° C). The tubes were subsequently centrifuged (2000 · g, 1 min), and aliquots of the supernatant stored for 1–2 days at 4° C (measurement of lactate dehydrogenase activity) or − 20° C (measurement of glycerol concentration). Half a ml of cell suspension was transferred to prechilled (0° C) plastic tubes containing 37.5 μl 5.8 N perchloric acid (ATP determination) and 0.5 ml was fixed in McDowels fixative for determination of cell morphology (% rod shaped cells).

b) Hypothermic/rewarmed myocytes
Cells undergoing hypothermia and rewarming, as well as their normothermic controls, were preincubated for 60 min on a linear shaking waterbath at 5° C and 37° C, respectively. The incubation conditions were otherwise as described above. At the end of the preincubation period one hypothermic and one normothermic incubation flask were removed for analysis of zero time values. Immediately thereafter PC-PLC (10 mU/ml) was added to one half of the incubation flasks, whereas the other half received an equivalent volume of buffer (determination of basal glycerol output). Incubation flasks subjected to rewarming were then moved from the cold (5° C) to the warm (37° C) waterbath and incubation continued for another 60 min. Cells kept constantly at 37° C served as controls.

c) Energy depleted myocytes
Energy depletion of myocytes was achieved by preincubation for 60 min in glucose free Krebs-Henseleit bicarbonate buffer under an atmosphere of 95% N_2-5% CO_2 (pO_2 = 2 mmHg). This resulted in a rapid decline in ATP, as well as hypercontracture (rounding up) of the myocytes (see results). Following energy depletion the incubation were continued for 60 more min under normoxic conditions in the presence or absence (determination of basal glycerol output) of PC-PLC. Myocytes preincubated under normoxic conditions served as controls.

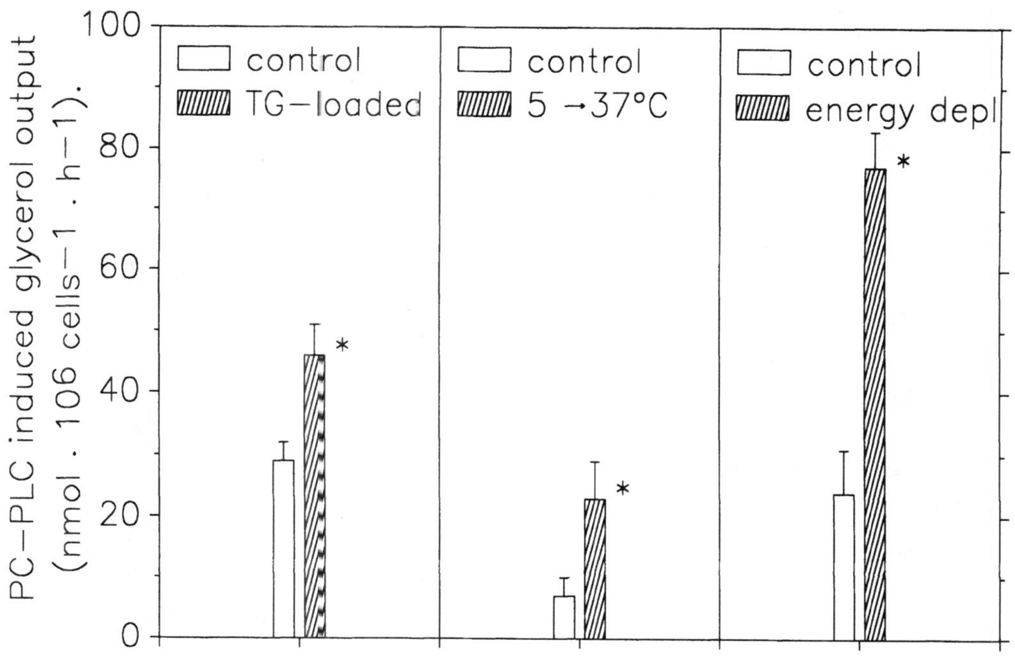

Fig. 1. PC-PLC induced glycerol output (i.e. glycerol output in excess of basal glycerol output, as measured in incubations without addition of PC-PLC) in TG-loaded, hypothermic/rewarmed and energy depleted myocytes, as well as in their respective controls. In each case PC-PLC (phosphatidyl choline preferring phospholipase C from Bacillus cereus) was added to the incubation flasks at a final concentration of 10 mU/ml. Incubation conditions were otherwise as described in 'Methods'.

Chemical measurements

Glycerol concentration was measured spectrofluorometrically (Kontron SFM 23 Spectrofluorometer) in the deproteinized and neutralized incubation medium according to a modification of the coupled enzymatic method described by Chernick [9]. The activity of lactate dehydrogenase (LDH) in the incubation medium was measured spectrofluorometrically (Hewlett Packard Spectrophotometer 8451 A) as recommended by the Committee on Enzymes of The Scandinavian Society for Clinical Physiology [10]. The cellular content of adenosine triphosphate (ATP) was measured in neutralized perchloric acid extracts of cells plus medium by means of HPLC technique (Waters HPLC unit, Waters Inc.) according to Sellevold *et al.* [11].

Materials

Collagenase (Worthington type II) was purchased from Cooper Biomedical Inc. (Malvern P.A. 19355, USA) and Joklik-MEM from Gibco Ltd. (Trident House, Paisley, Scotland). The following biochemicals were obtained from Sigma Chemical Company (St. Louis, M.O., U.S.A.): fatty acid free bovine serum albumin, 1-lactic dehydrogenase (1-lactate:NAD oxidoreductase, EC 1.1.1.27) and glycerol (approx. 99%). HPLC-grade methanol and acetonitrile were obtained from Rathburn Chemicals Ltd. (Walkenburn, Scotland). Glycerokinase (ATP:glycerol 3-phosphotransferase; EC 2.7.1.30), glycerol-3-phosphate dehydrogenase (sn-glycerol-3-phosphate:NAD 2-oxidoreductase; EC 1.1.1.8) were obtained from Boehringer Mannheim GmbH. All other chemicals used were from Sigma Chemical Company or from Merck a/s, Oslo. The PC-PLC used was isolated from Bacillus cereus [12] and appeared homogeneous as confirmed both by native – and SDS polyacrylamide gel electrophoresis followed by silver staining. One Unit of PC-PLC activity is defined as liberating 1 μmol of titratable H^+ per minute at pH 7.5 and 23° C in an egg yolk lipoprotein incubation medium [13]. Under these conditions this can be taken as representing the hydrolysis of 1 μmol of phospholipid per minute.

Statistics

Data are given either as mean ± SEM of 6–8 experiments in each group. Statistical analysis of the data was

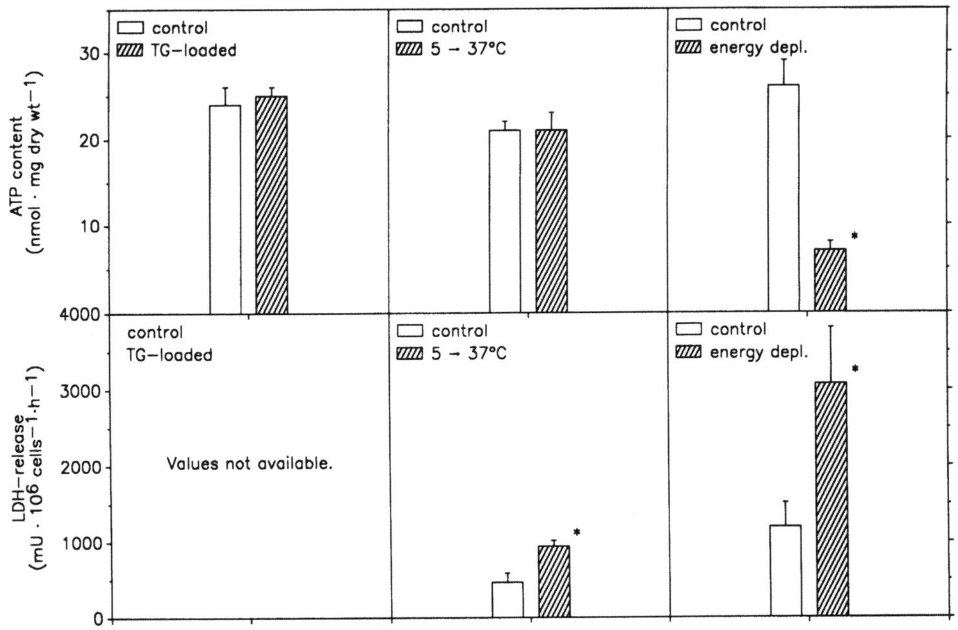

Fig. 2. Upper panel: cellular content of ATP of TG-loaded hypothermic/rewarmed and energy depleted myocytes, as well as of their respective control groups, at the onset of PC-PLC incubation. Lower panels: release of lactate dehydrogenase (LDH) from the same myocyte groups (except TG-loaded myocytes) during one hour incubation without addition of PC-PLC.

performed using Student's t-test for paired data. $P < 0.05$ was taken as limit of significance.

Results

Figure 1 shows PC-PLC induced glycerol output in TG-loaded hypothermic/rewarmed and energy depleted myocytes, as well as in their respective controls. In all three cases PC-PLC induced glycerol output was significantly ($p < 0.05$) higher in the intervention group than in the controls. The difference in PC-PLC induced glycerol output between controls and intervention group was three times higher for energy depleted myocytes than for hypothermic/rewarmed and TG-loaded myocytes. In the experimental series involving hypothermic/rewarmed myocytes the control and intervention group both showed lower PC-PLC induced glycerol output than in the other series.

Figure 2 (upper panels) shows the cellular content of ATP in the three different groups of myocytes and in their controls. The values refer to the content measured at the onset of PC-PLC incubation, demonstrating that neither TG-loading nor hypothermia/rewarming influenced the ATP content in these cells. On the other hand, ATP levels were markedly depressed in energy-depleted myocytes.

To further illustrate the condition of the cells inclined to PC-PLC exposure, the basal release (i.e. in the absence of PC-PLC) of lactate dehydrogenase (LDH) from hypothermic/rewarmed and energy depleted myocytes has been diagrammed in the lower panels of Fig. 2. Unfortunately, values of LDH release from TG-loaded myocytes were not available. A significantly higher LDH-release was observed both in hypothermic/rewarmed and in energy depleted myocytes, as compared to their controls. But, in the same way as for PC-PLC induced glycerol output, absolute values of LDH-release was considerably lower in the experiments involving hypothermic/rewarmed myocytes than those involving energy depleted myocytes.

Table 1 shows the percentage of rod shaped myocytes

Table 1. Morphology (% rod shaped cells) of freshly prepared TG-loaded myocytes and myocytes preincubated for 60 min under hypothermic (5° C) or hypoxic glucose free (energy depleting) conditions. Morphology of their appropriate controls are also included

Myocyte preparation	% rod shaped myocytes
TG-loaded	86 ± 3
Control	85 ± 2
Hypothermic/rewarmed	78 ± 2
Control	76 ± 2
Energy depleted	2 ± 1
Control	77 ± 4

in the various myocyte preparations at the onset of PC-PLC incubation. As can be seen, cell morphology was well maintained both in TG-loaded and hypothermic/rewarmed myocytes. In contrast, energy depleted preparations contained almost exclusively hypercontracted (rounded up) myocytes. Preincubation of the hypothermic/rewarmed and energy depleted series probably explains the somewhat lower fraction of rod shaped myocytes of their control groups, as compared to the controls of TG-loaded myocytes which were not subjected to preincubation.

Discussion

The present study demonstrates an elevated PC-PLC induced glycerol production in triacylglycerol (TG) loaded rat cardiomyocytes, as well as in myocytes subjected to hypothermia followed by rewarming. The amount of glycerol released from these myocyte preparations, however, was considerably less than that measured in parallel energy depleted myocytes.

One of the factors allowing the exogenously added PC-PLC to attack energy depleted myocytes may be loss of shielding of the membrane phospholipids by proteins and other membrane constituents. Rottem *et al.* [14] reported that intact cells of *Mycoplasma hominis* became readily accessible to *B. cereus* phospholipase C when part of the membrane proteins were removed by proteolytic digestion. The packing of the lipids in the membrane is another factor that could largely determine whether a phospholipase can interact with its substrate. It has previously been shown that Bacillus cereus phospholipase C fail to hydrolyze their substrates in the intact human erythrocyte, but that they will do so in a monolayer of sphingomyelin and lecithin at lateral surface pressures below 31 dynes/cm [15]. As demonstrated in the present study, rat cardiomyocytes undergoing energy depletion through hypoxic, substrate free incubation show hypercontraction of the myofilaments, causing rounding up of myocytes with bleb formation along the sarcolemma, as well as leakage of intracellular enzymes. This dramatic loss of membrane integrity may include both unmasking of phospholipids and irregularities in the packing of the lipid molecules, turning the membrane phospholipids into a favorable substrate for the exogenously added PC-PLC.

Hypothermic/rewarmed myocytes showed a slightly higher leakage of lactate dehydrogenase (LDH) than their normothermic controls, indicating a defect in the permeability barrier function of the cell membrane. On the other hand, the cellular content of high energy phosphates as well as cell morphology (% rod shaped myocytes) was not affected by hypothermia and rewarming. These findings, therefore, indicated less dramatic changes in the membrane structure of hypothermic/rewarmed myocytes than those of energy depleted myocytes. Earlier studies by Gordon *et al.* [6] and Verkleij *et al.* [7] have demonstrated phase separations and clustering of intramembranous proteins in myocardial tissue undergoing hypothermia. Furthermore, in experiments on synthetic liposomes Op den Kamp *et al.* [16] found that pig pancreatic phospholipase A_2 was able to attack these structures only at a temperature corresponding to the transition temperature of their lipid constituents. At the transition temperature, where the liquid crystalline and gel phase are in equilibrium with each other, irregularities in the packing of lipid molecules may occur at the border zone of the two phases. Temperature induced lesions like those discussed above may represent potentially attacking sites for phospholipases and explain the presently observed attack on hypothermic rewarmed myocytes by exogenous PC-PLC.

TG-loaded myocytes did not differ from their controls with respect to either microscopic appearance or leakage of LDH. However, previous studies by Kenno *et al.* [17] and Larsen *et al.* [18] have shown that these myocytes express a higher basal lipolytic rate than normal myocytes. Accordingly, TG-loaded myocytes have elevated intracellular levels of non esterified fatty acids (Larsen and Forsdahl, unpublished) which, through their amphiphilic nature, may distort the normal bilayer configuration of the membrane lipids [5, 19]. At any rate, the presence of excess fatty acids did not cause any change in cell morphology in the present study, as evaluated by light microscopic examination, nor did they influence the cellular content of ATP. Therefore, the fatty acid-membrane interaction must represent a subtle damage to the membrane. Nevertheless, the present results demonstrate that even such a mild damage can give access to exogenously added phospholipases.

In conclusion, this study favors the view that the physical state of the membrane phospholipids is the principal determinant of whether phospholipases will attack or not. It seems obvious that an irreversible injury (cell lysis) is not a prerequisite for exogenous phospholipase attack in rat myocardial cells, since myocytes with an intact sarcolemma and preserved energy homeostasis can, under certain circumstances, be the targets of PC-PLC. Further studies are required to find out whether these conclusions can be extrapolated to include also endogenous lipases.

Acknowledgements

This study was supported in part by the Norwegian Council on Cardiovascular Diseases (T.K.S.) and the Norwegian Research Council for Science and the Humanities (T.S.L.).

References

1. Van der Vusse GJ, van Bilsen M, Reneman RS: Is phospholipid degradation a critical event in ischemia- and reperfusion-induced damage? News in Physiological Sciences 4: 49–53, 1989
2. Stenbergen C, Jennings RB: Relationship between lysophospholipid accumulation and plasma membrane injury during total *in vitro* ischemia in dog heart. J Mol Cell Cardiol 16: 605–621, 1984
3. Van der Vusse GJ, Roemen THM, Prinzen FW, Coumans WA, Reneman RS: Uptake and tissue content of fatty acids in dog myocardium under normoxic and ischemic conditions. Circ Res 50: 538–546, 1982
4. Myrmel T, Larsen T, Forsdahl K, Skulberg A, Little C: Phospholipase C-evoked glycerol release in energy depleted rat myocardial cells. Mol Cell Biochem 88: 139–144, 1989
5. Katz AM, Messineo FC: Lipid-membrane interactions and the pathogenesis of ischemic damage in the myocardium. Circ Res 48: 1–16, 1981
6. Gordon LM, Sauerheber RD, Esgate JA: Spin label studies on rat liver and heart plasma membranes: Effects of temperature, calcium, and lanthanum on membrane fluidity. J Supramol Struct 9: 283–310, 1978
7. Verkleij AJ, Post JA: Physico-chemical properties and organization of lipids in membranes: their possible role in myocardial injury. In: H Stam, GJ van der Vusse, (eds.) 1987, pp 85–91 Lipid metabolism in the normoxic and ischemic heart
8. Larsen TG, Myrmel T, Skulberg A, Severson DL, Mjøs OD: Effects of hypoxia on lipolysis in isolated rat myocardial cells. Mol Cell Biochem 88: 139–144, 1989
9. Chernick SS: Determination of glycerol in acyl glycerols. Methods in enzymology 14: 627–630, 1969
10. Moss DW: The measurements of enzymes. In: DJ Hearse and de Leiris (eds.), 1979, Enzymes in cardiology, John Wiley and Sons, pp 154–198, 1979
11. Sellevold OFM, Jynge P, Aarstad K: High performance liquid chromatography: a rapid isocratic method for determination of creatine compounds and adenine nucleotides in myocardial tissue. J Mol Cell Cardiol 18: 517–527, 1986
12. Myrnes BJ, Little C: A simple purification scheme yielding crystalline phospholipase C from Bacillus cereus. Acta Chem Scand B 34: 375–377, 1980
13. Little C: Phospholipase C from Bacillus cereus. Methods Enzymol 71: 725–730, 1981
14. Rottem S, Hasin M, Razin S: Differences in susceptibility to phospholipase C of free and membrane-bound phospholipids of Mycoplasma hominis. Biochim Biophys Acta 323: 520–531, 1973
15. Roelofsen B: Phospholipases as tools to study the localization of phospholipids in biological membranes. A critical review. J Toxicol Toxin Reviews 1: 87–197, 1982
16. Op den Kamp JAF, de Gier J, van Deenen LLM: Hydrolysis of phosphatidylcholine liposomes by pancreatic phospholipase A_2 at the transition temperature. Biochim Biophys Acta 345: 253–256, 1974
17. Kenno KA, Severson DL: Lipolysis in isolated myocardial cells from diabetic rat hearts. Am J Physiol 249: H1024–H1030, 1985
18. Larsen TS, Severson DL: Influence of exogenous fatty acids and ketone bodies on rates of lipolysis in isolated ventricular myocytes from normal and diabetic rats. Can J Physiol Pharmacol 68: 177–182, 1990
19. Corr PB, Gross RW, Sobel BE: Amphiphatic metabolites and membranes dysfunction in ischemic myocardium. Circ Res 55: 135–154, 1984

Molecular and Cellular Biochemistry **116**: 53–58, 1992.

Phosphatidylcholine metabolism in ischemic and hypoxic hearts

Patrick C. Choy, Monroe Chan, Grant Hatch and Ricky Y.K. Man
Departments of Biochemistry & Molecular Biology and Pharmacology, Faculty of Medicine, University of Manitoba, Winnipeg, Canada

Abstract

The rates of phosphatidylcholine biosynthesis in the isolated hamster hearts under ischemic and hypoxic conditions were examined. Global ischemia was produced by perfusion of the heart with a reduced flow, whereas hypoxia was produced by perfusion with a N_2-saturated buffer. A 51% reduction in the biosynthesis of phosphatidylcholine was observed in the ischemic heart. The reduction was caused by a severe decrease in ATP level which resulted in a diminished conversion of choline into phosphocholine. A 22% reduction in the biosynthetic rate of phosphatidylcholine was also detected in the hypoxic heart. The reduction was caused by a diminished level of CTP which resulted in a decreased conversion of phosphocholine to CDP-choline. No compensatory mechanism was triggered during ischemia, but the CTP : phosphocholine cytidylyltransferase activity was enhanced in the hypoxic heart. Our results demonstrate the possible rate-limiting role of choline kinase and reconfirm the regulatory role of the cytidylyltransferase in the biosynthesis of phosphatidylcholine. (Mol Cell Biochem **116**: 53–58, 1992)

Key words: phosphatidylcholine, biosynthesis, ischemia, hypoxia, hamster heart

Introduction

Phospholipids are essential building blocks of all biological membranes [1]. The maintenance of the appropriate composition and content of phospholipids in the membrane are essential to cellular integrity and proper membrane function [2]. Beyond their role as structural components of the membrane, phospholipids are required for the full activities of certain enzymes [3]. In addition, membrane phospholipids have been shown to participate in the transmission of biological signals across the membrane, and their metabolic products may act as secondary messengers for cellular regulation [4, 5]. Existing evidence indicates that irrespective of etiological factors, cardiac disorders and heart failures are always associated with some defects in the cardiac membrane systems, and often with detectable alterations in the membrane phospholipid content and composition [6]. Since phospholipids are actively metabolized in the heart, changes in phospholipid contents must be caused by alterations in their biosynthesis and/or degradation.

Phosphatidylcholine is the principle phospholipid in the mammalian heart [7]. Studies on the isolated hamster heart have shown that over 90% of phosphatidylcholine is synthesized via the CDP-choline pathway [8]. Although the phosphorylation of choline is the first committed step in the CDP-choline pathway, the con-

Address for offprints: P.C. Choy, Department of Biochemistry and Molecular Biology, Faculty of Medicine, The University of Manitoba, 770 Bannatyne Avenue, Winnipeg, Manitoba, Canada R3E 0W3

version of phosphocholine to CDP-choline catalyzed by CTP : phosphocholine cytidylyltransferase is regarded as the rate-limiting step [9]. Using experimental models, several mechanisms for the control of phosphatidylcholine biosynthesis in the mammalian heart have been identified [10]. In general, the uptake of choline, the energy status of the tissue, the modulation of the rate-limiting enzyme are some of the mechanisms that regulate phosphatidylcholine biosynthesis in the heart [10]. Since ischemia would inevitably reduce the energy production capacity of the organ, it was not clear if short-term ischemia might affect the rate of phosphatidylcholine biosynthesis. In this study, the effect of ischemia on phosphatidylcholine biosynthesis was examined. As a corollary, the effect of hypoxia on phosphatidylcholine biosynthesis was also investigated.

Materials and methods

Materials

[*Me*-^{3}H]Choline and CDP-[*Me*-^{14}C]choline were obtained from New England Nuclear. Phospho[*Me*-^{3}H]choline was synthesized enzymically from [*Me*-^{3}H]choline by the procedure of Paddon and Vance [11]. Aqueous counting scintillant was obtained from Amersham Corporation. Thin-layer chromatographic plates (Sil-G25) were purchased from Brinkman Inc. All choline containing metabolites were obtained from Sigma Chemical Company. Other biochemical and solvents were of analytical grade and were purchased locally from the Canlab division of Baxter Company.

Animals

Syrian golden hamsters weighing 100 ± 15 g were used throughout the study. The hamsters were sustained on Purina rodent chow and tap water, *ad libitum*, in a light and temperature controlled room.

Perfusion of hamster heart

Hamster hearts were perfused in the Langendorf mode with 95% O_2-5% CO_2 saturated Krebs-Henseleit buffer as described previously [8]. The flow of the perfusate was maintained at a constant rate of 10 ml/g heart. Under these conditions, the aortic pressure was 75 ± 5 mm Hg and the pO_2 in the buffer was 522 ± 15 mm Hg. The viability of the heart was sustained for at least 6 h of perfusion [12]. In the ischemic study, the heart was perfused (pulse-labelled) at the normal flow rate for 15 min with 95% O_2-saturated buffer containing 0.01 mM [*Me*-^{3}H]choline and then perfused (chased) for 5 min with the same buffer containing 0.01 mM non-radioactive choline. Global ischemia was induced by reducing the flow rate of the chase buffer to 1.0 ml/g heart for 60 min whereas the flow rate of the chase buffer was maintained at 10 ml/g for the control. In the hypoxic study, the hamster heart was perfused (pulse-labelled) at the normal flow rate for 30 min with 95% O_2 saturated buffer containing 0.01 mM [*Me*-^{3}H]choline. Subsequently, hypoxia in the heart was produced by perfusion at normal flow rate for 60 min with the 95% N_2-saturated buffer containing 0.01 mM non-radioactive choline for 60 min. The pO_2 in this buffer was 22 ± 5 mm Hg. The hearts perfused under identical conditions with the 95% O_2-satured buffer were used as controls.

The viability of the hearts during perfusion was monitored by electrocardiac recording as described previously [13]. Perfusion of the heart under ischemic or hypoxic conditions caused a complete atrioventricular block. Histological analysis of the ventricular tissue by electron microscopy revealed some degree of mitochondrial swelling in both models, and the mitochondrial swelling was more severe in the ischemic heart. Subsequent perfusion with fully oxygenated buffer at normal flow rate resulted in an electrocardiogram which resembled that obtained from the control heart, indicating that the electrophysiological abnormalities generated under ischemia or hypoxia were reversible [13, 14].

Table 1. Radioactive choline uptake and its incorporation into phosphatidylcholine in hamster hearts perfused under ischemic and hypoxic conditions

	Total choline uptake	Phosphatidylcholine
	(dpm × 10^{-6})	
Control	2.00 ± 0.20 (4)	0.85 ± 0.05 (4)
60 min ischemic treatment	2.04 ± 0.16 (4)	0.42 ± 0.05 (4)*
Control	9.43 ± 0.73 (6)	5.21 ± 0.48 (6)
60 min hypoxic treatment	10.5 ± 1.69 (6)	4.05 ± 0.61 (6)*

$p < 0.05$ when compared with control.

Determination of radioactivity incorporated into choline-containing metabolites

Subsequent to perfusion, the hearts were homogenized in 20 ml chloroform/methanol (1 : 1, v/v). A portion of the homogenate was taken for the determination of total radioactivity. The homogenate was then allowed to separate into two phases by the addition of chloroform and 0.05 M KCl. The phosphatidylcholine fraction in the organic phase was separated from the other phospholipids by thin-layer chromatography with a solvent system containing chloroform/methanol/water/NH_4OH (70 : 30 : 4 : 2, v/v). The choline-containing metabolites in the aqueous phase were separated by thin-layer chromatography with a solvent system containing methanol/0.6% NaCl/NH_4OH (10 : 10 : 1, v/v) [8].

Determination of enzyme activities

After perfusion, the hearts were weighed and a 20% homogenate was prepared in 0.9% NaCl-5 mM Tris-HCl (pH 7.4). The post-mitochondrial fraction was obtained by centrifugation of the homogenate at 12,000 g for 10 min, and the supernatant obtained was centrifuged at 100,000 g for 60 min. The cytosolic fraction was stored at 4° for 4 h prior to use. The microsomal fraction was washed once with the same buffer and the pellet was resuspended in 1 ml of the buffer. Protein concentrations in the microsomal and cytosolic fractions were determined by the procedure of Lowry *et al.* [15]. Choline kinase in the cytosolic fraction was determined by the method of Brophy *et al.* [16]. CTP : phosphocholine cytidylyltransferase activities in the cytosolic and microsomal fractions were determined by the method of Vance *et al.* [17]. CDP-choline: 1,2-diacylglycerol cholinephosphotransferase was assayed in the microsomal fraction by previously described procedures [18].

Determination of pool sizes of metabolites

The pool sizes of the choline-containing metabolites were determined by a previously described procedure [19]. Briefly, choline, phosphocholine and CDP-choline were extracted from the heart and separated by column chromatography. Phosphocholine and CDP-choline were converted to choline by enzymatic hydrolysis. The amount of choline in each fraction was determined by enzymatic conversion into phosphocholine with labelled ATP. The pool sizes of ATP and CTP in the tissue extract were determined by HPLC equipped with a Partisil 10-SAX column [20].

Results

Effect of ischemia and hypoxia on phosphatidylcholine biosynthesis

The total choline uptake and the incorporation of labelled choline into phosphatidylcholine in hamster hearts perfused in ischemic and hypoxic conditions were determined. As depicted in Table 1, there was no significant difference in choline uptake between the control and ischemic or hypoxic hearts. However, there was a 51% reduction of labelled material incorporated into phosphatidylcholine in the ischemic heart. The reduction of labelled material incorporated into phosphatidylcholine was less pronounced (22%) in the hypoxic heart.

Analysis of choline-containing metabolites in the ischemic heart (Table 2) revealed that there was an accumulation of labelled choline with a corresponding reduction in phosphocholine labelling. The results suggest that the phosphorylation of choline was curtailed during ischemia. No change in the labelling of the CDP-choline fraction was detected between the control and ischemic hearts.

Table 2. Radioactivity incorporated into choline-containing metabolites in the control and ischemic hearts

	Control	Ischemic
	(dpm × 10^{-6}/g heart)	
Choline	0.28 ± 0.14 (4)	0.62 ± 0.10 (4)*
Phosphocholine	0.85 ± 0.09 (4)	0.73 ± 0.04 (4)*
CDP-choline	0.04 ± 0.02 (4)	0.04 ± 0.01 (4)

*$p < 0.05$.

Table 3. Radioactivity incorporated into choline-containing metabolites in the control and hypoxic hearts

	Control	Hypoxic
	(dpm × 10^{-6}/g heart)	
Choline	0.42 ± 0.09 (4)	0.53 ± 0.14 (4)
Phosphocholine	4.01 ± 0.32 (4)	6.23 ± 0.55 (4)*
CDP-choline	0.48 ± 0.06 (4)	0.31 ± 0.08 (4)*

*$p < 0.05$.

In the hypoxic heart, the distribution of radioactivity in choline-containing metabolites was different from that found in the ischemic heart (Table 3). No difference in the labelling of choline was detected, but an accumulation of labelled phosphocholine was observed during hypoxia. The labelling of CDP-choline was also reduced. It appears that the conversion of phosphocholine to CDP-choline catalyzed by CTP : phosphocholine cytidylyltransferase was inhibited during hypoxic treatment.

Changes in the labelling of the choline-containing metabolites might result from alterations in the pool sizes of these metabolites during ischemic or hypoxic treatments. Hence, the pool sizes of choline, phosphocholine and CDP-choline in the ischemic and hypoxic hearts were determined. The concentrations of choline, phosphocholine and CDP-choline were not changed in the ischemic or hypoxic hearts (data not shown).

Effects of ischemia and hypoxia on the enzymes in the CDP-choline pathway

The specific activities of the enzymes in the CDP-choline pathway in the ischemic and hypoxia hearts were determined and compared with the control (Table 4). No significant change in the activities of choline kinase, CTP : phosphocholine cytidylyltransferase and CDP-choline : 1,2-diacylglycerol choline phosphotransferase were detected in the heart perfused under ischemic conditions. However, the activity of CTP : phosphocholine cytidylyltransferase was enhanced in the microsomal fraction during hypoxic treatment.

Table 4. Activities of the enzymes in the CDP-choline pathway in ischemic and hypoxic hearts

Enzyme	Activity (nmol/min/mg protein)		
	Control**	Ischemic**	Hypoxic**
Choline kinase	0.52 ± 0.09	0.61 ± 0.11	0.47 ± 0.09
P-choline cytidylyltransferase			
(cytosolic)	0.57 ± 0.21	0.52 ± 0.15	0.48 ± 0.24
(microsomal)	0.63 ± 0.24	0.78 ± 0.27	1.31 ± 0.52*
P-choline transferase	1.04 ± 0.04	0.96 ± 0.07	1.01 ± 0.09

**n = 4. *P < 0.05.

Effects of ischemia and hypoxia on intracellular ATP and CTP concentrations

ATP is a cofactor required for the phosphorylation of choline whereas CTP is required by the cytidylyltransferase for the conversion of phosphocholine to CDP-choline. Hence, the pool sizes of ATP and CTP were determined in the heart during ischemia or hypoxia. As depicted in Table 5, the levels of ATP and CTP were severely and significant reduced in the ischemic and hypoxic hearts. The ATP concentration in the ischemic and hypoxic heart were 20% and 37% of the control hearts, respectively. The CTP concentration in the ischemic and hypoxic hearts were 39% and 28%, respectively. The reduction in ATP and CTP levels might reduce the products formed in the choline kinase and the cytidylyltransferase reactions.

Discussion

It is clear from this study that the energy status of the tissue is important for the regulation of phosphatidylcholine biosynthesis in the heart. In the ischemic heart, the severe reduction in ATP level (20% of the control) caused a reduction in the phosphorylation of choline and subsequently, phosphatidylcholine biosynthesis. Although the conversion of phosphocholine to CDP-choline has been regarded as the rate-limiting step in the CDP-choline pathway [9, 21], the reduction of CTP level (39% of control) was apparently not low enough to cause any reduction in the conversion of phosphocholine to CDP-choline. Our study demonstrates that the reaction catalyzed by choline kinase can also become rate-limiting under a highly diminished ATP pool.

In the hypoxic heart, the level of ATP and CTP were significantly reduced, but the reduction in ATP level (37% of control) was not severe enough to cause a significant change in the phosphorylation of choline. However, the reduction in CTP level (28% of control) caused a decrease in the conversion of phosphocholine

Table 5. ATP and CTP concentrations in ischemic and hypoxic hearts at 60 min of perfusion

	Control**	Ischemic**	Hypoxic**
ATP (μmol/g heart)	2.68 ± 0.45	0.54 ± 0.18*	0.99 ± 0.09*
CTP (nmol/g heart)	11.50 ± 2.13	4.57 ± 0.98*	3.25 ± 1.12*

**n = 4. *p < 0.05.

to CDP-choline, and subsequently, phosphatidylcholine. Interestingly, the cytidylyltransferase activity in the microsomal fraction was enhanced during hypoxic treatment. In earlier studies, the microsomal form of the cytidylyltransferase was shown to be the active form of the enzyme [21, 22]. The activation of the microsomal enzyme by lipids or by translocation of the cytosolic form into the microsomal form were the postulated mechanisms for the regulation of the enzyme activity [21]. Since no translocation of enzyme activity was detected in the hypoxic heart, the enhancement of enzyme activity probably caused by lipid activators [23, 24]. The activation of the microsomal enzyme could be regarded as the compensatory mechanism for maintaining phosphatidylcholine biosynthesis under energy depletion. The enhancement of cytidylyltransferase activity under low CTP supply was also observed in the heart of the myopathic hamsters [19]. The ability to trigger a compensatory mechanism by the cytidylyltransferase further confirms the regulatory role of this enzyme in the biosynthesis of phosphatidylcholine.

The reason for the differential reduction in ATP and CTP levels between the ischemic and hypoxic hearts is not explained. In view that the high energy phosphate of CTP originates from ATP, a reduction in ATP level should cause a proportional change in the CTP level in the organ. The differential reduction in ATP and CTP levels in between these two experimental models may arise from the fact that metabolites are normally accumulated in the ischemic heart [25]. We postulate that such accumulation would produce physiological changes in the cell (e.g. pH) which might affect the transfer of the high energy phosphate between the nucleotide groups.

Acknowledgement

This study was supported by the Medical Research Council of Canada.

References

1. White DA: The phospholipid composition of mammalian tissues. In: Ansell GB, Hawthorne JN, Dawson RMC (eds) Form and function of phospholipids. Elsevier Scientific Publishing Amsterdam, 1973, pp 441–483
2. Cullis PR, Hope MJ: Physical properties and functional roles of lipids in membranes. In: Vance DE, Vance JE (eds) Biochemistry of lipids and membranes. Benjamin Cummings Publishing, Menlo Park, CA, 1985, pp 25–72
3. Coleman R: Membrane-bound enzymes and membrane ultrastructure. Biochem Biophys Acta 300: 1–30, 1973
4. Berridge MJ, Irvine RF: Inositol triophosphate, a novel second messenger in cellular signal transduction. Nature (London) 312: 315–321, 1984
5. Exton JH: Signalling through phosphatidylcholine breakdown. J Biol Chem 265: 1–4, 1990
6. Dhalla NS, Ziegelhoffer A, Harrow JAC: Regulatory role of membrane systems in heart function. Can J Physiol Pharm 55: 1211–1234, 1977
7. Vance DE: Phospholipid metabolism in eucaryotes. In: Vance DE, Vance JE (eds) Biochemistry of lipids and membranes. Benjamin Cummings Publishing, Menlo Park, CA, 1985, pp 25–72
8. Zelinski TA, Savard JD, Man RYK, Choy PC: Phosphatidylcholine biosynthesis in isolated hamster hearts. J Biol Chem 255: 11423–11428, 1980
9. Vance DE, Choy PC: How is phosphatidylcholine biosynthesis regulated? Trends Biochem Sci 4: 145–148, 1979
10. Hatch GM, O K-M, Choy PC: Regulation of phosphatidylcholine metabolism in mammalian hearts. Biochem Cell Biol 67: 67–77, 1989
11. Paddon HB, Vance DE: The relationship between cholinephosphate phosphatase alkaline phosphatase and phosphatidylcholine biosynthesis in HeLa cells. Biochim Biophys Acta 488: 181–189, 1977
12. Arthur G, Choy PC: Acyl specificity of hamster heart CDP-choline : 1,2-diacylglycerol phosphocholine transferase in phosphatidylcholine biosynthesis. Biochim Biophys Acta 795: 221–229, 1984
13. Hatch GM, Choy PC: Effect of hypoxia on phosphatidylcholine biolsynthesis in the isolated hamster heart. Biochem J 268: 47–54, 1990
14. Piper HM, Schwartz P, Hutter J, Spieckermann PG: Energy metabolism and enzyme release of cultured adult rat heart muscle cells during anoxia. J Mol Cell Cardiol 16: 995–1007, 1984
15. Lowry OH, Rosebrough NJ, Farr AL, Randall RJ: Protein measurement with the Folin phenol reagent. J Biol Chem 193: 265–275, 1951
16. Brophy PJ, Choy PC, Toone JR, Vance DE: Choline kinase and ethanolamine kinase are separate soluble enzymes in rat liver. Eur J Biochem 78: 491–495, 1977
17. Vance DE, Pelech SD, Choy PC: CTP : phosphocholine cytidylyltransferase from rat liver. Methods Enzymol 71: 576–581, 1981
18. O K-M, Choy PC: Solubilization and partial purification of cholinephosphotransferase in hamster tissues. Lipids 25: 122–124, 1990
19. Choy PC: Control of phosphatidylcholine biosynthesis in myopathic hamster hearts. J Biol Chem 257: 10928–10933, 1982
20. Choy PC, Paddon HB, Vance DE: An increase in cytoplasmic CTP accelerates the reaction catalyzed by CTP : phosphocholine cytidylyltransferase in poliovirus-infected HeLa cells. J Biol Chem 255: 1070–1073, 1980
21. Vance DE: CTP: cholinephophsate cytidylyltransferase. In: Vance DE (ed) Phosphatidylcholine metabolism, CRC Press, Boca Raton, FL. 1989, pp 33–46
22. Slight R, Kent CR: Regulation of phosphatidylcholine biolsynthesis in mammalian cells. I. Effects of phospholipase C treat-

ment on phosphatidylcholine metabolism in chinese hamster ovary cells and LM mouse fibroblasts. J Biol Chem 258: 824–830, 1983
23. Choy PC, Vance DE: Lipid requirements for activation of CTP : phosphocholine cytidylyltransferase from rat liver. J Biol Chem 253: 5163–5167, 1978
24. Vance DE, Pelech SL: Enzyme translocation in the regulation of phosphatidylcholine biosynthesis. Trends Biochem Sci 9: 17–20, 1984
25. Katz AM, Messineo FC: Lipid-membrane interactions and the pathogenesis of ischemic damage in the myocardium. Circ Res 48: 1–16, 1981

Molecular and Cellular Biochemistry **116**: 59–67, 1992.

Occurrence and functions of the phosphatidylinositol cycle in the myocardium

Jos M.J. Lamers, Dick H.W. Dekkers, Karel Bezstarosti, Johanna T.A. Meij[1] and Han A.A. van Heugten
Department of Biochemistry, Faculty of Medicine and Health Sciences, Erasmus University Rotterdam, 3000 DR Rotterdam, The Netherlands; [1] Department of Physiology, Division of Cardiovascular Sciences, Faculty of Medicine, University of Manitoba, St. Boniface General Hospital Research Centre, Winnipeg, Manitoba, Canada R2H 2A6

Abstract

In the last decade a great deal of attention was awarded to a signal transduction pathway which is utilized primarily by 'Ca^{2+} mobilizing' signal molecules and which involves the hydrolysis of a quantitatively minor phospholipid, phosphatidylinositol 4,5-bisphosphate (PtdIns(4,5)P_2) by a PtdIns-specific phospholipase C (PLC). The evidence for the existence of receptor-mediated GTP binding protein-coupled PLC in myocardium and its possible functions are briefly summarized. The minireview is concentrated on the following aspects: 1) cellular localization and synthesis of polyphospho-PtdIns from PtdIns, 2) desensitization of the α_1-adrenergic agonist and endothelin-1 mediated PtdIns responses, 3) oscillatory Ca^{2+} transients initiated by PtdIns(4,5)P_2 hydrolysis, 4) polyunsaturated fatty acids as constituents of polyphospho-PtdIns and of the protein kinase C activator 1,2-diacylglycerol (DAG), 5) source other than PtdIns(4,5)P_2 contributing to the stimulated DAG, 6) role of the PtdIns pathway in cardiomyocyte growth and gene expression during the hypertrophic response. (Mol Cell Biochem **116**: 59–67, 1992)

Key words: heart, phosphatidylinositol cycle, alpha$_1$-adrenergic agonists, endothelins, polyunsaturated fatty acid, desensitization, hypertrophy

[1] *Abbreviations:* Phosphatidylinositol 4,5-bisphosphate, PtdIns(4,5)P_2; Phosphatidylinositol 4-monophosphate, PtdIns(4)P; Phosphatidylinositol, PtdIns; Inositol 1,4,5-trisphosphate, Ins(1,4,5)P_3; Inositol 1,3,4,5-tetrakisphosphate, Ins(1,3,4,5)P_4; Inositol 1-monophosphate, Ins(1)P; Inositol 1,4-bisphosphate, Ins(1,4)P_2; Inositol, Ins; Inositolphosphates, $InsP_n$; Guanine 5′-triphosphate, GTP; GTP binding protein, G-protein; Phosphatidylinositol-specific phospholipase C, PLC: Protein kinase C, PKC; 1,2-Diacylglycerol, DAG; Monoacylglycerol, MAG; cytidyldiphoshate-diacylglycerol, CDP-DAG; Sarcolemma, SL; Sarcoplasmic reticulum, SR; Stearic acid, 18:0; Polyunsaturated fatty acids, PUFA; Arachidonic acid, 20:4n-6; Linoleic acid, 18:2n-6; Eicosapentaenoic acid, 20:5n-3; Docosahexaenoic acid, 22:6n-3, Phosphatidic acid, PtdOH; Phospholipase D, PLD; Phosphatidylcholine, PtdChol

Address for offprints: J.M.J. Lamers, Department of Biochemistry, Faculty of Medicine and Health Sciences, Erasmus University Rotterdam, P.O. Box 1738, 3000 DR Rotterdam, The Netherlands

Introduction

The discovery of the second-messenger functions of inositol 1,4,5-trisphosphate (Ins(1,4,5)P_3) and 1,2-diacylglycerol (DAG), the products of receptor-stimulated phosphatidylinositol 4,5-bisphosphate (PtdIns(4,5)P_2) hydrolysis, marked a turning point in the studies on the mechanisms of mediation of functional hormone and neurotransmitter responses. The historical background of this discovery and the extensive bibliography of the enormously expanding knowledge in this field was recently presented by Rana and Hokin [1], the latter author who first observed the 'phosphoinositide' effect 35 years ago. It was, however, the Ca^{2+} gating hypothesis proposed by Michell [2] on basis of a survey of observations on the phosphoinositide turnover in a large number of tissues about 20 years later and the discovery of the enzyme protein kinase C (PKC) by Nishizuka *et al.* [3] in 1977, that evoked widespread interest in the role of polyphospho-PtdIns breakdown in cell signalling. PtdIns(4,5)P_2 hydrolysis catalyzed by PtdIns-specific phospholipase C (PLC) in response to activation of guanine-triphosphate (GTP) protein-coupled receptors, produces two compounds [4]. The first of these is Ins(1,4,5)P_3, a water-soluble molecule, whose formation triggers the mobilization of Ca^{2+} into the cytoplasm of stimulated cells. Ins(1,4,5)P_3 is metabolized by two routes, a part is quickly dephosphorylated to Ins(1,4)P_2 causing its immediate inactivation, but the remainder is phosphorylated to Ins(1,3,4,5)P_4, another short-lived compound that seems likely to control Ca^{2+} fluxes [5]. Ca^{2+} causes changes in the activity of a variety of proteins including protein kinases (calmodulin-dependent protein kinase and PKC). The second, membrane-associated product of PtdIns(4,5)P_2 hydrolysis is DAG. This compound activates one or more of the several isoenzymes of PKC [6]. Ins(1,4,5)P_3 and in several tissues Ins(1,3,4,5)P_4 as well, by mediating Ca^{2+} mobilization, may act synergistically with DAG to activate PKC. Activation of PKC in many cells is accompanied by its apparent translocation from the cytoplasm to the membrane compartment, where the catalytic activity is dependent on DAG, phosphatidylserine and Ca^{2+}. DAG is rapidly metabolized to phosphatidic acid (PtdOH), cytidyldiphosphate-DAG and/or monoacylglycerol (MAG) and exists only transiently after transmembrane signalling. PKC and Ca^{2+}-calmodulin-dependent protein kinase activation elicit a variety of cellular responses by phosphorylation of target proteins on serine and threonine residues. Although there is general agreement that PKC plays an important role in the initiation and/or modulation of receptor-linked responses, the precise nature or molecular details of this involvement remain elusive. There are several suggestions of possible functions of PKC, including involvement in modulation of ionconductance, regulation of receptor interaction with components of (other) signal transduction pathways, modulation of Ca^{2+} sensitivity of contractile proteins and gene expression [6].

The receptors involved in the activation of the adenylate cyclase and PtdIns cascade pathways have one common feature. They are present in the plasmamembrane as complexes with GTP binding proteins (G-proteins). The first evidence for GTP requirement of PLC [7] was obtained around 1985, almost 15 years after the discovery of GTP requirement of adenylate cyclase [8]. In spite of the clear functional evidence for a socalled G_p protein in the coupling of receptors to PLC, there is no direct information of any of its structural properties [4]. The present study is intended to organize available data on the α_1-adrenergic agonist- and endothelin-1 mediated PtdIns cycle in the myocardium and to consider possible sites at which this signalling pathway might regulate metabolic and physiological responses of the myocardium to hormones and neurotransmitters.

Early studies on the polyphospho-PtdIns turnover in the myocardium

The initial studies on the PtdIns cycle followed the α_1-adrenergic and muscarinic stimulation of ^{32}P incorporation into the total polyphospho-PtdIns fraction which response was called the 'phosphoinositide effect'. This kind of evidence was first obtained for myocardium (rat) in 1966 by Gaut and Huggins, although it was not clearly recognized as an adrenoceptor-mediated effect [9]. Many years later several reports definitely demonstrated that α_1-adrenergic and muscarinic cholinergic agonists caused an increased ^{32}P incorporation into polyphospho-PtdIns and PtdOH and not into the other phospholipids of myocardium [10–13]. In general the α_1-adrenergic effect was equally strong in atria and ventricles whereas the cholinergic effect on the ^{32}P

labelling of PtdIns was predominantly present in atria (rat, cat, guinea pig and dog). Actual progress in myocardial phosphoinositide research came around the early 1980's when the first reports appeared describing the occurrence of receptor-mediated hydrolysis of PtdIns(4,5)P_2 in non-myocardial tissues [14]. This led to a general revival of interest in the hormone-sensitive phosphoinositide turnover after it was recognized that the increased ^{32}P incorporation into the polyphospho-PtdIns and PtdOH was, in fact, a reaction secondary to receptor-stimulated PtdIns(4,5)P_2 hydrolysis resulting information of DAG. The rapid dephosphorylation of Ins(1,4,5)P_3, resynthesis of PtdIns from free inositol and CDP-DAG and the stepwise phosphorylation of PtdIns are all part of the PtdIns cycle. The turnover rate of the cycle will increase in response to the sudden acceleration of the PtdIns(4,5)P_2 hydrolysis. Therefore a more direct methodology for measurement of the rate of PtdIns cycle became fashion: the formation of ^{3}H-InsP_n from ^{3}H-Ins prelabelled polyphospho-PtdIns. However, millimolar amounts of Li^+ that blocks Ins (1)P phosphatase, have to be added to cell or tissue incubations in order to prevent rapid Ins(1)P and Ins (4)P breakdown [15]. This methodology was first applied on atria and ventricles to confirm the cholinergic stimulation of the PtdIns(4,5)P_2 breakdown [16]. Unlike the previous studies no difference in stimulation was found between atria and ventricles [12, 13]. Predominantly ^{3}H-Ins(1)P is formed and the amounts of ^{3}H-Ins(1,4)P_2, ^{3}H-Ins(1,4,5)P_3 and ^{3}H-Ins(1,3,4,5)P_4 formed are usually extremely low, making quantification difficult. The whole heart or the atrial and ventricular tissues examined in the earlier studies not only contain cardiomyocytes but also vascular smooth muscle cells, fibroblasts, vascular endothelial cells, nerve endings, etc., cells that all might contribute to the ^{32}P incorporation into polyphospho-PtdIns or PtdOH, or to the ^{3}H-InsP_n formation from ^{3}H-PtdIns(4,5)P_2. More recent studies that used almost homogeneous preparations of cultured or freshly isolated cardiomyocytes have conclusively shown the presence of muscarinic- and α_1-adrenergic receptor-coupled PtdIns cycle activity in heart muscle cells [17–20]. Besides the muscarinic- and α_1-adrenergic receptor agonists, angiotensin II [21], endothelin-1 [22], thrombin [23] and adenine nucleotides [24] have been reported to induce hydrolysis of PtdIns(4,5)P_2 in cardiac cells. Even non-receptor events such as atrial dilatation or mechanical stretching of cultured cardiac myocytes stimulate the PtdIns cycle [25, 26]. There is no doubt that G-proteins are involved in the myocardial α_1-adrenergic agonist and endothelin-1 evoked PtdIns responses, but the evoked identification of these G-proteins is complicated by the contradictory findings of its pertussis toxin sensitivity [27, 28]. Indirect evidence for a G_p-protein coupled to myocardial PtdIns-specific PLC comes from the observations that the rate of PtdIns(4,5)P_2 hydrolysis by saponin-permeabilized cardiomyocytes or cardiomyocyte microsomes is stimulated by stable GTP analogues [29, 30]. For more information on the PtdIns cycle of myocardium is referred to other reviews [28, 31, 32].

Synthesis of polyphospho-PtdIns from PtdIns

In plasmamembranes PtdIns phosphorylation at the 4-position of the inositol ring is catalyzed by a membrane-localized PtdIns 4-kinase and this reaction is followed by phosphorylation of the 5-position by Ptd(4)P 5-kinase resulting in the formation of PtdIns(4,5)P_2. These rapid phosphorylation processes can be reversed by PtdIns(4,5)P_2 and PtdIns(4)P phosphomonoesterases. The phosphorylated species of PtdIns constitute *in vivo* about 8% of the total sarcolemmal (SL) PtdIns pool and the activities of the kinases are sufficiently high for the maintenance of this relative composition. The PtdIns 4-kinase and PtdIns(4)P 5-kinase have been studied in membranes isolated from dog [33, 34] and rabbit hearts [35, 36]. Not in accordance with the generally accepted view that the PtdIns kinases are localized predominantly in the plasmamembrane is the presence of significant activities in cardiac sarcoplasmic reticulum (SR) [33, 34, 36]. On the other hand, the enzymes involved in the *de novo* synthesis of PtdIns are highly enriched in SR [36]. A PtdIns transfer protein which has been purified from the cytosol, may mediate the transport of the *de novo* synthetized PtdIns from the SR to the SL [36]. *In vitro* incubation of cardiac SR with γ-^{32}P-ATP and cyclic AMP-dependent or Ca^{2+}-calmodulin-dependent protein kinase induces not only phosphorylation of the Ca^{2+} pump regulating protein phospholamban but also that of PtdInsP and PtdIns(4)P [37]. Recently it has been shown that isoproterenol stimulation of isolated hearts is associated with changes in basal PtdIns turnover and these effects were not blocked by α_1- but by a β-adrenergic antagonist [38, 39]. Cyclic AMP- and Ca^{2+}-calmodulin-mediated control of SR- and possibly also SL levels of polyphospho-PtdIns may represent other examples of the socalled 'crosstalk' between the

PtdIns and adenylate cyclase cascade pathways [40]. Moreover, it can not be excluded that changes in the polyphospho-PtdIns levels have a direct regulatory influence on Ca^{2+} transport systems in SL and/or SR [41, 42].

In the *in vitro* experiments with isolated cardiac SL or SR, ^{32}P incorporation into PtdIns(4)P far exceeded that into PtdIns(4,5)P_2 [34, 35, 43]. However, when ^{3}H-inositol or $^{32}P_i$ are offered to cardiomyocytes the label is equally distributed between PtdIns(4)P and PtdIns (4,5)P_2 fractions [20]. Likely the endogenous PtdIns (4)P is enzymatically degraded during the isolation of SL or SR membrane fragments, because the amount of ^{32}P-PtdIns(4,5)P_2 formed in the *in vitro* experiments can be markedly increased by adding exogenous PtdIns(4)P [34, 35, 43]. An interesting observation that was recently described by Mesaeli *et al.* [43] for rat heart SL, is the sensitivity of the PtdIns kinases to inhibition by μmolar concentrations of free Ca^{2+}. Another report on canine heart SL, however, did not describe any effect of free Ca^{2+} [34]. A feedback inhibition mechanism acutely operating at the SL controlled by emptying of the Ins (1,4,5)P_3 releasable Ca^{2+} pools of SR may be considered, but this idea is difficult to reconcile with the proposed forward activation of PtdIns-specific PLC by Ca^{2+} [44, 45]. Although depletion of cellular PtdIns (4,5)P_2 does not occur, Ca^{2+} inhibition of PtdIns(4,5)P_2 synthesis is not in conflict with the desensitization of the α_1-adrenoceptor- or endothelin-1 mediated PI response observed in cultured cardiomyocytes [20, 46].

Desensitization of the receptor-mediated response in cardiomyocytes despite maintained levels of PtdIns(4,5)P_2

Previously we observed that after prolonged α_1-adrenergic agonist stimulation (over 30 min) there is an attenuation of the formation of $InsP_n$ [20]. A similar result was obtained by Otani *et al.* [47]. However, other reports described $InsP_n$ productions that remained linear up to 40 min of incubation [17, 19]. At present we have no suitable explanation for this discrepancy, but in our study we have been able to exclude factors like depletion of the PtdIns(4,5)P_2 pool that could have been the cause of the attenuation of the rate of $InsP_n$ production [20]. Recently we discovered that the desensitization occurs much faster in cultured rat neonatal cardiomyocytes with the endothelin-1 mediated PtdIns response [46]. In this work we followed the time course of changes of all ^{3}H-Ins containing cell substances upon endothelin-1 stimulation. Proof for the extremely fast resynthesis of polyphospho-PtdIns during endothelin-1 (10^{-8} M) stimulation of PtdIns response comes from the observation that the free intracellular Ins level decreased dramatically, while PtdIns(4)P and PtdIns (4,5)P_2 remained relatively constant. This occurred at a time (45 min) where more than 30% of total cellular Ins was released as $InsP_n$. Even the blockade of Ins(1)P conversion into free Ins by LiCl, that is always present in the incubation media for trapping the formed $InsP_n$, had no consequences for the maintenance of PtdIns (4,5)P_2 levels. Another, often overseen, disadvantage of using LiCl is that its presence might lead to unphysiologically high levels of DAG by limiting the supply of Ins for recycling of DAG to PtdIns. Despite the constant levels of PtdIns(4,5)P_2, the generation of $InsP_n$ nearly came to a halt after 30 min of continuous stimulation of myocardial cells with 10^{-8} M endothelin-1 [46]. That the desensitization was endothelin-1 receptor specific is proven by the finding that the α_1-adrenergic agonist response could still be evoked in endothelin-1 desensitized cells. This homologous desensitization of the receptor-mediated PtdIns response has not been described before in myocardium. There is one report on homologous desensitization of calcitonin receptor-mediated cyclic AMP response in cultured rat heart cells [48].

Based upon studies on α_1-adrenergic receptors in other tissues we proposed previously that PKC is involved in the receptor desensitization [20, 49]. Our recent findings showing homologous receptor desensitization, however, makes the PKC involvement unlikely, although receptor-specific activation of certain PKC isoenzymes can not be excluded. Indeed, cultured cardiomyocytes express at least 3 different PKC isoenzymes α-PKC, β-PKC and a PKC isoform recognized by the monoclonal antibody CK 1.4 [50, compare also 6]. What has created so much confusion is that phorbolesters, synthetic activators of PKC, efficiently counteract e.g. α_1-adrenergic agonist and endothelin-1 mediated PtdIns responses [20, 51, 52]. The explanation seems to lie in the differential activation of PKC by phorbolester and by Ins(1,4,5)P_3/DAG pathway as has been clearly shown in neonatal rat cardiomyocytes [53]. Phorbolester-induced activation of PKC is more pronounced and persistent and is accompanied by loss of PKC activity and generation of Ca^{2+}/DAG-independent kinase activity. These results suggest caution in extrapolating phorbolester effects to those of natural PKC activating agonists such as norepinephrine.

The possibility of oscillatory Ca^{2+} transients initiated by PtdIns(4,5)P_2 hydrolysis

The beating activity of the excitable cardiomyocytes is driven by regular oscillations in cytosolic Ca^{2+} levels. These Ca^{2+} oscillations arise sequentially from the voltage-dependent Ca^{2+} entry through the slow channels inducing Ca^{2+} release from the SR, pumping of Ca^{2+} back into the SR and out of the cell. The repolarization of the membrane and closing of slow channels are the signals for return to resting Ca^{2+} levels. Activation of the adenylate cyclase cascade pathway by e.g. β-adrenergic agonists increases the probability of opening of slow Ca^{2+} channels and the capacity of the SR and SL in pumping of Ca^{2+} and reduces the Ca^{2+} affinity of troponin C [54]. This results in higher peak levels of Ca^{2+} (positive inotropy) and increased frequency of Ca^{2+} oscillation (positive chronotropy). Unlike the β-adrenergic agonist response, myocardial α_1-adrenergic agonist stimulation does not affect slow Ca^{2+} channels and even lengthens the action potential duration [55]. The α_1-adrenergic agonist-induced positive inotropic effect is thought to involve Ins(1,4,5)P_3 induced release of Ca^{2+} from the SR and increased Ca^{2+} sensitivity of the myofibrillar proteins [28, 30, 31, 55]. Assuming these mechanisms of α_1-adrenergic agonist action, oscillating levels of Ins(1,4,5)P_3 are required for a pulsatile Ca^{2+} release from SR. There is now overwhelming evidence that transient Ca^{2+} oscillations found in non-excitable cells are controlled by activation of the PtdIns pathway [44, 45]. Vigne *et al.* [52] studied α_1-adrenergic agonist-, angiotensin II-, and endothelin-1-induced positive inotropy and changes in cytosolic Ca^{2+} in newborn rat atrial cells. Endothelin-1 appeared the most potent inotropic agent. Ca^{2+} transients were elicited lasting not more than 4 min. The Ca^{2+} store that is mobilized in response to receptor-mediated PtdIns(4,5)P_2 breakdown was insensitive to caffeine and ryanodine, which properties distinguishes the InsP(1,4,5)P_3- from the Ca^{2+}-inducable SR Ca^{2+} release. Since many investigations have shown that Ins(1,4,5)P_3 is able to release Ca^{2+} from cardiac SR [56], either oscillatory Ins (1,4,5)P_3 levels or Ins(1,4,5)P_3 itself may cause Ca^{2+} oscillations on top of elevated Ca^{2+} levels during a heart beat. There are indications that Ins(1,3,4,5)P_4 plays a role in the induction of Ca^{2+} oscillations in myocardium [5], but its formation is not clearly demonstrated [28, 31, 32, 57].

Fatty acids as constituents of polyphospho-PtdIns and DAG

It has been shown both *in vivo* and *in vitro* that one of the fatty acid chains of DAG must be unsaturated for optimal PKC activity [6, 58]. DAG has also been suggested to be an important source of arachidonic acid (20:4n-6). The reports that the polyphospho-PtdIns in mammalian tissues are rich in 20:4n-6 support both functional aspects of DAG, but most of the data are confined to PtdIns [59]. While in myocardium PtdIns, PtdIns(4)P and PtdIns(4,5)P_2 represent not more than respectively 8, 0.2 and 0.2% of the total phospholipid fraction, we set up an isolation procedure for accurate determination of their fatty acid composition [60]. PtdIns(4)P and PtdIns(4,5)P_2 fractions were separated from the other negatively charged phospholipid fractions using neomycin-coated glassbead columns. In rat as well as in porcine heart it could be shown that PtdIns (4)P and PtdIns(4,5)P_2 contained much more stearic acid (18:0) and less 20:4n-6 that PtdIns [60]. These results suggest interesting possibilities, that only a localized membrane pool of PtdIns is accessible to the PtdIns kinases in the sarcolemma. On the other hand, they might suggest that fatty acid compositions of polyphospho-PtdIns in SL and SR are different.

Dietary n-3 polyunsaturated fatty acids (PUFA) have been shown to result in a reduction of the α_1- but not of the β-adrenoceptor mediated changes in cardiac inotropy [61] and so the question arose whether the difference in the α_1-adrenergic response after incorporation of n-3 PUFA into heart membranes is caused by alterations in the rate of PtdIns(4,5)P_2 hydrolysis. A massive incorporation of eicosapentaenoic acid (20:5n-3) and docosahexaenoic acid (22:6n-3) at the expense of linoleic acid (18:2n-6) and 20:4n-6 was observed in the cardiac total phospholipid fraction from rats fed with n-3 PUFA. We showed that 20:5n-3 as well as 18:2n-6 treatment of cultured heart cells had no effect on the basal and α_1-adrenoceptor-mediated PLC activity [62]. However, in our study only minor changes occurred in fatty acid composition of the phospholipids. In a recent study, by increasing the duration and quantity of added 20:5n-3, we succeeded in changing 20:5n-3 content of the phospholipids of cardiomyocyte similar to the changes observed in the previous report of Reibel *et al.* [61], and we could observe a decrease in the rate of PtdIns(4,5)P_2 hydrolysis (unpublished results). In conclusion, variations in the PUFA composition of membrane phospholipids do affect the α_1-adrenoceptor stimulated rate of PtdIns(4,5)P_2 hydrolysis as we suggested before [63].

Source other than PtdIns(4,5)P_2 contributing to the stimulated DAG

There is growing appreciation that PtdIns(4,5)P_2 is not the only phospholipid whose breakdown is stimulated by extracellular signals. Recent studies clearly indicate that 'Ca^{2+} mobilizing' signal molecules also cause rapid hydrolysis of the bulk phospholipid fraction, phosphatidylcholine (PtdChol), by phospholipase D (PLD) to generate PtdOH and choline [64]. PtdOH can be degraded further by PtdOH phosphomonoesterase to DAG. PLD activation is like the PtdIns-specific PLC activation a widespread phenomenon, suggesting its critical involvement in signal transduction [64]. Convincing evidence that the activation of PLD by phorbolesters is causally dependent on activation of PKC has been presented [64]. Thus, there is little doubt that activation of PKC would result in stimulation of PLD, but does it really mediate the agonist-induced activation by PLD? The fact that PLD is activated by 'Ca^{2+} mobilizing' hormones known to activate PLC raises the attractive hypothesis that both PLD and PLC are regulated by the same G-protein [65]. It is interesting to note that PKC has a negative modulatory effect on PLC [20, 30] while exerting a positive modulatory effect on PLD. Thus it has been speculated that, if indeed both PLD and PLC are regulated by the same G-protein, PKC might serve to switch this G-protein from DAG production from the initial transient activation of PLC to the more persistent activation of PLD [65]. This would accomplish the redirection of DAG production from the quantitatively minor PtdIns(4,5)P_2 source to the more plentiful source, PtdChol, and might represent a feed-forward mechanism in the activation of PKC.

Not only DAG is second messenger in the PLD pathway but likely also its precursor PtdOH. PtdOH accumulation may cause transient increase of negative charge density at the cytoplasmic side of the membrane of specific cytosolic proteins with a net positive charge. However, at present just one paper describes muscarinic receptor-mediated PLD activation in myocardium [66]. No information is currently available regarding the fatty acid composition of receptor-mediated DAG products in myocardium. The identification of the molecular species of the stimulated DAG will be crucial in establishing a) the source(s) and b) the role(s) of DAG. In this respect attention should be paid to a report on analysis of molecular species of DAG formed during α-thrombin stimulation of cultured fibroblasts [67]. The PtdIns(4,5)P_2 contributed a significant proportion of DAG generated 15 s after α-thrombin addition, while PtdChol contributed most of the DAG generated after 5 min and up to 1 h.

Role of the PtdIns pathway in cardiomyocyte growth and gene expression during the hypertrophic response

The specific signals that initiate and maintain different types of cardiomyocyte hypertrophy, the intracellular mechanisms that transduce these signals, and the reason for postnatal failure of cardiomyocyte hyperplasia have resisted elucidation. However, the development of model systems employing isolated or cultured myocytes and the application of new techniques in cell and molecular biology seem to provide the foundation to begin to resolve these issues [68]. It appears that e.g. α_1-adrenergic agonist [68], endothelin-1 [69] and phorbolester [70] stimulation or mechanical stretch [71] of neonatal rat myocardial cells produce several features of the hypertrophic response including induction of proto-oncogenes, expression of contractile protein isogenes characteristic of fetal development (isogene switching), accumulation of contractile proteins and their corresponding mRNAs and increase in cell size. Evidence from many other systems suggests a common role of the PtdIns cycle in the regulation of gene expression by hormones, neurotransmitters and growth factors. Moreover, proto-oncogene encoded proteins appear to fall within a relatively limited number of functional groups: growth factors, receptors, protein kinases (PKC and protein-tyrosine kinases) and DNA transcription factors [72]. Evidence is accumulating that PKC is one of the important transducers connecting the α_1-adrenergic endothelin-1 and angiotensin II receptors at the cellsurface or the mechanical loading stimulus to transcription in the nucleus. Recently, Simpson *et al.*

[68] observed by immunofluorescence that upon α_1-adrenergic stimulation of cardiomyocytes the PKC isoform recognized by monoclonal antibody CK 1.4 is translocated to the myofibrils, whereas the β-PKC is translocated to surface membrane and perinuclear area. Although less work has been done in this field it is not excluded that the Ins(1,4,5)P_3/Ca^{2+} pathway, aside from Ca^{2+}/DAG activation of PKC, plays a role in the transduction of growth factor responses. Some caution should also be exercised in extrapolating the operation of these basic mechanisms in isolated cardiomyocytes to the *in vivo* situation. In particular, cell culture is an artificial system and the properties of the cells vary with culture conditions. On the other hand, reports that show acute α_1-adrenergic effects on protein synthesis in freshly isolated cardiac myocytes and perfused hearts have already appeared [73].

Acknowledgements

Parts of this work were supported by the Netherlands Heart Foundation and NATO, Belgium.

References

1. Rana RS, Hokin LE: Role of phosphoinositides in transmembrane signalling. Physiol Rev 70: 115–164, 1990
2. Michell RH: Inositol phospholipids and cell surface receptor function. Biochim Biophys Acta 415: 81–147, 1975
3. Inoue M, Kishimoto A, Takai Y, Nishizuka Y: Studies on a cyclic nucleotide-independent protein kinase and its proenzyme in mammalian tissues. II Proenzyme and its activation by Ca^{2+} dependent protease from rat brain. J Biol Chem 252: 7610–7616, 1977
4. Fain JN: Regulation of phosphoinositide-specific phospholipase C. Biochim Biophys Acta 1053: 81–88, 1990
5. Berridge MJ, Irvine RF: Inositol phosphates and cell signalling. Nature 341: 197–205, 1989
6. Nishizuka Y: The molecular heterogeneity of protein kinase C and its implication for cellular regulation. Nature 334: 661–665, 1988
7. Cockroft S, Gomperts BD: Role of guanine nucleotide binding protein in the activation of polyphosphoinositide phosphodiesterase. Nature 314: 534–536, 1985
8. Rodbell M, Krans HM, Pohl SL, Birnbaumer L: The glucagon sensitive adenyl cyclase system in plasmamembranes of rat liver. J Biol Chem 246: 1872–1876, 1971
9. Gaut ZN, Huggins CG: Effect of epinephrine on the metabolism of the inositolphosphatides in rat heart *in vivo*. Nature 212: 612–613, 1966
10. Kiss Z, Farkas T: The effect of isoproterenol on the metabolism of phosphatidylinositol by rat heart *in vitro*. Biochem Pharmacol 24: 999–1002, 1975
11. Quist E, Sanches M: Alpha-adrenergic drugs induce a phospholipid effect in heart. Proc West Pharmacol Soc 26: 333–335, 1983
12. Quist E: Evidence for a carbachol-stimulated phosphatidylinositol effect in the heart. Biochem Pharmacol 31: 3130–3133, 1982
13. Sekar MC, Roufogalis BD: Comparison of muscarinic and α-adrenergic receptors in rat atria based on phosphoinositide turnover. Life Sci 35: 1527–1533, 1984
14. Michell RH, Kirk CJ, Jones LM, Downes CP, Creba JA: The stimulation of inositol lipid metabolism that accompanies calcium mobilization in stimulated cells defined characteristics and unanswered questions. Philos Trans R Soc Lond 296: 123–137, 1981
15. Berridge MJ, Downes CP, Hanley MR: Lithium amplifies agonist-dependent phosphatidylinositol response in brain and salivary glands. Biochem J 206: 587–595, 1982
16. Brown SL, Brown JH: Muscarinic stimulation of phosphatidylinositol metabolism in atria. Mol Pharmacol 24: 351–356, 1983
17. Brown JH, Buxton IL, Brunton LL: α_1-Adrenergic and muscarinic cholinergic stimulation of phosphoinositide hydrolysis in adult rat cardiomyocytes. Circ Res 57: 532–537, 1985
18. Meij JTA, Lamers JMJ: Alpha-1-adrenergic stimulation of phosphoinositide breakdown in cultured neonatal rat ventricular myocytes. Mol Cell Biochem 88: 73–75, 1989
19. Steinberg SF, Chow YK, Robinson RB, Bilezikian JP: A pertussis toxin substrate regulates $alpha_1$-adrenergic dependent phosphatidylinositol hydrolysis in cultured rat myocytes. Endocrinology 120: 1889–1895, 1987
20. Meij JTA, Lamers JMJ: Phorbolester inhibits $alpha_1$-adrenoceptor mediated phosphoinositol breakdown in cardiomyocytes. J Mol Cell Cardiol 21: 661–668, 1989
21. Baker KM, Singer HA: Identification and characterization of guinea pig angiotensin II ventricular and atrial receptors: coupling to inositolphosphate production. Circ Res 62: 896–904, 1988
22. Vigne P, Lazdunski M, Frelin C: The inotropic effect of endothelin-1 on rat atria involves hydrolysis of phosphatidylinositol. FEBS Lett 249: 143–146, 1989
23. Chien WW, Mohabir R, Clusin WT: Effect of thrombin on the calcium homeostasis in chick embryonic heart cells. Receptor-operated calcium entry with inositol trisphosphate and a pertussis-toxin sensitive G-protein as second messengers. J Clin Invest 85: P1436–P1443, 1990
24. Leggsyer A, Poggioli J, Renard D, Vassort G: ATP and other adenine compounds increase mechanical activity and inositol trisphosphate production in rat heart. J Physiol 401: 185–199, 1988
25. Komuro J, Kaida T, Shibazaki T, Kurabayashi M, Katoh Y, Hoh E, Takaku F, Yazaki Y: Stretching cardiac myocytes stimulates proto-oncogen expression. J Biol Chem 265: 3595–3598, 1990
26. Von Harsdorf R, Lang R, Fullerton M, Smith AI, Woodcock EA: Right atrial dilatation increases inositol-(1,4,5)trisphosphate accumulation. FEBS Lett 233: 201–205, 1988
27. Robishaw JD, Foster KA: Role of G proteins in the regulation of the cardiovascular system. Annu Rev Physiol 51: 229–244, 1989
28. Brown JH, Jones LG: Phosphoinositide metabolism in the heart. In: Putney JW (ed) Phosphoinositides and receptor mechanisms. Alan R. Liss Inc. New York, 1986, pp 245–270
29. McDonough PM, Goldstein D, Brown JH: Elevation of cytoplasmic calcium concentration stimulates hydrolysis of phosphatidyli-

nositol biphosphate in chick heart cells: Effect of sodium channel activators. Mol Pharmacol 33: 310–315, 1988
30. Meij JTA, Bezstarosti K, Panagia V, Lamers JMJ: Phorbolester and the actions of phosphatidylinositol 4,5-bisphosphate specific phospholipase C and protein kinase C in microsomes prepared from cultured cardiomyocytes. Mol Cell Biochem 105: 37–47, 1991
31. De Chaffoy de Courcelles D: Is there evidence of a role of the phosphoinositol-cycle in the myocardium? Mol Cell Biochem 88: 65–72, 1989
32. Meij JTA, Panagia V: Catecholamines and Heart Disease: status of phospholipid signalling pathways. In: Ganguly PK (ed) Catecholamines and heart disease. CRC Press, Boca Raton, in press
33. Quist E, Satumtira N, Powell P: Regulation of polyphosphoinositide synthesis in cardiac membranes. Archs Biochem Biophys 271: 21–32, 1989
34. Kasinathan C, Xu ZC, Kirchberger MA: Polyphosphoinositide formation in isolated cardiac plasmamembranes. Lipids 24: 818–823, 1989
35. Varsanyi M, Messer M, Brandt NR, Heilmeyer LMG: Phosphatidylinositol 4,5-bisphosphate formation in rabbit skeletal and heart muscle membranes. Biochem Biophys Res Commun 138: 1395–1404, 1986
36. Wolf RA: Synthesis, transfer and phosphorylation of phosphoinositides in cardiac membranes. Am J Physiol 259: C987–C994, 1990
37. Farkas G, Enyedi A, Sarkadi B, Gardos G: Cyclic AMP-dependent protein kinase and Ca^{2+}-calmodulin stimulate the formation of polyphosphoinositides in a sarcoplasmic reticulum preparation of rabbit heart. FEBS Lett 176: 235–238, 1984
38. Edes I, Solaro RJ, Kranias EG: Changes in phosphoinositide turnover in isolated guinea pig hearts stimulated with isoproterenol. Circ Res 65: 989–996, 1989
39. Jakab G, Rapundalo S, Solaro RJ, Kranias EG: Phosphorylation of phospholipids in isolated guinea pigs stimulated with isoprenaline. Biochem J 251: 189–194
40. Houslay MD: 'Crosstalk': a pivotal role for protein kinase C in modulating relationships between signal transduction pathways. Eur J Biochem 195: 9–27, 1991
41. Lamers JMJ: Cardiac sarcolemmal calcium transport systems and their modulation by the second messengers cyclic AMP, calcium and phosphoinositide products. In: AM Kidway (ed) Sarcolemmal Biochemistry. CRC Press Inc., Boca Raton, Florida, 1987, vol. II, pp 67–98.
42. Pierce GN, Panagia V: Role of phosphatidylinositol in cardiac sarcolemmal membrane sodium-calcium exchange. J Biol Chem 264: 15344–15350, 1989
43. Mesaeli N, Lamers JMJ, Panagia V: Characteristics and Ca^{2+} sensitivity of polyphosphoinositide synthesis in rat heart sarcolemma. J Mol Cell Cardiol 23(Suppl III): 566, 1991
44. Peterson OH, Wakui M: Oscillating Ca^{2+} signals evoked by activation of receptors linked to inositol lipid hydrolysis: Mechanism of generation. J Membrane Biol 118: 93–105 (1990)
45. Berridge MJ: Calcium oscillations. J Biol Chem 265: 9583–9586, 1990
46. Van Heugten HAA, Bezstarosti K, Lamers JMJ: Endothelin-1 induced phosphoinositide breakdown in cardiomyocytes is subject to homologous desensitization. J Mol Cell Cardiol, 23 (Suppl V): 547, 1991
47. Otani H, Otani H, Das DK: Evidence that phosphoinositide response is mediated by α_1-adrenoceptor stimulation and not linked with excitation-contraction coupling in cardiac muscle. Biochem Biophys Res Commun 136: 863–869, 1986
48. Fisher RA, Robertson SM, Olson MS: Stimulation and homologous desensitization of calcitonin gene-related peptide receptors in cultured beating rat heart cells. Endocrinology 123: 106–122, 1988
49. Bouvier M, Leeb-Lundberg LMF, Benovic JL, Caron MG, Lefkowitz RJ: Regulation of adrenergic receptor function by phosphorylation. II Effects of agonist occupancy on phosphorylation of α_1- and β_2-adrenergic receptors by protein kinase C and the cyclic AMP-dependent protein kinase. J Biol Chem 262: 3106–3112, 1987
50. Simpson PC, Kariya K, Karns LR, Long CS, Karliner JS: Adrenergic hormones and control of cardiac myocyte growth. Mol Cell Biochem 104: 35–43, 1991
51. Endoh M, Otomo J, Norota I: Phorbol-12,13-dibutyrate antagonizes the α_1-adrenoceptor-mediated positive inotropic effect in the rabbit ventricular myocardium. Br J Clin Pharmac 30: 115S–117S
52. Vigne P, Breittmayer J-P, Marsault R, Frelin C: Endothelin mobilizes Ca^{2+} from a caffeine- and ryanodine-insensitive intracellular pool of rat atrial cells. J Biol Chem 265: 6782–6787, 1990
53. Henrich CJ, Simpson PC: Differential acute and chronic response of protein kinase C in cultured neonatal rat heart myocytes to α_1-adrenergic and phorbolester stimulation. J Mol Cell Cardiol 20: 1081–1085, 1988
54. Lamers JMJ: Calcium transport system in cardiac sarcolemma and their regulation by the second messengers cyclic AMP and calcium-calmodulin. Gen Physiol Biophys 4: 143–154, 1985
55. Benfey BG: Minireview. Function of myocardial α-adrenoceptors. Life Sci 46: 743–757, 1990
56. Kentish JC, Barsotti RJ, Trevor JL, Mulligan IA, Patel JR, Ferenczi MA: Calcium release from cardiac sarcoplasmic reticulum induced by photorelease of calcium or Ins(1,4,5)P_3. Am J Physiol 258, H610–H615, 1990
57. Heathers GP, Corr PB, Rubin LJ: Transient accumulation of inositol (1,3,4,5)-tetrakisphosphate in response to alpha$_1$-adrenergic stimulation in adult cardiac myocytes. Biochem Biophys Res Commun 156: 485–492, 1988
58. Bell RM, Burns DJ: Lipid activation of protein kinase C. J Biol Chem 266: 4661–4664
59. Holub BJ: The cellular forms and functions of the inositol phospholipids and their metabolic derivatives. Nutr Rev 45: 65–71, 1987
60. Lamers JMJ, Dekkers DHW, Mesaeli N, Meij JTA, Panagia V, Van Heugten HAA: The fatty acid composition of (poly)phosphoinositides from rat and pig myocardium. J Mol Cell Cardiol, 23 (Suppl V): 5100, 1991
61. Reibel DK, Holahan MA, Hock CE: Effects of dietary fish oil on cardiac responsiveness to adrenoceptor stimulation. Am J Physiol 254: H494–H499, 1988
62. Meij JTA, Bordoni A, Dekkers DHW, Guarnieri C, Lamers JMJ: Alterations in polyunsaturated fatty acid composition of cardiac membrane phospholipids and α_1-adrenoceptor mediated phosphatidylinositol turnover. Cardiovasc Res 24: 94–101, 1990
63. Lamers JMJ, Hartog JM, Verdouw PD, Hülsmann WC: Dietary fatty acids and myocardial function. Bas Res Cardiol 82: 209–221, 1987

64. Exton JH: Signaling through phosphatidylcholine breakdown. J Biol Chem 265: 1–4, 1990
65. Liscovitch M: Signal-dependent activation of phosphatidylcholine hydrolysis: role of phospholipase D. Biochem Soc Transact 19: 402–407, 1991
66. Lindmar R, Löffelholz K, Sandman J: On the mechanisms of muscarinic hydrolysis of choline phospholipids in the heart. Biochem Pharmacol 37: 4689–4695, 1988
67. Pessin MS, Raben DM: Molecular species analysis of 1,2-diglycerides stimulated by α-thrombin in cultured fibroblasts. J Biol Chem 264: 8729–8738, 1989
68. Simpson PC, Kariya K-I, Karns LR, Long CS, Karliner JS: Adrenergic hormones and control of cardiac myocyte growth. Mol Cell Biochem 104: 35–43, 1991
69. Shubeita HE, McDonough PM, Harris AN, Knowlton KU, Glembotski CC, Brown JH, Chien KR: Endothelin induction of inositol phospholipid hydrolysis, sarcomere assembly, and cardiac gene expression in ventricular myocytes. A paracrine mechanism for myocardial mechanism for hypertrophy. J Biol Chem 265: 20555–20562, 1990
70. Dunnmon PM, Iwaki K, Henderson SA, Sen A, Chien KR: Phorbolesters induce immediate-early genes and activate cardiac gene transcription in neonatal rat myocardial cells. J Mol Cell Cardiol 22: 901–910, 1990
71. Komuro I, Katoh Y, Kaida T, Shibazaki Y, Kurabayashi M, Hoh E, Takaku F, Yazaki Y: Mechanical loading stimulates cell hypertrophy and specific gene expression in cultured rat cardiac myocytes. Possible role for protein kinase C activation. J Biol Chem 266: 1265–1268, 1991
72. Simpson PC: Proto-oncogenes and cardiac hypertrophy. Ann Rev Physiol 51: 189–202, 1988
73. Fuller S, Gaitanaki CJ, Sugden PH: Effects of catecholamines on protein synthesis in cardiac myocytes and perfused hearts isolated from adult rats. Stimulation of translation is mediated through the α_1-adrenoceptor. Biochem J 266: 727–736, 1990

Molecular and Cellular Biochemistry **116**: 69–73, 1992.

Modulation of phosphatidylethanolamine biosynthesis by exogenous ethanolamine and analogues in the hamster heart

Christopher R. McMaster, Paul G. Tardi and Patrick C. Choy
Department of Biochemistry and Molecular Biology, Faculty of Medicine, University of Manitoba, 770 Bannatyne Ave., Winnipeg, MB., Canada, R3E 0W3

Abstract

In the hamster heart, exogenous ethanolamine is taken up by the heart and utilized for the biosynthesis of phosphatidylethanolamine. The role of the exogenous supply of ethanolamine on phosphatidylethanolamine biosynthesis was examined by perfusing hamster heart with various concentrations of labelled ethanolamine. Analysis of the radioactivity distributed in the ethanolamine-containing metabolites indicated that at low exogenous ethanolamine concentrations ($\leq 0.1\ \mu M$), the conversion of phosphoethanolamine to CDP-ethanolamine was rate-limiting for phosphatidylethanolamine biosynthesis. However, perfusion with higher concentrations of ethanolamine ($\geq 0.4\ \mu M$) resulted in the phosphorylation of ethanolamine becoming rate-limiting. Since the intracellular ethanolamine levels remained unchanged, the alterations in radioactivity distribution suggested that the newly imported ethanolamine was preferentially utilized for phosphatidylethanolamine biosynthesis. The effects of ethanolamine analogues on ethanolamine uptake and subsequent conversion to phosphatidylethanolamine at physiological concentrations of exogenous ethanolamine were examined. Monomethylethanolamine was found to inhibit ethanolamine uptake, the conversion of ethanolamine to phosphoethanolamine and incorporation of radioactivity into phosphatidylethanolamine. The accumulation of radioactivity in the ethanolamine fraction by monomethylethanolamine, despite of the inhibition of ethanolamine uptake, further confirms the rate-limiting role of ethanolamine kinase in the biosynthesis of phosphatidylethanolamine. (Mol Cell Biochem **116**: 69–73, 1992)

Key words: ethanolamine uptake, ethanolamine kinase, phosphatidylethanolamine, hamster heart

Introduction

Phosphatidylethanolamine is a major phospholipid in mammalian hearts [1]. In the hamster heart, over 90% of the phosphatidylethanolamine is synthesized *de novo* via the CDP-ethanolamine pathway [2]. Ethanolamine taken up by the heart is rapidly phosphorylated to phosphoethanolamine which is then converted to CDP-ethanolamine. The CDP-ethanolamine is condensed with diacylglycerol to form phosphatidylethanolamine [3]. Although the first committed step for the *de novo* synthesis of phosphatidylethanolamine is catalyzed by ethanolamine kinase, the reaction catalyzed by CTP: phosphoethanolamine cytidylyltransferase for the conversion of phosphoethanolamine to CDP-ethanolamine is generally regarded as the rate-limiting step in this path-

Address for offprints: C.R. McMaster, Department of Biochemistry and Molecular Biology, Faculty of Medicine, University of Manitoba, 770 Bannatyne Ave., Winnipeg, MB, Canada, R3E 0W3

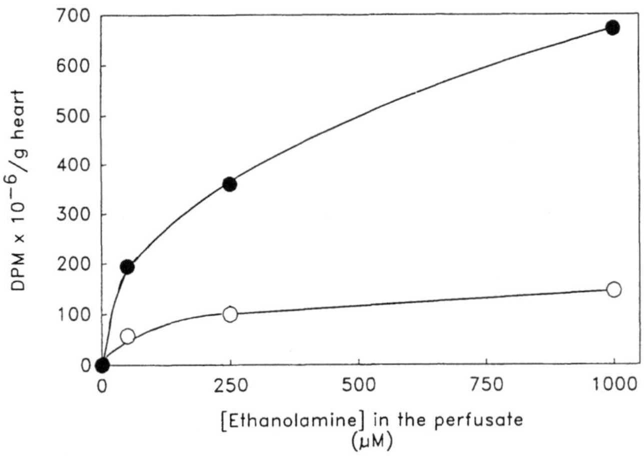

Fig. 1. Hamster hearts were perfused in Krebs-Henseleit buffer containing 0.04–1,000 μM [1-^{3}H] ethanolamine. Total uptake of radioactivity (●) and incorporation into phosphatidylethanolamine (○) are indicated. The values are the average of at least 4 different experiments done in duplicate.

way [3, 4]. In the last two decades, several control mechanisms for the regulation of phosphatidylethanolamine biosynthesis in mammalian tissues were identified [2–9], but it was not known whether the exogenous supply of ethanolamine would affect the *de novo* biosynthesis of phosphatidylethanolamine. Perfusion of the isolated hamster hearts with various concentrations of exogenous radioactive ethanolamine and the determination of the distribution of labelled metabolites appear to be a viable approach to study the role of exogenous ethanolamine on phosphatidylethanolamine biosynthesis. The effect of ethanolamine analogues on ethanolamine uptake and subsequent metabolism to phosphatidylethanolamine was examined at a physiological concentration of exogenous ethanolamine.

Materials and methods

Materials

[1-^{3}H] Ethanolamine hydrochloride (8.8 Ci/mmol) was purchased from Amersham International Limited. Ethanolamine, choline, glycine, alanine and activated charcoal were obtained from Sigma Chemical Company. Monomethylethanolamine and dimethylethanolamine were the products of Aldrich Chemical Company, Inc. Phenylisothiocyanate was purchased from Pierce Chemical Company. Pyridine, triethylamine and thin layer chromatography plates were purchased from Fisher Scientific. Celite was obtained from Supelco Inc. Acetonitrile (HPLC-grade) and all other chemicals were acquired from the Canlab division of Baxter Diagnostics Corporation.

Hamster heart perfusions

Hearts from Syrian golden hamsters (120 ± 20 g) were perfused in the Langendorff mode with Krebs-Henseleit buffer containing 0.1–1,000 μM labelled ethanolamine at an aortic pressure of 80 mm Hg [2]. Subsequent to perfusion, the heart was homogenized in chloroform/methanol (1/1; v/v) and the homogenate was centrifuged at 2,000 × g to pellet the tissue. Chloroform and water were added to cause phase separation and the aqueous and organic phases were removed. The ethanolamine-containing metabolites in the aqueous phase were separated by thin-layer chromatography (TLC) in a solvent containing ethanol/2% ammonium hydroxide (1/2; v/v) and visualized with ninhydrin spray. The orga-

Table 1. Radioactivity incorporation into ethanolamine-containing metabolites

[Ethanolamine] in the perfusate	Ethanolamine	Phosphoethanolamine	CDP-ethanolamine
	(dpm × 10^{-6}/g heart)		
0.04 μM #	0.019 ± 0.005 (30%)	0.035 ± 0.004 (56%)	0.009 ± 0.002 (14%)
0.1 μM #	0.051 ± 0.015 (38%)	0.071 ± 0.010 (53%)	0.013 ± 0.002 (9%)
250 μM +	1.275 ± 0.105 (51%)	0.929 ± 0.132 (37%)	0.294 ± 0.008 (12%)
1000 μM +	3.257 ± 0.832 (58%)	1.797 ± 0.133 (32%)	0.562 ± 0.090 (10%)

Hearts were perfused as described in Materials and methods. Specific radioactivities of the perfusate were 3.33 × 10^6 dpm/nmol (#) or 3.33 × 10^4 dpm/nmol (+).

nic phase metabolites were also separated by TLC in a solvent containing chloroform/methanol/water/acetic acid (70/30/4/2; v/v) and visualized with iodine stain. The corresponding bands were removed from the TLC place and radioactivity was determined by scintillation counting.

Pool size determinations

Ethanolamine pool size determinations were determined by taking an aliquot of the aqueous phase and applying it to a charcoal/celite column (1/2; w/w). Ethanolamine was eluted from the column with 10 mM sodium phosphate buffer, pH 7.4, containing 2% ethanol. An aliquot of this sample was reacted with phenylisothiocyanate [10] and the resulting derivative was separated from other metabolites by reverse phase HPLC with a Pharmacia Spherisorb ODS (3 μm) column. A gradient from 0–40% acetonitrile in 10 mM potassium phosphate buffer, pH 6.4, was used to elute the derivatives. Peak area analysis was performed by a Beckman 450 Data System. Phosphatidylethanolamine was separated from other phospholipids by TLC as described in the preceding section, and the lipid fraction was eluted from the silica gel. The pool size of phosphatidylethanolamine was determined by analysis of the lipid phosphorus content [11].

Table 2. Effect of ethanolamine analogues on ethanolamine uptake

Ethanolamine analogue in the perfusate	Uptake of ethanolamine
	(dpm × 10^{-6}/g heart)
Control	2.06 ± 0.38
0.5 mM Monomethylethanolamine	1.26 ± 0.19*
0.5 mM Dimethylethanolamine	1.96 ± 0.10
1.0 mM Glycine	2.24 ± 0.34
1.0 mM Alanine	2.00 ± 0.24

Hamster hearts were perfused with 50 μM [1-^{3}H] ethanolamine in the presence of ethanolamine analogues for 30 min. The specific radioactivity of ethanolamine in the perfusate was 3.33 × 10^4 dpm/nmol. Values are the mean ± standard deviation of three separate experiments done in duplicate. * $p < 0.02$.

Results

Effect of exogenous ethanolamine on ethanolamine uptake and its incorporation into phosphatidylethanolamine

Ethanolamine uptake was linear from 0.04–100 μM ethanolamine in the perfusate. The efficiency of incorporation of the newly imported ethanolamine into phosphatidylethanolamine was not proportional to ethanolamine uptake but decreased progressively over the linear range of ethanolamine uptake. The reduction in efficiency was more pronounced at high rates of ethanolamine uptake (Fig. 1). For example, over 70% of the newly imported ethanolamine was incorporated into phosphatidylethanolamine when the heart was perfused with 0.1 μM ethanolamine, but the efficiency was reduced to 29% at 1,000 μM ethanolamine in the perfusate. The distribution of radioactivity in the metabolites of the CDP-ethanolamine pathway was examined (Table 1). When ethanolamine uptake was low (0.04–0.1 μM ethanolamine in the perfusate) the majority of the radioactivity was associated with the phosphoethanolamine fraction. At higher rates of ethanolamine uptake (0.4–1000 μM ethanolamine in the perfusate) the radioactivity was found to accumulate in the ethanolamine fraction with a corresponding decrease in radioactivity in the phosphoethanolamine fraction. The radioactivity associated with the CDP-ethanolamine fraction was substantially less than those found in the ethanolamine or phosphoethanolamine fractions, and remained relatively constant under all experimental conditions.

Effect of ethanolamine analogues on ethanolamine uptake

Hamster hearts were perfused with 50 μM [1-^{3}H] ethanolamine in the presence of various ethanolamine analogues. The presence of 0.5 mM monomethylethanolamine in the perfusate significantly inhibited ethanolamine uptake. However, the ethanolamine uptake in the heart was not affected by similar concentrations of dimethylethanolamine, glycine, or alanine in the perfusate (Table 2). In view of the inhibition of ethanolamine uptake by monoethylethanolamine, analysis of ethanolamine-containing metabolites in control hearts and hearts perfused with monomethylethanolamine were conducted. The analysis revealed that the radioactivity associated with ethanolamine fraction was not significantly changed but the radioactivity in the phosphoethanolamine fraction was decreased to 33% of control value, together with a 63% decrease in radioacticity in the CDP-ethanolamine fraction.

Pool size analysis

Changes in radioactivity in the ethanolamine and phosphoethanolamine fractions might be caused by changes in the pool size of ethanolamine. However, analysis of the pool size of ethanolamine in hamster hearts perfused with different concentrations (0.04–1,000 μM) of ethanolamine revealed that there was no significant alteration in the intracellular pool of ethanolamine. The total phosphatidylethanolamine content in the heart was not changed under all experimental conditions. The ethanolamine pool size ranged from 1.07–1.27 μmol/g heart while the phosphatidylethanolamine pool size was 11.41–12.43 μmol/g heart.

Discussion

It is clear from this study that exogenous ethanolamine modulates phosphatidylethanolamine biosynthesis via the CDP-ethanolamine pathway in the hamster heart. However, the efficiency with which exogenous ethanolamine was incorporated into phosphatidylethanolamine was not proportional with ethanolamine uptake. This discrepancy was further investigated by analyzing the ethanolamine-containing metabolites when hearts were perfused with various concentrations of ethanolamine. When ethanolamine import was low ($\leq$0.1 μM ethanolamine in the perfusate), the majority of the radioactivity was associated with phosphoethanolamine which confirmed the previous finding that the reaction catalyzing the conversion of phosphoethanolamine to CDP-ethanolamine was rate-limiting [3, 5]. As the rate of ethanolamine uptake increased, the percentage of radioactivity found in the ethanolamine fraction increased, with a corresponding decrease in phosphoethanolamine. This indicated that the ethanolamine kinase reaction had become rate-limiting for phosphatidylethanolamine biosynthesis. The shift in the rate-limiting step was not caused by a change in the overall ethanolamine concentration in the tissue. An alternative explanation is that the newly imported ethanolamine would constitute a separate pool from the existing cellular ethanolamine pool. The newly imported ethanolamine pool was preferentially utilized for phosphatidylethanolamine biosynthesis. When the rate of ethanolamine uptake saturated the capacity of the heart to phosphorylate ethanolamine, the phosphorylation of ethanolamine became rate-limiting. In light of the reported serum ethanolamine levels (10–50 μM) [12, 13], the conversion of ethanolamine to phosphoethanolamine seem to be an important factor for the regulation of phosphatidylethanolamine biosynthesis in the hamster heart.

The effects of various ethanolamine analogues on ethanolamine uptake and phosphatidylethanolamine biosynthesis by the hamster heart were examined. Monomethylethanolamine and dimethylethanolamine are ethanolamine analogues with modified amino terminals while glycine and alanine contain modified carboxy moieties. In this study, only monomethylethanolamine significantly inhibited ethanolamine uptake. As the majority of the radioactivity in the ethanolamine-containing metabolites was associated with the ethanolamine fraction in the monomethylethanolamine perfused heart, the rate-limiting role of ethanolamine kinase was not changed by the presence of monomethylethanolamine. Interestingly, the label associated with phosphoethanolamine was decreased to 33% by monomethylethanolamine perfusion, which suggests that the ethanolamine kinase was also inhibited by monomethylethanolamine.

In conclusion, this study has highlighted the rate-limiting role of the ethanolamine kinase reaction in

Table 3. Radioactivity incorporated into ethanolamine-containing metabolites when perfused with monomethylethanolamine

Ethanolamine analogue	Ethanolamine	Phosphoethanolamine	CDP-ethanolamine
	(dpm $\times$ 10^{-6}/g heart)		
Control	1.46 $\pm$ 0.14	0.40 $\pm$ 0.03	0.08 $\pm$ 0.01
Monomethylethanolamine	1.34 $\pm$ 0.14	0.13 $\pm$ 0.02	0.05 $\pm$ 0.01

Isolated hamster hearts were perfused as described in Table 2.

phosphatidylethanolamine biosynthesis at physiological concentrations of ethanolamine. It is clear that the uptake of ethanolamine and its subsequent phosphorylation are tightly coupled. Our results demonstrate the intimate relationship between ethanolamine uptake and its subsequent synthesis into phosphatidylethanolamine via the CDP-ethanolamine pathway.

Acknowledgements

The study was supported by the Medical Research Council of Canada. CRM is the recipient of a Medical Research Council Studentship. PGT is the recipient of a Canadian Heart and Stroke Foundation Studentship.

References

1. Ansell GB, Spanner S: Phosphatidylserine, phosphatidylethanolamine and phosphatidylcholine. In: JN Hawthorne and GB Ansell (eds) Phospholipids. Elsevier Biomedical Press. Amsterdam, 1982, pp 1–49
2. Vance DE: Phospholipid metabolism in eucaryotes. In: DE Vance and JE Vance (eds) Biochemistry of Lipids and Membranes. The Benjamin/Cummings Publishing Company, Inc., Menlo Park, California, 1985, pp 242–270
3. Zelinski TA, Choy PC: Phosphatidylethanolamine biosynthesis in isolated hamster heart. Can J Biochem 60: 817–823, 1982
4. Sundler R, Åkesson B: Regulation of phospholipid biosynthesis in isolated rat hepatocytes. J Biol Chem 250: 3359–3367, 1975
5. Sundler R: Biosynthesis of rat liver phosphatidylethanolamines from intraportally injected ethanolamine. Biochim Biophys Acta 306: 218–226, 1973
6. Yorek MA, Rosario RT, Dudley DT, Spector AA: The utilization of ethanolamine and serine for ethanolamine phosphoglyceride synthesis by human Y79 retinoblastoma cells. J Biol Chem 260: 2930–2936, 1985
7. Lipton BA, Davidson EP, Ginsburg BH, Yorek MA: Ethanolamine metabolism in cultured bovine aortic endothelial cells. J Biol Chem 265: 7195–7201, 1990
8. Tijburg LBM, Geelen MJH, Van Golde LMG: Biosynthesis of phosphatidylethanolamine via the CDP-ethanolamine route is an important pathway in isolated rat hepatocytes. Biochem Biophys Res Comm 160: 1275–1280, 1989
9. Arthur G, Page L: Synthesis of phosphatidylethanolamine and ethanolamine plasmalogen by the CDP-ethanolamine and decarboxylase pathway in rat heart, kidney and liver. Biochem J 273: 121–125, 1991
10. Heinrikson RL, Meridith SC: Amino acid analysis by reverse-phase high-performance liquid chromatography: precolumn derivatization with phenylisothiocyanate. Anal Biochem 136: 65–74, 1984
11. Bartlett GR: Phosphorus assay in column chromatography. J Biol Chem 234: 466–468, 1959
12. Baba S, Watanabe Y, Gejyo F, Arakawa M: High-performance liquid chromatographic determination of serum aliphatic amines in chronic renal failure. Clin Chim Acta 136: 49–56, 1984
13. Milakofsky L, Hare TA, Miller JM, Vogel WH: Rat plasma levels of amino acids and related compounds during stress. Life Sci 36: 753–761, 1985

Molecular and Cellular Biochemistry **116**: 75–78, 1992.

Eicosapentaenoic and docosahexaenoic acids in cultured rat ventricular myocytes and hypoxia-induced alterations of phospholipase-A activity

Alain Grynberg,[1] Gilles Nalbone,[2] Jeannie Leonardi,[2] Huguette Lafont[2] and Pierre Athias[3]

[1] Station de Recherches sur la Qualité des Aliments de l'Homme, INRA, 21034 Dijon, [2] INSERM U130, 13009 Marseille and [3] Laboratoire d'électrophysiologie, Faculté de Médecine, 21034 Dijon, France

Abstract

Hypoxia was reported to induce a decrease in phosphatidylcholine-hydrolyzing phospholipase activity (PC-PLA) in cultured rat cardiomyocytes. This work was intended to compare the influence of the presence of either eicosapentaenoic acid (EPA) or docosahexaenoic acid (DHA) in the phospholipids on the PC-PLA activity in normoxic and hypoxic conditions. The enrichment of the medium with EPA or DHA resulted in cell phospholipids containing about 2% or 22% DHA, respectively. These cells were then submitted for 3.5 h to either normoxia or hypoxia and the PC-PLA activities were assayed using [1-^{14}C] dioleoyl-PC (pH 8.4 for PC-PLA2 and 4.9 for PC-PLA1). The results show that both enzymic activities are significantly higher in DHA-rich cardiomyocytes. Hypoxia induced a significant decrease in PC-PLA2 (about 25%) which was not statistically different between the two groups of cells. The hypoxia-induced decrease in PC-PLA1 was not found significant. In conclusion, the nature of the long chain n-3 polyunsaturated fatty acids in the phospholipids appears to contribute to the regulation of PC-PLA activity but not to influence its decrease during hypoxia. (Mol Cell Biochem **116**: 75–78, 1992)

Key words: heart, cell culture, hypoxia, phospholipase, EPA, DHA

Introduction

Fish oil administration to human has been recommended to lower the incidence of cardiovascular diseases. This beneficial effect essentially concerns a reduced platelet aggregation which results in a lower thrombotic risk [see ref 1 for review] and has been mainly attributed to eicosapentaenoic acid (C20 : 5 n-3, or EPA). But the effect of the other major fatty acids (FA) of fish oils, such as docosahexaenoic acid (C22 : 6 n-3, or DHA) remains largely unknown. Numerous studies demonstrated that a modification of the fatty acid intake results in large alterations in the heart membrane fatty acid composition [2–6]. On the basis of studies performed on animals [7], human [8] or cultured cells [5], a direct influence of such alterations on cardiac function has been proposed in physiologic as well as in hypoxic conditions. Heart membrane stability is essential to ensure a correct electrophysiological activity under normoxic conditions as well as a satisfactory recovery after ischemic insult and reperfusion. Phospholipases A are supposed to play a determinant role in the membrane stability process [9]. The activation of phospholipases has been proposed to explain the enhanced degradation

Address for offprints: A. Grynberg, Unité de Nutrition Lipidique, I.N.R.A., 17 rue Sully, 21034 Dijon, France of cardiac membranes after a long period of ischemia or during reperfusion [10]. Recently, it was reported that in rabbit heart, a highly plasmalogen selective phospholipase A2 was activated as early as 15 min of ischemia [11]. Conversely, other results obtained from either rat heart [12, 13] or cultured cardiomyocytes [14] showed a reduction of diacylphospholipid-hydrolyzing phospholipases. Recently, we showed that in standard culture conditions, the activity of phospholipases in the cardiac cell could be influenced by its phospholipid fatty acid composition. Increasing the n-3/n-6 polyunsaturated fatty acid ratio lead to a reduced phosphatidylcholine-hydrolyzing phospholipase A2 (PC-PLA2) activity, although lysosomial acid phosphatidylcholine-hydrolyzing phospholipase A1 (PC-PLA1) was not affected [15].

This work was intended to compare the influence of the presence of either eicosapentaenoic acid or docosahexaenoic acid in cardiomyocyte phospholipids on the PC-PLA activities during hypoxia.

Material and methods

Culture media

The standard culture medium was composed of Ham's F10 medium (Seromed) supplemented with 10% fetal calf serum and 10% human serum [16]. Two experimental media (EPAMed and DHAMed) were obtained by supplementation of the standard medium with either EPA or DHA (0.1 mM), respectively. Sterilization was achieved by filtration through 0.22 µm filters (Millex GS, Millipore). The calcium activity was standardized to 1.2 mM in all the media. The final fatty acid composition of these 2 experimental media is presented in Table 1.

Culture preparation and handling

The cultures of rat ventricular cardiomyocytes were prepared from 2 to 4 days-old Wistar rats. After mincing the hearts, the cardiac cells were dissociated through seven proteolytic treatments in 0.1% trypsin (Difco) at 30° C. Myocyte enrichment was performed using a two-step (30 and 120 min) differential attachment procedure [5]. The cells were resuspended in standard culture medium. The cell suspension was seeded in 60 × 15 mm petri dishes (Falcon 3802) at a final density of 2.10^6 cells per dish, and incubated at 37° C in a humidified atmosphere containing 5% CO_2, 19% O_2 and 76% N_2. The cells were grown for 1 day (i.e. till confluency) in standard culture medium and then for 4 days in either EPAMed or DHAMed.

Hypoxic treatment

The medium was withdrawn and the cells were rinced 3 times with Puck's F balanced salt solution. The cultures were then incubated in oxidizable substrate free Puck's F solution covered with a layer of paraffin oil [17]. The culture dishes werer placed in a 24-well gaz-controlled incubation chamber [18] continuously flushed (600 ml/min) with either air (control treatment) or N_2 (hypoxic treatment). For each medium, 6 N_2-treated dishes were compared to 6 control (air-treated) dishes.

Lipid analysis

In the media, the fatty acids were transmethylated without previous extraction, according to Lepage and Roy [19]. In the cells, the lipids were extracted by the method of Folch *et al.* [20] and the phospholipids were separated from non-phosphorous lipids using silica cartridges [21]. The fatty acid were transmethylated with BF_3-methanol and the methyl esters were analysed by gas chromatography on a Carbowax 20 M capillary column.

Table 1. Fatty Acid Composition (%) of the EPA or DHA supplemented media (EPAMed and DHAMed, respectively)

	EPAMed	DHAMed
16 : 0	23.7	23.3
18 : 0	8.5	7.6
18 : 1 n-9	17.2	16.3
18 : 1 n-7	3.8	2.7
18 : 2 n-6	26.2	25.5
18 : 3 n-6	0.4	0.3
18 : 3 n-3	0.3	0.4
20 : 3 n-6	1.4	1.5
20 : 4 n-6	5.8	6.1
20 : 5 n-3	7.0	0.7
22 : 6 n-3	1.5	10.3
minor FA	4.2	4.9

Protein and enzyme assay

At the end of the control or hypoxic treatment, the cells were washed 3 times in Puck's F solution, harvested with a rubber policeman and homogenized in 0.01 M Tris-HCl buffer (pH 7.0). Proteins were assayed according to Lowry *et al.* [22] using serum albumin as the standard. PC-PLA was assayed as previously described [14, 15], using [1-^{14}C]-dioleoyl-phosphatidylcholine (NEN) with a specific activity of 50 Bq/nmol. Alkaline PC-PLA2 (EC 3.1.1.4) was assayed at pH 8.4 in the presence of calcium (5 mM), and acid PC-PLA1 (EC 3.1.1.32) at pH 4.9 without added calcium, both for 20 min. The enzyme assays were done in duplicate with 40 μg of proteins which was in the linear range of the protein concentration range reaction. The specific activity was expressed as nmol per hr per mg protein. The data were submitted to a 2-way analysis of variance (ANOVA) to take into account the EPA vs DHA effect, the control vs hypoxia effect and the cross-interaction of these 2 factors.

Results and discussion

The fatty acid composition of the phospholipid fraction of cultured cardiomyocytes grown in EPA or DHA supplemented conventional medium is presented in Table 2. Supplementation of the medium with DHA resulted in a high level of this fatty acid in the phospholipids (22.1%). Conversely, supplementation with EPA induced a high level in EPA (12.9%) but also in C22 : 5 n-3 (14.9%), indicating a rapid elongation of C20 : 5 n-3. Interestingly, the DHA level remained low in this cell group, underlining previous reports which showed that the delta-4 desaturase activity is low in heart muscle cells [5, 6, 23]. No other major difference in fatty acid composition appeared between the EPAMed or DHAMed grown cardiomyocytes.

Table 2. Fatty Acid Composition (%) of the Phospholipid Fraction (PL) of cultured cardiomyocytes grown in the EPA or DHA supplemented media (EPAcell and DHAcell, respectively)

	EPAcell	DHAcell
16 : 0	17.2 ± 1.57	15.5 ± 1.10
18 : 0	18.1 ± 1.80	19.1 ± 1.71
18 : 1	13.3 ± 0.99	13.5 ± 0.95
18 : 2 n-6	7.3 ± 0.11	10.1 ± 0.33
20 : 3 n-6	0.8 ± 0.02	0.9 ± 0.09
20 : 4 n-6	9.6 ± 0.67	11.9 ± 1.00
20 : 5 n-3	12.9 ± 1.00	1.6 ± 0.23
22 : 4 n-6	1.1 ± 0.10	0.2 ± 0.03
22 : 5 n-3	14.9 ± 3.29	1.1 ± 0.32
22 : 6 n-3	1.8 ± 0.55	22.1 ± 3.17
minor FA	3.0 ± 0.31	3.0 ± 0.26
∑SFA	36.8	35.5
∑MUFA	14.7	14.5
∑PUFA	48.8	43.2

(SFA, MUFA and PUFA = Saturated, Monounsaturated and Polyunsaturated Fatty Acids, respectively. 18 : 1 represents the sum of the n-7 and n-9 isomers.)

In control conditions both PC-PLA1 and PC-PLA2 specific activities were significantly higher in DHA rich cells than in EPA rich cells (+ 23% for PC-PLA1 and + 17% for PC-PLA2). This confirms that the activity of phospholipase A in cardiomyocytes is affected by the nature of membrane polyunsaturated fatty acids. This was demonstrated before by the influence of the n-6 vs n-3 PUFA balance on PC-PLA2 [15] and here by the influence of the EPA vs DHA balance on PC-PLA1 and PC-PLA2.

Our results confirm and extend our previous observations [14] and others [12, 13] that hypoxia induces a significant decrease in PC-PLA2 activity. This could be considered as a beneficial event since it may contribute to the limitation of membrane degradation. In this study, the decrease was significant for PC-PLA2

Table 3. Comparison of Hypoxic Treatment (HT, 3h) and Normoxic Treatment (NT) on Phospholipase A1 (PLA1) and Phospholipase A2 (PLA2) Specific Activity (nmol/hr/mg protein) in cultured Cardiomyocytes Grown on EPA or DHA Supplemented Media. (EPAcell and DHAcell, respectively, Mean ± SEM, n = 6.)

		EPAcell		DHAcell
PLA1	Control	2.6 ± 0.11		3.2 ± 0.28
	Hypoxia	2.3 ± 0.20		2.6 ± 0.21
	a		$p < 0.05$	
ANOVA	b		n.s.	
PLA1	c		n.s.	
	i		n.s.	
PLA2	Control	8.6 ± 0.24		10.0 ± 0.44
	Hypoxia	6.1 ± 0.94		7.8 ± 0.61
	a		$p < 0.05$	
ANOVA	b		$p < 0.05$	
PLA2	c		$p < 0.05$	
	i		n.s.	

(a, b, and c = Fatty Acid Factor in Control Cells, Control/Hypoxia Factor in EPA Cells and Control/Hypoxia Factor in DHA Cells, respectively; i = Cross-interaction of the Fatty Acid Factor and the Hypoxia Factor; n.s. = not significant.)

(– 29% in EPA cells and – 22% in DHA cells) but not for PC-PLA1 (– 12% in EPA cells and – 18% in DHA cells).

The cross-interaction of the fatty acid factor and the hypoxia factor was not significant. This suggests that the amplitude of hypoxia-induced decrerases was not significantly different between the 2 cell groups, whatever the enzyme activity considered. Since the two cell groups displayed n-6/n-3 PUFA ratio very close (0.65 and 0.92), and similar contents of each saturated, monounsaturated and polyunsaturated fatty acids, the present study allows a comparison of the specific effects of EPA and DHA. The results point out that this balance alone does not influence the hypoxia-induced phospholipase decrease. However, the lower activity in EPA cells in physiologic conditions resulted in a lower activity in hypoxic conditions which, in turn, could be related to a decreased membrane degradation in such a pathological situation.

References

1. Kinsella JE, Lokesch B, Stone RA: Dietary n-3 polyunsaturated fatty acids and amelioration of cardiovascular diseases. Am J Clin Nutr 52: 1–28, 1990
2. Charnock JS, Abeywardena M, McLennan PL: Comparative changes in the fatty acid composition of rat cardiac phospholipids after long term feeding of sunflower seed oil or tuna fish oil supplemented diet. Ann Nutr Metab 30: 393–406, 1986
3. Hartog JM, Verdouw PD, Klompe M, Lames JMJ: Dietary mackerel oil in pigs: effect on plasma lipids, cardiac sarcolemmal phospholipids and cardiovascular parameters: J Nutr 117: 1361–1378, 1988
4. Nalbone G, Leonardi J, Termine E, Portugal H, Lechene P, Pauli AM, Lafont H: Effects of fish oil, corn oil and lard on lipid peroxidation status and glutathione peroxidase activities in rat heart. Lipids 24: 179–186, 1989
5. Grynberg A, Fantini E, Athias P, Degois M, Guenot L, Courtois M, Khatami S: Modification of the n-6/n-3 fatty acid ratio in the phospholipids of rat ventricular myocytes in culture by the use of synthetic media: functional and biochemical consequences in normoxic and hypoxic conditions. J Mol Cell Cardiol 20: 863–874, 1988
6. Liautaud S, Grynberg A, Mourot J, Athias P: Contribution to the use of cultured cardiomyocytes as a model in nutritional studies: Fatty acid composition of the hearts of rats fed linseed oil or sunflower oil and cardiomyocytes grown on their sera. Cardioscience 2: 55–61, 1991
7. Charnock JS: Antiarrhythmic effect of fish oils. In: AP Simopoulos (Ed) World Review of Nutrition and Dietetics, Washington D.C., 1991, pp 278–291
8. Burr ML, Gilbert JF, Holliday RM, Elwood PC, Fehily AM, Rogers S, Sweetnam PM, Deadman NM: Effect of changes in fat, fish and fiber intakes on death and myocardial infarction: diet and reinfarction trial (DART). Lancet ii: 757–761, 1989
9. Van der Vusse GJ, van Bilsen M, Reneman RS: Is phospholipid degradation a critical event in ischemia-induced damage? NIPS 4: 49–53, 1989
10. Weglicki WB: Degradation of phospholipids in myocardial membranes. In: Wildenthal K (Ed) Degradative processes in heart and skeletal muscle, Elsevier, Amsterdam, 1980, pp 377–388
11. Hasen SL, Ford DA, Gross RW: Activation of a membrane-associated phospholipase A2 during myocardial ischemia which is highly selective for plasmalogen substrate. J Biol Chem 266: 5629–5633, 1991
12. Bentham JM, Higgins AJ, Woodward B: Phospholipase A2 activity during ischemia in the isolated rat heart. Basic Res Cardiol 82 (S1): 127–136, 1987
13. Schwertz D, Halverson J, Levitsky S: Differential alterations in phospholipase C and phospholipase A2 activity in ischemic rat heart. J Mol Cell Biol 21 (S2): S136, 1989
14. Grynberg A, Nalbone G, Degois M, Leonardi J, Athias P, Lafont H: Activities of some enzymes of phospholipid metabolism in cultured rat ventricular myocytes in normoxic and hypoxic conditions. Biochim Biophys Acta 958: 24–30, 1988
15. Nalbone G, Grynberg A, Chevalier A, Leonardi J, Termine E, Lafont H: Phospholipase A activity of cultured rat ventricular myocytes is affected by the nature of cellular polyunsaturated fatty acids. Lipids 25: 301–306, 1990
16. Grynberg A, Athias P, Degois M (1986): Effect of change in growth environment on culture myocardial cells investigated in a standardized medium. *In vitro*, 22: 44–50
17. Athias P, Grynberg A: Electrophysiological studies on heart cells in culture. In: 'The Heart Cell in Culture', Pinson A, Ed. CRC Press, Boca Raton, 1987, pp 125–158
18. Chevalier A, Demaison L, Grynberg A, Athias P: Influence of phospholipid polyunsaturated fatty acid composition on some metabolic disorders induced in rat cardiomyocytes by hypoxia and reoxygenation. J Mol Cell Cardiol 22: 1177–1186, 1990
19. Lepage C, Roy C: Direct transesterification of all classes of lipids in a one step resection. J Lip Res 27: 114–120, 1986
20. Folch J, Lees M, Sloane-Stanley GH: A simple method for the isolation and purification of total lipids from animal tissues. J Biol Chem 226: 487–509, 1957
21. Juaneda P, Rocquelin G: Rapid and convenient separation of phospholipids from non phosphorus lipids in rat heart using silica cartridges. Lipids 20: 40–41, 1985
22. Lowry DH, Rosebrough NJ, Farr AL, Randall RJ: Protein measurement with the Folin phenol reagent. J Biol Chem 193: 265–275, 1957
23. Mohammed BS, Hagve TA, Sprecher H: The metabolism of 20- and 22-carbon unsaturated acids in rat heart and myocytes as mediated by feeding fish oil. Lipids 25: 854–858, 1990

Molecular and Cellular Biochemistry **116**: 79–87, 1992.

Incorporation of radioiodinated fatty acids into cardiac phospholipids of normoxic canine myocardium

G.W. Sloof, F.C. Visser, T. Teerlink,[1] E.F.I. Comans, M.J. Eenige van, G.J. van der Vusse,[2] and F.F. Knapp Jr.[3]

[1] *Dept. of Cardiology, Metabolic Laboratory, Free University Hospital Amsterdam, Amsterdam, The Netherlands;* [2] *Dept. of Physiology, University of Limburg, The Netherlands;* [3] *Nuclear Medicine Group, Oak Ridge National Laboratory (ORNL), Oak Ridge, Tennessee, USA*

Abstract

The aim of this study was to assess the phospholipid distribution of radioiodinated 17-iodoheptadecanoic acid (IHDA), 15-(p-iodophenyl)pentadecanoic acid (p-IPPA) and 15-(p-iodophenyl)-3,3-dimethylpentadecanoic acid (DMIPPA) under normoxic conditions and to compare these data with the fatty acid composition of the phospholipid classes. After simultaneous i.v. injection of the radioiodinated fatty acids (I-123-IHDA; I-131-p-IPPA; I-125-DMIPPA) in open-chest dogs seven myocardial biopsies were taken over 40 min (n = 26). After lipid extraction of the biopsies the organic phase was analyzed for both neutral and polar lipids by two different TLC systems. The following polar lipid fractions were analyzed: lysophopshatidylcholine (LPC), sphingomyelin (SPH), phosphatidylcholine (PC; lecithin), phosphatidylinositol (PI), phosphatidylserine (PS), phosphatidylethanolamine (PE), diphosphatidylglycerol (DPG; cardiolipin) and neutral lipids. Fractions were counted in a gamma well counter and corrected for cross-over and recovery. Results of the polar phospholipids analysis showed that IHDA has the highest incorporation into the phospholipids. The IHDA was mainly incorporated into PI (45.6%) followed by PC (30.9%), PE (14.0%) and PS (5.6%). The p-IPPA was predominantly incorporated incorporated into PC (37.2%), followed by PS (20.1%) and PE (13.7%). In contrast to IHDA, incorporation of p-IPPA into PI was small (6.4%). The DMIPPA analogue was incorporated into phopsholipids to only a very small degree, compared to IHDA and p-IPPA. PS (27.4%) was the only considerable phospholipid fraction into which DMIPPA was incorporated.

The results clearly demonstrated that these radioiodinated fatty acid analogues have entirely different patterns of phospholipid incorporation.Major resemblances have been found between the incorporation into phospholipids of IHDA and the phospholipid distribution of the natural counterpart: stearic acid. The p—IPPA phospholipid incorporation only partly resembles the phospholipid distribution of palmitic acid. DMIPPA is because of its modified structure, incorporated into phospholipids to a low extent, mainly into PS. (Mol Cell Biochem **116:** 79–87, 1992)

Key words: normal canine myocardium, radioiodinated fatty acids, phospholipid distribution

Address for offprints: G.W. Sloof, Dept of Cardiology, Free University Hospital, De Boelelaan 1117, 1007 MB Amsterdam, The Netherlands

Introduction

Phospholipids are essential components of myocardial tissue and play an important role in structural, metabolic and membrane functions of the cell. Distrubances in phospholipid metabolism may therefore jeopardize cell integrity [1]. In fact, reversible or irreversible myocardial damage due to ischemia correlates with altered membrane functions [2–8]. For example, elevated levels of arachidonic acid, a fatty acid predominantly incorporated into phospholipids, have been found in ischemic myocardium, suggesting phospholipid degradation [9–11]. Moreover, disturbances in phospholipid metabolism in membranes seem to play an important role in the pathogenesis of arrhythmias [5, 12].

The oxidation of fatty acids is the major energy source for normoxic myocardium [13] and fatty acid metabolism can be traced in patients by labelling the fatty acid chain with radioiodine. The radioiodine can be monitored with conventional gamma cameras, yielding noninvasive information about regional uptake and metabolism in normal and pathological myocardium. A variety of different radioiodinated fatty acid analogues (IFAs) have been developed for patient studies. These include:

17-Iodoheptadecanoic acid (IHDA), an aliphatic fatty acid radiolabelled with iodine in the terminal (omega) position, which has been studied intensively over the past decade [14–21]. This agent is avidly taken up in normal myocardium, whereafter the acyl chain is rapidly oxidized, splitting off the radioiodine. The free iodide leaves the cell and enters the vascular compartment resulting in background activity.

Because of the high background activity of IHDA, Machulla *et al.* [22] proposed terminal phenyl-substituted fatty acids and developed 15-(p-iodophenyl—pentadecanoid acid (p-IPPA). The p-IPPA analogue is catabolized to para-iodobenzoic acid and is either directly excreted or transformed in the liver to para-iodohippuric acid and then excreted by the kidneys [23]. Rapid excretion would lower levels of radioactivity in the blood, resulting in more favourable heart to blood ratios.

However, rapid clearance of both IHDA and p-IPPA from the heart at rest may limit their use for assessment of myocardial uptake by single photon emission computerized tomography (SPECT). Knapp *et al.* [24] therefore designed a type of 'molecular microsphere' by synthesis of 15-(p-iodophenyl)-3,3-dimethylpentadecanoic acid (DMIPPA) analogue. Geminal dimethyl-substitution at the 3-position of the carbon chain prevents β-oxidation. Thus DMIPPA is mainly trapped in the lipid pool, resulting in prolonged retention of the tracer in the myocardium [21, 24, 25]. Therefore this relatively new agent may be useful to study regional differences in fatty acid uptake using SPECT.

Although the addition of an heteroatom or phenyl ring may change the biological properties of IHDA and p—IPPA, respectively, several studies have indicated that the kinetics of the tracer correlates with fatty acid metabolism of the non-iodinated homologues [17, 18, 20, 26, 27]. These studies focused on myocardial uptake, oxidation and lipid incorporation in normal myocardium or alterations of these parameters under various pharmacologic, metabolic or ischemic interventions.

Because the radioiodinated fatty acids are incorporated into phospholipids and their phospholipid distribution is unknown, the aim of this study was to assess the phospholipid distribution of these IFAs in the normoxic dog heart. Furthermore, by analyzing the distribution over an experimental period of 40 min the time-course of the radioiodinated phospholipids were studied. Finally, we compared these distribution profiles with the distribution of the natural fatty acids.

Materials and methods

Radiolabelled fatty acids

17-[^{123}I]Iodoheptadecanoic acid (IHDA) and 15-(p-[^{131}I]iodophenyl)pentadecanoic acid (p-IPPA) were obtained commercially (Cygne, Eindhoven, The Netherlands), and had a specific activity of 4.0 mCi/mg and 15.0 mCi/mg, respectively. The 15-(p-[^{125}I]iodophenyl)-3,3-dimethylpentadecanoic acid (DMIPPA) was prepared by the thallation-iodination route [24, 28] and had a specific activity of 8.0 mCi/mg. Fatty acids were complexed to 20% human serum albumin, prior to administration [24].

Animal experiments

Twenty-six male mongrel dogs (17 to 34 kg) were studied after overnight fasting. After premedication with 2.0–4.0 ml Thalamonal[r] i.m. (droperidol 2.5 mg/fentanyl 50 μg per ml), anaesthesia was induced by fentanyl i.v. (20 μg/kg) immediately followed by etomidate i.v. (0.4 mg/kg). Before thoracotomy, droperidol (0.5 mg/kg) and fentanyl (40 μg/kg) were administered by bolus injection, followed by continuous infusion of fentanyl (10–20 μg/kg/hr). Dogs were ventilated with N_2O/O_2 (2/1, by vol.) and muscle relaxation was maintained with pancuronium i.v. (0.2 mg/kg prior intubation, followed by 0.01 mg/kg/kr). Arterial blood gas samples were taken every 30 min and ventilation was adjusted in such a way that proper blood gas levels were maintained i.e. pO_2 about 13.3 kPa (100 mmHg) and pCO_2 about 5.0 kPa (38 mmHg). The pH was kept at 7.4 by adminstering sodium hydrogen carbonate ($NaHCO_3$ 4.2%) when required. To prevent arrhythmias during biopsy retrieval, lidocain 1.2 mg/kg/hr was administered i.v.

A catheter was introduced into the carotid artery and used for blood sampling and arterial pressure measurements. For measurements of cardiac output and central body temperature a Swan Ganz catheter was introduced. Heparin 200 IU/kg i.v. was given to prevent blood clotting in the catheters. A thoracotomy was performed through the fifth left intercostal space and the heart was suspended in a pericardial cradle. The electrocardiogram and pressure curves were continuously monitored. Central body temperature was maintained between 37 and 38 degrees C using a water heated pad.

Biopsies and blood samples

Via a cephalic vein, 4.2 ± 0.9 mCi IHDA (n = 26), 2.3 ± 0.9 mCi p-IPPA (n = 24) and 0.9 ± 0.2 mCi DMIPPA (n = 18) were injected simultaneously in one bolus. Thereafter, seven myocardial biopsy specimens (weighing 32.4 ± 12.7 mg) were taken with a fast spinning hollow needle at 6, 10, 15, 20, 25, 30 and 40 min after injection. Immediately after rinsing in ice-cold saline, specimens were frozen in liquid nitrogen. The time between taking the biopsy samples and freezing required less than 20 seconds. Before (t = 0) and at 35 min after IFA injection, the cardiac output was measured by the thermodilution method. At 0, 18 and 35 min, arterial blood samples were drawn for determination of unlabelled FFA, lactate and glucose levels.

Counting procedure

After weighing, the biopsy specimens were submerged in vials filled with fluid nitrogen and counted for 1 min in a gamma well counter (LKB-Wallac 1282 Compu Gamma, multi-isotope option), with window settings appropriate for ^{125}I, ^{123}I and ^{131}I. Decay and cross-over were corrected.

Biochemical analysis

Lipid extraction of heart biopsy specimens was performed using the method of Bligh and Dyer [29]. Frozen biopsies were transferred to a Potter-Elvehjem glass-teflon homogenizer. Ice-cold water (0.6 ml) was added and the tissue was homogenized for 20 seconds at 0° C. Thereafter 2.25 ml of chloroform/methanol (1/2, by vol.) was added and homogenization was continued for 1 min at 0° C. The homogenate was transferred to a glass tube and the tissue residue remaining in the Potter tube was reextracted twice with 1.0 ml of chloroform/methanol/water (1/2/0,8, by vol.). The two extracts were combined with the homogenate and centrifuged for 5 min at 1500 × g. The supernatant was pipetted off and the remaining pellet was washed once by suspension in 1.0 ml of chloroform/methanol/water (1/2/0.8, by vol.). After centrifugation this supernatant was added to the supernatant obtained earlier. Water (1.4 ml) and chloroform (1.4 ml) were added to the supernatant to initiate phase separation. After thorough mixing and centrifugation, the aqueous phase, containing the oxidation products, was separated from the chloroform phase, containing the lipids. The chloroform fraction was dried under a stream of nitrogen (40° C). The three fractions (lipid phase, aqueous phase and pellet) were counted for 1 min in the gamma well counter, correcting for cross-over and decay.

Thin layer chromatography (TLC)

Lipid fractions were redissolved in 70 μl of chloroform/methanol (2/1, by vol.). To analyze the various lipid classes two separate TLC systems were employed. Neutral lipids were separated on plastic backed silica gel TLC plates (Merck nr. 5748) using hexane/diethyl ether/acetic acid (65/35/1, by vol.) as developing solvent. In this system polar lipids remain at the origin. Polar lipids were separated on glass backed silica gel

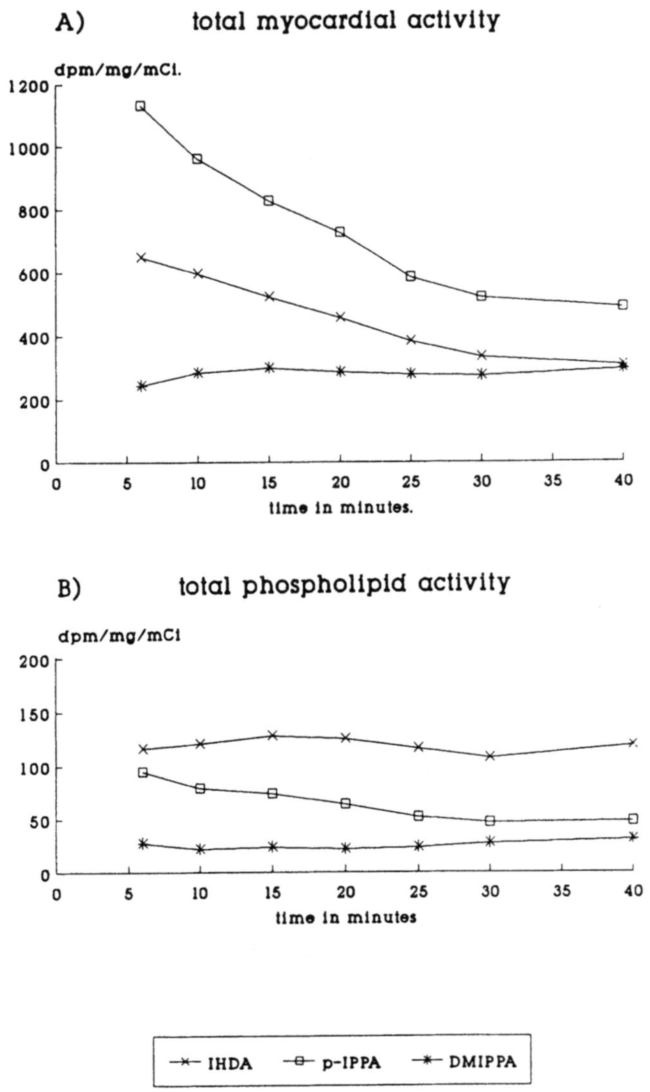

Fig. 1. Time-activity curves of total myocardial radioactivity (A) and total phospholipid activity (B) in biopsy specimens of normoxic canine myocardium after simultaneous i.v. injection of the radioiodinated fatty acids at t = 0 min. Data are mean values and presented as disintegrations per minute per milligram tissue per milliCurie injected dose.

high-performance TLC plates (Merck nr. 5642) with chloroform/hexane/methanol/acetic acid/water (48/28/16/12/1.2, by vol.) as developing solvent. Thirty μl aliquots of the lipid extracts were spotted on both TLC plates. Standards were obtained from Sigma and were also applied. After chromatography the plates were air-dried and individual lipids were detected by exposing the plates to iodine vapour. Individual polar lipids were recovered by scraping the silica gel from the glass backed TLC plates. The following fractions were obtained: 1. lysophosphatidylcholine (LPC, $R_f = 0.04$); 2. sphingomyelin (SPH, $R_f = 0.11$); 3. phosphatidylcholine (PC, $R_f = 0.21$); 4. phsophatidylinositol (PI, $R_f = 0.29$); 5. phosphatidylserine (PS, $R_f = 0.40$); 6. phosphatidylethanolamine (PE, $R_f = 0.55$); 7. diphosphatidylglycerol (DPG, $R_f = 0.68$); 8. neutral lipids (NL, $R_f > 0.80$). The lipid fractions were counted for 5 min in the gamma well counter, correcting for cross-over and decay. Because polar lipids consist predominantly of phospholipids, this fraction will further be mentioned as phospholipids.

Further, we assessed the relative composition of phospholipids in 2 dogs. After TLC and perchloric acid destructions of the phospholipids, lipid phosphorus was determined by the procedure of Rouser *et al.* [30]. We found, data expressed as percentage of total phospholipid: LPC: 0.9%, SPH: 3.6%, PC: 48.0%, PI: 5.2%, PS: 2.5%, PE: 30.9%, DPG: 6.2% and not identified: 2.7%. This composition correlates with the phospholipid composition in normoxic dog heart found by others [6, 31].

Data analysis

To compare relative cross-over of ^{125}I, ^{123}I and ^{131}I from the three different free fatty acids, the count rates (already corrected for cross-over and decay) were also corrected for 1) the efficiency of the gamma well counter for the three different radioisotopes used, 2) the weight of the biopsy specimen and 3) the radioactivity injected. Correction for recovery was made for the lipid extraction and for both TLC procedures as well. The average recovery of the lipid extraction was 95% and the balance of recovery of the TLC's was 91%. The activity is presented as disintegrations per minute per milligram tissue per milliCurie injected fatty acid (dpm/mg/mCi). All data are expressed as mean ± standard error of mean. Slight differences in sampling time of biopsy specimen were corrected by linear interpolation of the two adjacent sampling times.

Results

Haemodynamics and substrate levels

Before myocardial biopsy specimens were taken, mean arterial blood pressure and heart rate in the dogs were 104 ± 2 mmHg and 123 ± 3 beats per minute, respectively. At the end of the experiment values averaged 97 ± 2 mmHg and 133 ± 5 beats per minute. The average cardiac output was 4.6 ± 0.2 l/min at the start and 4.3 ± 0.1 l/min at the end of the experiment. Just before the injection of the radioiodinated fatty acids, plasma arterial substrate levels for free fatty acids, glucose and lactate were 0.84 ± 0.08 mmol/l, 7.6 ± 0.3 mmol/l and 2.9 ± 0.4 mmol/l, respectively. These values were consistent during the entire experimental period.

Total myocardial activity and aqueous phase activity

p-IPPA had the highest total myocardial activity: 1132 ± 102 dpm/mg/mCi vs 647 ± 48 dpm/mg/mCi for IHDA (Fig. 1A). At peak activity (t = 6 min), the aqueous phase activity was 65 ± 3% and 55 ± 3% of total myocardial activity for IHDA and p-IPPA, respectively. Both p-IPPA and IHDA activity decreased over time. DMIPPA myocardial activity remained nearly constant during the experimental period with a maximal value of 299 ± 30 dpm/mg/mCi at the 15th min. The aqueous phase of DMIPPA activity was maximally 31 ± 3% at t = 10 min. Pellet activity averaged 1.1% and never exceeded 3% of total tissue activity.

Total activity of phospholipids

IHDA showed the highest activity in phospholipids (Fig. 1B): 116 ± 17 dpm/mg/mCi at t = 6 min and remained constant (19 ± 2% of total activity at t = 6 min). p-IPPA phospholipid activity decreased from 101 ± 14 dpm/mg/mCi at 6 min to 48 ± 6 dpm/mg/min at t = 40 min (8.6 ± 0.6% of total activity at t = 6 min). Phospholipid incorporation of DMIPPA was low and remained constant: 28 ± 5 dpm/mg/mCi at t = 6 min (9.0 ± 1.0% of the total radioactivity).

Incorporation of IFAs into cardiac phospholipids

IHDA

The IHDA was mainly incorporated into PI: 53 ± 8 dpm/mg/mCi, followed by PC: 35 ± 5 dpm/mg/mCi, PE: 15 ± 2 dpm/mg/mCi and PS: 5.6 ± 0.8 dpm/mg/mCi (Fig. 2A). Small amounts of radioactivity were found in the DPG, SPH and LPC fractions and the mean values were 4.3, 2.6 and 0.8 dpm/mg/mCi, respectively. The radioactivity in all these fractions remained constant during the entire experimental period. There were no significant differences between values at t =

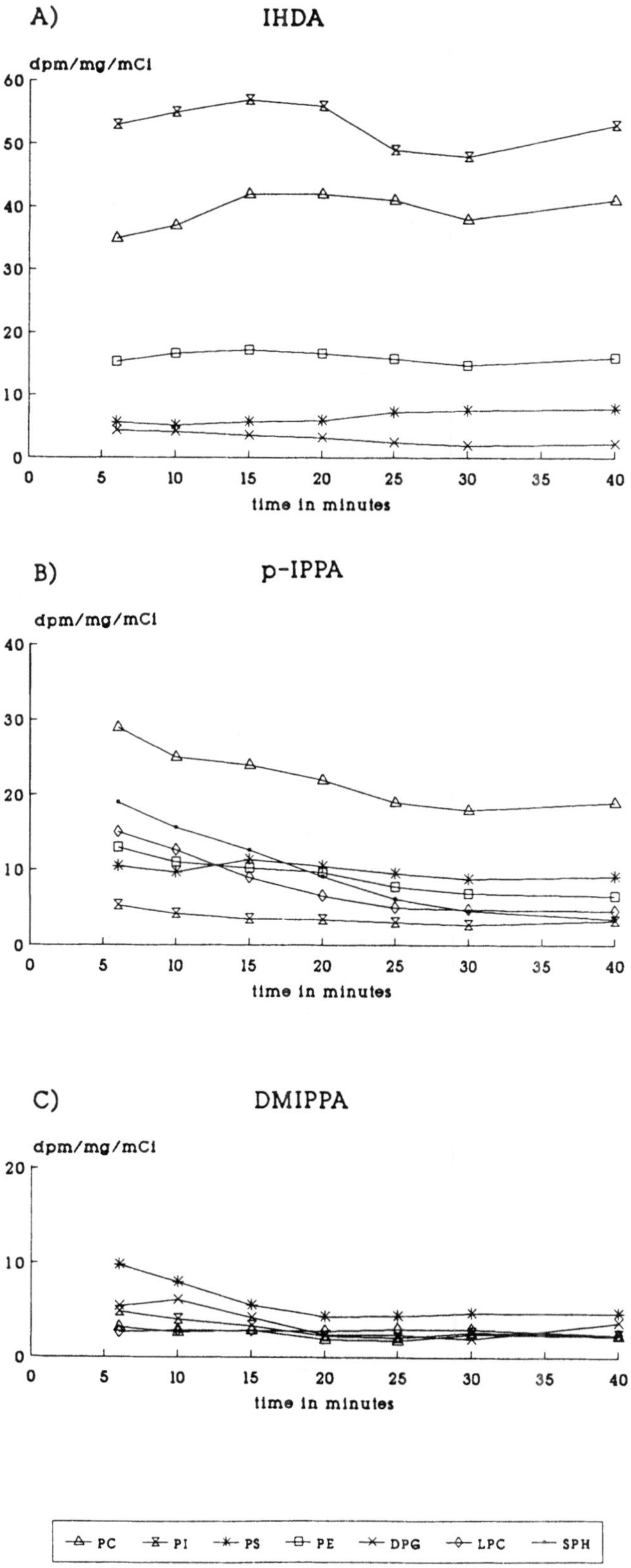

Fig. 2. Time-activity curves of radioactivity in phospholipid classes in biopsy specimens of normoxic canine myocardium after simultaneous i.v. injection of A) I-123-IHDA, B) I-131-p-IPPA and C) I-125-DMIPPA. Data are mean values and presented as disintegrations per minute per milligram tissue per milliCurie injected dose. Note that for each figure the scale on the Y-axis is different.

6 min versus t = 40 min for all phospholipid classes (paired t-test: $p > 0.05$).

p-IPPA

Throughout the experimental period the majority of p-IPPA was incorporated into PC with a value of 29 ± 4 dpm/mg/mCi at t = 6 min (Fig. 2B). Activity of p-IPPA in PC decreased significantly during the experimental period to 19 ± 3 dpm/mg/mCi: $p < 0.001$ paired t-test comparing values at t = 6 vs t = 40 min. At 6 min after administration, the activity in the SPH and LPC fractions were 23 ± 3 and 16 ± 2 dpm/mg/mCi, respectively. The radioactive content of the LPC and SPH fractions also decreased significantly over the experimental period (t = 6 vs t = 40 min: $p < 0.001$). The value of the PE fraction was 13 ± 1 dpm/mg/mCi at t = 6 min and decreased significantly to 6.6 ± 0.9 dpm/mg/mCi at t = 40 min: $p < 0.01$. The PS fraction was 11 ± 1 dpm/mg/mCi at 6 min and remained constant and incorporation into PI was low and remained essentially constant (5.3 ± 0.8 dpm/mg/min at t = 6 min).

DMIPPA

Because activity of DMIPPA in the total phospholipid was small (Fig. 1B), count rates in the various subfractions were low. PS was the only fraction in which a considerable amount of radioactivity was found with values of 11 ± 2 dpm/mg/mCi at t = 6 min and 4.7 ± 0.9 dpm/mg/mCi at t = 40 min ($p < 0.05$, Fig. 2C).

Relative distribution of IFAs in cardiac phospholipids

The relative distribution of the IFAs in the phospholipid classes are listed in Table 1. Values are presented as a percentage of the total phospholipid activity at 40 min. Taking the difference in tissue content of the various phospholipid classes into account, the preferential incorporation of a IFA for a specific phospholipid class can be demonstrated by the relative specific activity values (RSA), also listed in Table 1. RSA is the ratio of the relative activity of the IFA in a phospholipid class compared to the relative amount of that phospholipid class in the myocardium.

The data show that there was a preferential incorporation of IHDA into PI (RSA = 8.8) followed by PS (RSA = 2.2). p-IPPA was preferentially incorporated into LPC (RSA = 12.4) and PS (RSA = 8.0). Although p-IPPA was mainly incorporated into PC, the RSA is 0.8. DMIPPA was preferentially incorporated into PS (RSA = 11.0).

Discussion

The aim of this study was to assess the incorporation of radioiodinated IHDA, p-IPPA and DMIPPA into the various phospholipid classes in normal canine myocardium. The findings have demonstrated that the three IFAs have an entirely different extent and profile of incorporation into phospholipids. These differences will be discussed in relation to the distribution of the natural fatty acids in the phospholipid classes found in normoxic canine myocardium. Because previous reports suggested uptake, oxidation and lipid incorporation of p-IPPA resembled that of palmitic acid [26, 27], we compared p-IPPA phospholipid incorporation with the phospholipid distribution of palmitic acid (16 : 0).

Table 1. Relative distribution (%) and relative specific activity (RSA) of the IFAs in the phospholipids classes at 40 min. The relative distribution is expressed as percentage of total phospholipid activity (mean ± SEM). The relative specific activity is the ratio of the relative activity of the IFA in a phospholipid class compared to the relative amount of that phospholipid class in the myocardium. (-.- = inaccurate value, not presented)

	Phospholipid class						
	LPC	SPH	PC	PI	PS	PE	DPG
IHDA							
%	0.9 ± 0.9	1.6 ± 1.5	30.9 ± 7.9	45.6 ± 9.3	5.6 ± 2.2	14.0 ± 2.1	1.4 ± 1.2
RSA	1.0	0.4	0.6	8.8	2.2	0.4	0.2
p-IPPA							
%	9.3 ± 6.4	7.1 ± 3.9	37.2 ± 8.2	6.4 ± 2.7	20.1 ± 3.7	13.7 ± 2.7	6.2 ± 3.6
RSA	12.4	2.3	0.8	1.2	8.0	0.4	1.0
DMIPPA							
%	13.1 ± 9.0	9.8 ± 4.8	11.4 ± 6.4	10.5 ± 3.3	27.4 ± 7.6	10.3 ± 8.5	17.5 ± 15.9
RSA	-.-	2.7	0.2	2.0	11.0	0.3	2.8

Since in IHDA, the terminal methyl group of stearic acid has been replaced by radioiodine, we compared IHDA as an analogue of stearic acid (18 : 0).

Extent of phospholipid incorporation

IHDA showed the greatest incorporation into phospholipids (116 ± 17 dpm/mg/mCi) and DMIPPA the lowest (28 ± 5 dpm/mg/mCi). The low incorporation of DMIPPA into phospholipids is consistent with earlier studies of Knapp *et al.* [25] in rat hearts. Total phospholipid activity of both IHDA and DMIPPA remained constant during the 40 min period. This may indicate that phospholipid incorporation has already occurred soon after tracer administration and that turnover is small relative to the size of the lipid pool. The extent of p-IPPA incorporation into phospholipids was intermediate between IHDA and DMIPPA, but in contrast to IHDA and DMIPPA activity decreased over time. The finding that IHDA is incorporated more into the phospholipids than p-IPPA correlates well with the fatty acid content of phospholipids found in normoxic dog heart by van der Vusse *et al.* [9], who found that 16.4% of the fatty acids in total phospholipids was stearic acid and 8.6% was palmitic acid.

Phospholipid distribution

IHDA

IHDA was primarily incorporated into PI (45.6%) followed by PC (30.9%), PE (14.0%) and PS (5.6%) as summarized in Table 1. Taking the size of the phospholipid class into account, there was a preferential incorporation into PI and PS with RSA values 8.8 and 2.2, respectively. The RSA value for PC was 0.6 and 0.4 in case of PE. Roemen and van der Vusse [31] found that 50.6% of the fatty acids in PI in normal canine myocardium was stearic acid and 48.3% in the case of PS. For PC and PE, the two major phospholipids in the heart, these values were lower representing 8.3% and 30.8%, respectively. From their data we calculated the relative specific distribution (RSD) for stearic acid, which is the equivalent of the RSA for IHDA. The greatest RSD was found for PI (2.8) followed by PS (2.7), LPC (1.8), PE (1.7), PC (0.5) and DPG (0.1).

With the exception of PE, the sequence of the RSA values for IHDA corresponds well with the sequence of the RSD values for stearic acid. These data suggest that resemblances exist between IHDA and stearic acid distribution in phospholipids: particularly into the PI, PS and PC fractions. Yet the incorporation of IHDA into PE is lower than one would expect from stearic acid distribution in PE. Interestingly, in contrast to stearic acid, which is equally distributed between PI and PS, the high RSA of PI compared to PS (8.8 versus 2.2) suggest a specific incorporation into PI. Thus, this modified fatty acid may give insight into the factors that determine phospholipid incorporation.

During the 40 min experimental period no significant changes in distribution of IHDA were found. After rapid incorporation into the various phospholipid classes, the IHDA remained unaltered in the phospholipids during at least the 40 min, instead of being cleared. Thus, turnover of IHDA in the various phospholipid classes was small.

p-IPPA

p-IPPA was mainly incorporated into PC (37.2%), followed by PS (20.1%), PE (13.7%), LPC (9.3%) and SPH (7.1%): Table 1. In contrast to IHDA, the incorporation into PI was small (6.4%). Roemen and van der Vusse [31] found that 35.0% of the fatty acids in PC was palmitic acid, 21.1% in case of LPC and 9.1% in the case PE. Only 1.7% of the fatty acid in PI was palmitic acid and 1.2% in case of PS. The calculated RSD values for palmitic acid in the various phospholipid fractions were: PC (1.9), LPC (1.1), PE (0.5), PI (0.09), DPG (0.07) and PS (0.06).

The large p-IPPA incorporation into PC is consistent with the large content of palmitic acid in PC. Also the considerable radioactive content of p-IPPA in PE with a low RSA value (0.4) is in line with the distribution of palmitic acid in PE. Yet, comparing the RSA values for p-IPPA (Table 1) with the RSD values for palmitic acid, major differences are observed. In particular the finding that p-IPPA was preferentially incorporated into PS (RSA = 8.0) resulting in 20.1% of the radioactivity in phospholipids does not agree with palmitic acid distribution in PS. A possible explanation is that the phenyl ring induces preferential incorporation into PS (see DMIPPA discussion). A relatively large content of radioactivity was found in LPC (RSA: 12.4). Theoretically contamination of the TLC at the origin, where the LPC is found, with iodobenzoic acid might have happened. Still the same distribution pattern of p-IPPA over LPC with respect to the other phospholipids was reported by Kropp *et al.* [32]. They found 3 ± 1% of total lipid incorporation in the LPC fraction and 12 ± 6% in PC. Radioactivity from p-IPPA in the PC, SPH, LPC and PE fractions decreased significantly during the experimental period. The decline of these fractions are responsible for the decrease of p-IPPA activity in total phospholipid. This considerable turnover of p-IPPA might be due to the steric hindrance of the phenyl ring.

DMIPPA

In spite of its modified structure, DMIPPA is not only extracted by the myocardium from the blood plasma, but is also incorporated into phospholipids albeit at a low level: 28 ± 5 dpm/mg/mCi, which is 9.0 ± 1.0% of total myocardial activity. The largest amount of DMIPPA radioactivity in phospholipids was found in PS (27.1%). There was a considerable turnover of DMIPPA in PS, which, like p-IPPA, might be due to the steric hindrance of the phenyl ring. The mechanism why DMIPPA is mainly incorporated into this small phospholipid fraction is unclear. Both DMIPPA and p-IPPA, with the same terminal phenyl ring, were largely incorporated into this phospholipid class, suggesting a relation between the presence of the phenyl ring and the incorporation into PS.

Consequences for cardiac imaging

IHDA: radioactivity in IHDA-phospholipids contributes a constant value to the total myocardial activity, which is detected externally. Eenige van *et al.* [33] demonstrated that the best curve fitting procedure for IHDA scintigraphy is a monoexponential plus constant; the constant includes the background activity of the free iodide. The activity of IHDA found in the phospholipids will be part of this constant.

p-IPPA: the decrease of activity in phospholipids introduces another exponent to the time-activity curve of total myocardial activity. Curve fitting of p-IPPA scintigraphy is therefore more complex and must at least have two exponentials.

DMIPPA: the constant level of activity in phospholipids contributes to the constant level of total myocardial activity found after DMIPPA administration.

In conclusion, the three IFAs show an entirely different phospholipid incorporation. Major resemblances have been found between the incorporation into phospholipids of IHDA and the phospholipid distribution of the natural counterpart, stearic acid. p-IPPA phospholipid incorporation only partly resembles the phospholipid distribution of palmitic acid. DMIPPA is because of its modified structure, incorporated into phospholipids to a low extent, mainly into the PS fraction. These different profiles of the IFAs might serve as a tool for studies on fatty acid incorporation into specific phospholipid classes and on phospholipid homeostasis.

Acknowledgements

Animal preparations by B.J.H. van der Water, J. Grimbergen and G. Vink was greatly appreciated. We thank the staff of the Radionuclide Centre of the Free University for their hospitality and technical assistance. We also thank Viggo-Spectramed for the generous supply of medical equipment. Research at ORNL sponsored by the Office of Healthy and Environmental Research, U.S. Department of Energy, under contract with Martin Marietta Energy Systems, Inc. F.F. Knapp, Jr., also thanks the Alexander von Humboldt Foundation in Germany for support as a Senior American Scientist during the 1991–1992 period. Supported in part by the Netherlands Heart Foundation project number: 88.097.

References

1. Van der Vusse GJ, van Bilsen M, Reneman RS: Is phospholipid degradation a critical event in ischemia and reperfusion induced damage. News in physiological sciences (NIPS), 4: 949–953, 1989
2. Chien KR, Reeves JP, Buja LM, Bonte F, Parkey RW, Willerson JT: Phospholipid alterations in canine ischemic myocardium; Temporal and topographical correlations with Tc-99m-PPi accumulation and an *in vitro* sarcolemmal Ca^{2+} permeability defect. Circ Res 48: 711–719, 1981
3. Reimer KA, Jennings RB, Hill ML: Total ischemia in dog hearts. *In vitro.* II. High energy phosphate depletion and associated defects in energy metabolism, cell volume regulation, and sarcolemmal integrity. Circ Res 49: 901–911, 1981
4. Shaikh NA, Downar E: Time course of changes in porcine myocardial phospholipid levels during ischemia. A reassessment of the lysolipid hypothesis. Circ Res 49: 316–325, 1981
5. Corr PB, Gross RW, Sobel BE: Amphipathic metabolites and membrane dysfunction in ischemic myocardium. Circ Res 55: 135–154, 1984
6. Steenbergen C, Jennings RB: Relationship between lysophospholipid accumulation and plasma membrane injury during total *in vitro* ischemia in dog heart. J Mol Cell Cardiol 16: 605–621, 1984
7. Das DK, Engelman RM, Rousou JA, Breyer RH, Otani H, Lemeshow S: Role of membrane phospholipids in myocardial

injury induced by ischemia and reperfusion. Am J Physiol 251 (Heart Circ Physiol 20): H71–H79, 1986

8. Miyazaki Y, Gross RW, Sobel BE, Saffitz JE: Selective turnover of sarcolemmal phospholipids with lethal cardiac myocyte injury. Am J Physiol 259 (Cell Physiol 28): C325–C331, 1990
9. Van der Vusse GJ, Roemen ThHM, Prinzen FW, Coumans WA, Reneman RS: Uptake and tissue content of fatty acids in dog myocardium under normoxic and ischemic conditions. Circ Res 50: 538–546, 1982
10. Chien KR, Han A, Sen A, Buja LM, Willerson JT: Accumulation of unesterified arachidonic acid in ischemic canine myocardium. Relationship to a phosphatidylcholine deacylation-reacylation cycle and the depletion of membrane phospholipids. Circ Res 54: 313–322, 1984
11. Prinzen FW, van der Vusse GJ, Arts T, Roemen THM, Coumans WA, Reneman RS: Accumulation of nonesterified fatty acids in ischemic canine myocardium. Am J Physiol 247 (Heart Circ Physiol 16): H264–H272, 1984
12. Man RYK, Slater TL, Pelletier MP, Choy PC: Alterations of phospholipids in ischemic canine myocardium during acute arrhythmia. Lipids 18: 677–681, 1983
13. Opie LH: Metabolism of the heart in health and disease. Am Heart J 76: 685–698, 1968
14. Van der Wall EE, Den Hollander W, Heidendal GAK, Westera G, Majid PA, Roos JP: Dynamic myocardial scintigraphy with 123-I-labeled free fatty acids in patients with myocardial infarction. Eur J Nucl Med 6: 383–389, 1981
15. Visser FC, Eenige van MJ, Westera G, Den Hollander W, Duwel CMB, Van der Wall EE, Heidendal GAK, Roos JP: Metabolic fate of radioiodinated haptadecanoic acid in the normal canine heart. Circulation 72: 565–571, 1985
16. Visser FC, Eenige van MJ, Duwel CMB, Roos JP: Radioiodinated free fatty acids; can we measure myocardial metabolism? Eur J Nucl Med 12: S20–S23, 1986
17. Lerch RA: Effect of impaired fatty acid oxidation on myocardial kinetics of ^{11}C- and ^{123}I-labelled fatty acids. Eur J Nucl Med 12: S36–S38, 1986
18. Eenige van MJ, Visser FC, Duwel CMB, Karreman AJP, Lingen van A, Roos JP: Comparison of 17-iodine-131 heptadecanoic acid kinetics from externally measured time-activity curves and from serial myocardial biopsies in an open-chest canine model. J Nucl Med 29: 1934–1942, 1988
19. Eenige van MJ, Visser FC, Duwel CMB, Roos JP: Clinical value of studies with radioiodinated heptadecanoic acid in patients with coronary artery disease. Eur Heart J 11: 258–268, 1990
20. Duwel CMB, Visser FC, Eenige van MJ, Den Hollander W, Roos JP: The fate of ^{131}I-17-iodoheptadecanoic acid during lactate loading: its oxidation is strongly inhibited in favor of its esterification. Nucl Med 29: 24–27, 1990
21. Visser FC, Sloof GW, Eenige van MJ, Knapp FF: Radioiodinated heptadecanoic acid, phenylpentadecanoic acid and dimethylphenylpentadecanoic acid in the normal canine heart. Comparison of uptake, oxidation and lipid distribution. '(Abstr)' Eur Heart J 11: P524, 1990
22. Machulla HJ, Marsmann M, Dutschka K: Biochemical concept and synthesis of a radioiodinated phenylfatty acid for *in vivo* metabolic studies of the myocardium. Eur J Nucl Med 5: 171–173, 1980
23. Dudczak R: Myokardszintigraphie mit Jod-123 markierten Fettsäuren. Wien Klin Wochenschr 95: 1–38, 1983
24. Knapp FF Jr, Goodman MM, Callahan AP, Kirsch G: Radioiodinated 15-(p-iodophenyl)-3,3-dimethylpentadecanoic acid: a useful new agent to evaluate myocardial fatty acid uptake. J Nucl Med 27: 521–531, 1986
25. Knapp FF Jr, Ambrose KR, Goodman MM: New radioiodinated methyl-branched fatty acids for cardiac studies. Eur J Nucl Med 12: S39–S44, 1986
26. Reske SN, Sauer W, Machulla HJ, Winkler C: 15(p-[123I]Iodophenyl)pentadecanoic acid as tracer of lipid metabolism: comparison with [1-^{14}C]palmitic acid in murine tissues. J Nucl Med 25: 1335–1342, 1984
27. Reske SN, Schön S, Schmitt W, Machulla HJ, Knopp R, Winkler C: Effect of myocardial perfusion and metabolic interventions on cardiac kinetics of phenylpentadecanoic acid (IPPA) I123. Eur J Nucl Med 12: S27–S31, 1986
28. Goodman MM, Kirsch G, Knapp FF Jr: Synthesis and evaluation of radioiodinated terminal p-iodophenyl-substituted α- and β-methyl-branched fatty acids. J Med Chem 27: 390–397, 1984
29. Bligh EG, Dyer WJ: A rapid method of total lipid extraction and purification. Can J Biochem Physiol 37: 911–917, 1959
30. Rouser G, Fleischer S, Yamamoto A: Two-dimensional thin-layer chromatographic separation of polar lipids and determination of phospholipids by phosphorus analysis of spots. Lipids 5: 494–496, 1970
31. Roemen ThHM, van der Vusse GJ: Application of silica gel column chromatography in the assesssment of non-esterified fatty acids and phosphoglycerides in myocardial tissue. J of Chromatography 344: 304–308, 1985
32. Kropp J, Ambrose KR, Knapp FF Jr, Nissen HP, Biersack HJ: Incorporation of radioiodinated IPPA and BMIPPA fatty acid analogues into complex lipids from isolated rat hearts. J Nuc Med Biol. In press
33. Eenige van MJ, Visser FC, Duwel CMB, Bezemer PD, Westera G, Karreman AJP, Roos JP: Analysis of myocardial time activity curves of I-123 heptadecanoic acid; I curve fitting. Nucl Med 26: 241–247, 1987

Molecular and Cellular Biochemistry **116**: 89–93, 1992.

Kinetic changes of ethanolamine base exchange activity and increase of viscosity in sarcolemmal membranes of hamster heart during development of cardiomyopathy

Alba Vecchini,[1] Luciano Binaglia,[1] Paolo Di Nardo,[2] Manuela Bartoli,[2] Marilena Minieri[2] and Giuseppe Tallarida[2]

[1] *Department of Biochemistry, University of Perugia, Via del Giochetto, 06100 Perugia, Italy and*
[2] *Lab. Cellular and Molecular Cardiology (Fisiopatologia Medica) Dept. of Internal Medicine, II University of Roma 'Tor Vergata', Via O. Raimondo, 00173 Roma, Italy*

Abstract

The activity of the phospholipid base exchange enzyme specific for ethanolamine has been measured in cardiac sarcolemmal membrane preparations from Syrian golden and UM-X7.1 cardiomyopathic hamsters. In Syrian golden hamsters, the Km of the enzyme for ethanolamine does not change with age, whereas it almost doubles in membranes from cardiomyopathic animals, from the 30th to the 150th day of age. During the same period, the membrane cholesterol content increases by 68% in cardiomyopathic hamsters, whereas it does not change significantly in the Syrian golden hamster strain. As a consequence, in the adult animal, the cholesterol to phospholipid ratio and the viscosity of sarcolemmal membranes are higher in UM-X7.1 strain than in Syrian golden hamsters. A cause-consequence relationship between the enzymatic changes and the compositional modifications in the sarcolemma occurring in UM-X7.1 hamsters during the development of cardiomyopathy is proposed. (Mol Cell Biochem **116:** 89–93, 1992)

Key words: cardiomyopathic hamsters, base-exchange, phospholipids, sarcolemma, heart, cholesterol, fluidity

Introduction

Cardiac tissue of cardiomyopathic hamsters (CMPH) [1, 2] has been the subject of several clinical and physiological studies, as well as of morphological and biochemical investigations [3–10]. In fact, the hamster hereditary cardiomyopathy has been regarded as a useful animal model to investigate mechanisms leading to cardiac damage and ventricular failure.

Among the numberous abnormalities of CMPH heart, those regarding the membranes of cardiomyocytes deserve particular attention [10]. Indeed, an abnormal number of intramembranous particles was recently found in the sarcolemmal membranes of CMPH cardiomyocytes [11]. This morphological evidence could be somehow related to the high cholesterol content of both sarcolemmal and other intracellular membranes of CMPH heart [8, 12, 13]. Moreover, evidence

Address for offprints: L. Binaglia, Department of Biochemistry, School of Medicine, University of Perugia, Via del Giochetto, 06100 Perugia, Italy is emerging that the molecular species composition of membrane phospholipids is different in Syrian Golden hamsters (SGH) and CMPH [Vecchini A: unpublished results]. These experimental data raised the hypothesis that compositional and structural changes of sarcolemmal and intracellular membranes could be responsible for the morphological, physiological and clinical changes occurring during the development of the cardiomyopathic hamster. Therefore, a detailed study on the turnover of the myocardial lipids of the CMPH strain is relevant. Nevertheless, only little information is available on this field. Panagia *et al.* [7] have described a defect in the phospholipid N-methylation process in heart sarcolemma during the development of cardiomyopathy, whereas Vecchini *et al.* have measured the activity of the phospholipid base exchange enzymes [14] in cardiac sarcolemmal membranes from three months-old CMPH and SGH [15]. The sarcolemmal membranes from control and cardiomyopathic hamsters exhibited different Km values for the substrate of the base exchange enzyme involved in the synthesis of phosphatidylethanolamine. The high Km for ethanolamine of the enzyme in myopathic hamsters was paralleled by a higher content of cholesterol in sarcolemmal membranes of myopathic hamsters when compared to control animals. Taking into account the well known lipid-dependence of the ethanolamine exchange enzyme [16, 17], a direct relationship was hypothesized between high membrane cholesterol content and high Km of the enzyme.

Since no information is available on changes of sarcolemmal membrane composition during the development of cardiomyopathy, both cholesterol and phospholipid content of sarcolemmal membranes from UM-X7.1 cardiomyopathic hamster [18] and SGH strains were measured in the present study. In addition, the kinetic parameters of the ethanolamine base exchange enzyme in the same membranes were measured and a correlation between the enzymatic and the compositional changes was proposed.

Experimental

Materials

UMX 7.1 hamsters were kindly provided by Dr. G. Jasmin, Montreal, Canada. [1-^{3}H]ethanolamine (specific activity, 22.4 Ci/mmole) was purchased from Amersham International plc (Amersham, U.K.). Ethanolamine, N-2-hydroxyethylpiperazine-N′-2-ethanesulfonic acid (Hepes) and Etylene-glycol-bis(β-aminoethylether)-N,N,N′,N′-tetraacetic acid (EGTA) were from Sigma Chemical Co. (St. Louis, MO, U.S.A.). Sodium pyrophosphate, $CaCl_2$ and KCl were from E. Merck A.G. (Darmstadt, West Germany). 1,6-diphenylhexatriene (DPH) was obtained from Aldrich-Chemie GmbH (Steinheim, Germany). Other chemicals were from Carlo Erba (Milano, Italy).

Methods

Hamsters were studied at the age of 30 to 150 days. Sarcolemmal membranes were prepared from the cardiac ventricules of UM-X7.1 and Syrian Golden hamster according to Pitts [19] and characterized as previously described [15]. Proteins were determined according to Lowry *et al.* [20]. Cholesterol was quantitated in the Folch lipid extract [21] according to Mascini *et al.* [22]. Total phospholipid content was determined as inorganic phosphate, after mineralization, using the method described by Baykov *et al.* [23].

The fluorescence polarization measurements were performed according to Shinitzky and Barenholz [24]. Sarcolemmal membrane suspensions were supplemented under vigorous stirring with a 4 mM solution of DPH in tetrahydrofuran to reach the following concentrations: 80 μg protein/ml, 2.6 μM DPH, 0.59 μg tetra-

Table 1. Km for ethanolamine of the phospholipid base exchange enzyme activity in sarcolemmal membranes from UM-X7.1 and Syrian golden hamsters at various age

age (days)	SGH	UM-X7.1
30	19.2 ± 3.3	18.3 ± 2.8*
45	19.5 ± 0.6	26.1 ± 0.3#
60	19.4 ± 3.3	28.2 ± 2.4##
90	19.6 ± 2.7	30.4 ± 3.3##
150	27.3 ± 2.2	36.5 ± 0.5##

Sarcolemmal membranes isolated from Syrian golden hamsters (SGH) and from cardiomyopathic hamsters (UM-X7.1) were incubated with isotopically labelled ethanolamine as described in the Experimental-section. The Km value for ethanolamine was calculated from the experimental data by linear regression, according to the Lineweaver-Burk equation.

The data ± S.D. have been calculated from the results obtained from 4 separate determinations.

* non significantly different from control, # significantly different ($P < 0.05$) from control, ## significantly different ($P < 0.01$) from control.

hydrofuran per milliliter. The phospholipid/DPH molar ratio was about 37. The mixtures were left 1 h in the dark, at 20° C.

Fluorescence measurements were performed with a model RF 5000 Shimadzu spectrofluorophotometer (Kyoto, Japan), equipped with thermostated cell holder and magnetic microstirrer. The excitation and emission wavelengths were set at 360 nm and 420 nm, respectively. The temperature of the membrane suspension was measured with an accuracy of ± 0.1° C, using a termistor thermometer. The degree of fluorescence polarization (P) was calculated according to Shinitzky and Barenholz [24], using the equation

$$P = I_{\|} - I_{\perp} \times G / I_{\|} + I_{\perp} \times G$$

where $I_{\|}$ and $I_{\perp}$ are the emission intensities detected by the analyzer oriented parallel ($I_{\|}$) and perpendicular ($I_{\perp}$), respectively, to the direction of polarization of the excitation light. The grating correction factor G of the instrument was calculated according to Antunes-Madeira and Madeira [25].

The Michaelis-Menten constant (Km) of the ehtanolamine exchange enzyme for ethanolamine was measured in sarcolemmal membranes isolated from the cardiac ventricules of cardiomyopathic and control hamsters. Aliquots of the sarcolemmal membrane suspensions (1 mg protein per milliliter of 0.25 M sucrose) were incubated 5 min at 37° C in a medium containing a fixed amount of [1-^{3}H]ethanolamine and variable amounts of unlabelled ethanolamine. Each incubation mixture (final volume: 0.2 ml) contained 0.1 mg sarcolemmal protein, 50 mM Hepes buffer (pH 7.8), 2 mM $CaCl_2$, 400 nCi of [1-^{3}H]ethanolamine (specific activity 22.4 Ci/mmol) and cold ethanolamine at concentrations ranging between 2 μM to 0.4 mM. The enzymatic reactions were stopped by adding 3 ml of a chloroform/methanol mixture (2 : 1, by vol). Lipids were extracted according to Folch *et al.* [21] and isolated by thin layer chromatography on HPTLC Kieselgel 60 plates (Merck, Darmstadt, West Germany), using a chloroform/methanol/water mixture (65 : 25 : 4, by vol) as developing solvent [26]. Alternatively, the two-dimensional development procedure described by Horrocks [27] was used. After exposure of the plates to iodine vapour, the lipids co-chromatographing with an authentic phosphatidylethanolamine standard were scraped off and their radioactivity was measured on the silica scrapes as previously described [26].

Results

The Km values for ethanolamine of the sarcolemmal phospholipid base exchange enzyme are not statistically different in 30 day-old SGH and UM-X7.1 hamsters. As shown in Table 1, the Km does not change in the SGH strain during the following two months, whereas it increases by more than 50% in the UM-X7.1 hamster. A sustained increase of the Km is observed in membranes from the UM-X7.1 strain, being doubled at the end of the life span considered (30 to 150 days of age).

The phospholipid and cholesterol content of sarcolemmal membrane preparations from 30 day-old myopathic and control hamsters are not significantly different (Table 2). An identical phospholipid content was found when sarcolemmal membrane preparations from 90 and 150 days-old hamsters of both strains were analyzed. On the contrary, the cholesterol level at similar age is significantly higher in membranes from UM-X7.1 hamsters when compared to those from the SGH strain (Table 2). Moreover, the cholesterol content in UM-X7.1 hamsters show an age-related increase when compared to 30 days old animals (+ 36% at 90 days and + 68% at 150 days). As a consequence, the sarcolemmal cholesterol to phospholipid ratio increases with age in the cardiomyopathic UM-X7.1 hamster.

The increase of the cholesterol to phospholipid ratio

Table 2. Phospholipid and cholesterol content of heart sarcolemmal membranes from Syrian golden and from cardiomyopathic UM-X7.1 hamsters

	Total phospholipids (μmols/mg protein)		Cholesterol (μmols/mg protein)		Chol/PL (molar ratio)	
age (days)	SGH	UM-X7.1	SGH	UM-X7.1	SGH	UM-X7.1
30	1.20 ± 0.01	1.15 ± 0.01*	0.28 ± 0.03	0.25 ± 0.01*	0.233 ± 0.011	0.211 ± 0.013*
90	1.19 ± 0.02	1.16 ± 0.03*	0.30 ± 0.01	0.34 ± 0.01#	0.242 ± 0.009	0.283 ± 0.012#
150	1.16 ± 0.02	1.12 ± 0.03*	0.29 ± 0.03	0.42 ± 0.08##	0.251 ± 0.013	0.375 ± 0.023##

Phospholipids and cholesterol were quantitated according to Mascini *et al.* [22] and Baykov *et al.* [23], respectively. Mean values ± standard deviation were calculated from the results obtained from eight separate determinations. * not significantly different from the respective control, # significantly different ($P < 0.05$) from control, ## significantly different ($P < 0.01$) from control.

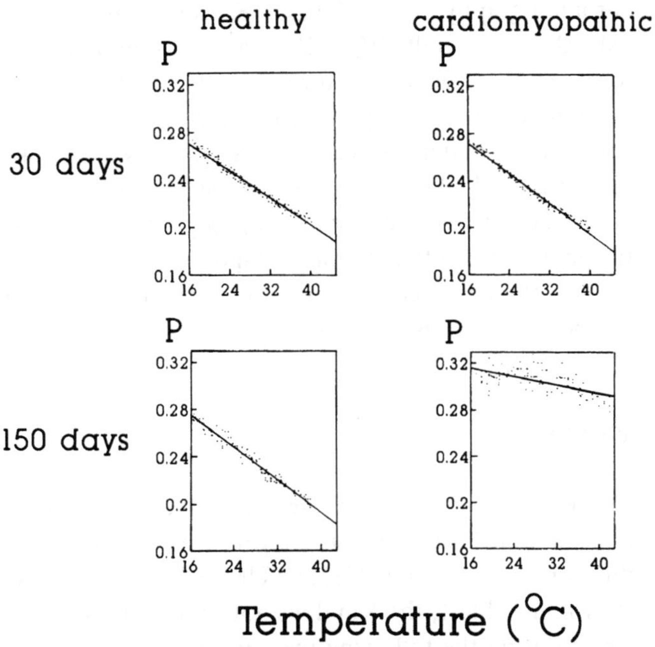

Fig. 1. Degree of fluorescence polarization (P) of diphenylhexatriene in sarcolemmal membranes isolated from Syrian golden hamsters (SGH) and cardiomyopathic hamsters (UM-X7.1) at 1 month and 5 months of age.

in sarcolemmal membranes from UM-X7.1 hamsters is reflected by an impressive decrerase of the membrane fluidity, as follows from the high degree of polarization fluorescence of the membrane-probe diphenylhexatriene, when compared to the fluorescence in membranes from age-matched SGH, in the temperature range from 16° C to 40° C (Fig. 1). In contrast, the degree of polarization of diphenylhexatriene in the same temperature range was not statistically significantly different when membranes from 30 days old animals were analyzed.

Discussion

In the present paper, the changes of the Km for ethanolamine of the heart sarcolemmal phospholipid base exchange enzyme during the development of cardiomyopathy in UM-X7.1 Syrian golden hamster strain are described. The kinetic changes are paralleled by an increase of the sarcolemmal membrane cholesterol content and of the cholesterol to phospholipid ratio. The phospholipid and cholesterol content of sarcolemmal membranes from the hamster strains that were used in the present study (Table 2) appear to be high when compared to previously published results from Panagia *et al.* [28]. Indeed, if the analytical data obtained from different authors are reconsiderd, it appears evident that the amount of sarcolemmal lipids present per miligram protein in the various membrane preparations largely depends on the method used for membrane isolation [28–33]. The procedure for purifying sarcolemmal membranes used in the present work appears to be reliable and reproducible, as far as quantitation of its lipid components is concerned. As a consequence of the increased cholesterol to phospholipid ratio a rigidification of sarcolemmal membranes, during development of cardiomyopathy, occurs, as follows from the increased degree of fluorescence polarization of membrane probe diphenylhexatriene.

Indeed, a relation between the kinetic parameters of the phospholipid base exchange enzyme and the cholesterol to phospholipid ratio in the sarcolemmal membranes can be expected, due to the well known dependence of the enzyme on the composition of the lipid phase of the host membrane [16, 17]. In fact, it was observed previously that the apparent activation energy of the phospholipid base exchange enzymes could be modified by changes in the fatty acid composition of membrane phospholipids of rat brain microsomes [16, 17]. The present data support the hypothesis that modification of the lipid environment of the phospholipid base exchange enzyme inducing changes of membrane fluidity modulates the conformation of the enzyme molecule and, as a consequence, its kinetic features [34].

It is conceivable that the changes in the sarcolemmal phospholipid base exchange activity of CMPH due to an increase in membrane viscosity also hold for membrane proteins involved in ion-transport, as previously hypothesized [4, 35].

However, the present results are not sufficient to recognize membrane modifications as the fundamental cause of cardiomyopathy in the hamster and further studies are needed to strengthen this hypothesis. In fact, it may well be that compositional changes of cardiac sarcolemma are a consequence of pathological processes that, among others, also involve membrane structure and function.

References

1. Homburger F, Baker JR, Nixon CW, Whitney R: Primary, generalized polymyopathy and cardiac necrosis in an inbred line of Syrian hamsters. Med Exp (Basel) 6: 339–345, 1962
2. Homburger F, Baker JR, Nixon CW, Welgram G: A new hereditary disease of Syrian hamsters: primary generalized polymyopathy and cardiac necrosis. Arch Intern Med 110: 660–662, 1962

3. Jasmin G, Proschek L: Hereditary polymyopathy and cardiomyopathy in the Syrian hamster. I. Progression of heart and skeletal muscle lesions. Muscle Nerve 5: 20–25, 1982
4. Proschek L, Jasmin G: Hereditary polymyopathy and cardiomyopathy in the Syrian hamster. II. Development of heart necrotic changes in relation to defective mitochondrial function. Muscle Nerve 5: 26–32, 1982
5. Paterson RA, Layberry RA, Nadkarni BB: Cardiac failure in the hamster. A biochemical and electron microscopic study. Lab Invest 26: 755–766, 1972
6. Whitmer JT: Energy metabolism and mechanical function in perfused hearts of Syrian hamsters with dilated or hypertrophic cardiomyopathy. J Mol Cell Cardiol 18: 307–317, 1986
7. Okumura K, Panagia V, Jasmin G, Dhalla NS: Sarcolemmal phospholipid N-methylation in genetically determined hamster cardiomyopathy. Biochem Biophys Res Commun 143: 31–37, 1987
8. Panagia V, Singh JN, Anand-Srivastava MB, Pierce GN, Jasmin G, Dhalla NS: Sarcolemmal alterations during the development of genetically determined cardiomyopathy. Cardiovasc Res 18: 567–572, 1984
9. Hano O, Mitsuoka T, Matsumoto Y, Ahmed R, Hirata M, Hirata T, Mori M, Yano K, Hashiba K: Arrhythmogenic properties of the ventricular myocardium in cardiomyopathic Syrian hamster, BIO 14.6 strain. Cardiovasc Res 25: 49–57, 1991
10. Factor SM, Sonnenblick EH: Microvascular spasm in the cardiomyopathic Syrian hamster as a cause of focal necrosis and myocardial failure. In: C Kawai and WH Abelmann (eds) Pathogenesis of myocarditis and cardiomyopathy. Recent experimental and clinical studies. University of Tokyo Press, 1987, pp 63–78
11. Berry B, Poulsen R, Yunge L, Bruneval P, Fitchett D, de Chastonay C, Gabibani G, Huttner I: Numerical densities of intramembrane particles in the cardiac sarcolemma of normal and myopathic Syrian hamsters. J Mol Cell Cardiol 15: 503–513, 1983
12. Singh JN, Dhalla NS, McNamara DB, Bajusz E, Jasmin G: Membrane alteration in failing hearts of cardiomyopathic hamsters. In: A Fleckenstein and G Rona (eds) Recent advances in studies on cardiac structure and metabolism. University Park Press, Baltimore, 1975, pp 259–268
13. Slack BE, Boegman RJ, Downie JW, Jasmin G: Cardiac membrane cholesterol in distrophic and verapamil-treated hamsters. J Mol Cell Cardiol 12: 179–185, 1980
14. Arienti G, Pirotta M, Giorgini D, Porcellati G: Base-exchange reactions of phospholipid metabolism in chick brain microsomal membranes. Biochem J 118: 3–4, 1970
15. Vecchini A, Binaglia L, Di Nardo P, Minieri M, Tallarica G: Phospholipid base exchange enzyme activity in sarcolemmal membranes from the heart of cardiomyopathic hamsters. Mol Cell Biochem (submitted)
16. Binaglia L, Roberti R, Vecchini A, Alunni-Bistocchi G, Porcellati G: Influence of fatty acid composition of membrane phospholipids on membrane-bound enzymic activities. Prog Fd Nutr Sci 4: 65–70, 1980
17. Binaglia L, Roberti R, Vecchini A, Porcellati G: The chemical composition of the lipid phase in biological membranes and their enzyme kinetics. Gazz Chim It 109: 437–440, 1979
18. Bajusz E, Jasmin G: Hereditary disease model of congestive cardiomyopathy: Studies on a new line of Syrian Hamsters. Fed Proc 31: 621–628, 1972
19. Pitts BJR: Stoichiometry of sodium-calcium exchange in cardiac sarcolemmal vesicles. J Biol Chem 254: 6232–6235, 1979
20. Lowry OH, Rosebrough NJ, Farr AL, Randall RJ: Protein measurement with the Folin phenol reagent. J Biol Chem 193: 265–275, 1951
21. Folch J, Lees M, Sloane-Stanley GM: A simplified method for the isolation and purification of total lipids from animal tissues. J Biol Chem 226: 497–509, 1957
22. Mascini M, Moscone D, Palleschi G: Determination of free and total cholesterol in human bile samples using enzyme electrodes. Clin Chim Acta 132: 7–15, 1983
23. Baykov AA, Evtushenko OA, Avaeva SM: A malachite green procedure for ortophosphate determination and its use in alkaline phosphatase-based enzyme immunoassay. Anal Biochem 171: 266–270, 1988
24. Shinitzky M, Barenholz Y: Fluidity parameters of lipid regions determined by fluorescence polarization. Biochim Biophys Acta 515: 367–394, 1978
25. Antunes-Madeira MC, Madeita VMC: Membrane fluidity as affected by the organochlorine insecticide DDT. Biochim Biophys Acta 1023: 469–474, 1990
26. Vecchini A, Roberti R, Freysz L, Binaglia L: Partial purification of ethanolaminephosphotransferase from rat brain microsomes. Biochim Biophys Acta 918: 40–47, 1987
27. Horrocks LA: The alk-1-enyl group content of mammalian myelin phosphoglycerides by quantitative two-dimensional thin layer chromatography. J Lipid Res 9: 469–472, 1968
28. Panagia V, Singh JN, Anand-Srivastava MB, Pierce GN, Jasmin G, Dhalla NS: Sarcolemmal alterations during the development of genetically determined cardiomyopathy. Cardiovasc Res 18: 567–572, 1984
29. Weglicki WB, Owens K, Kennet FF, Kessner A, Harris L, Wise RM: Preparation and properties of highly enriched cardiac sarcolemma from isolated adult myocytes. J Biol Chem 255: 3605–3609, 1980
30. Tibbits GF, Sasaki M, Ikeda M, Shimada K, Tsuruhara T, Nagatomo T: Characterization of rat myocardial sarcolemma. J Mol Cell Cardiol 13: 1051–1061, 1981
31. Panagia V, Lamers JMJ, Singal PK, Dhalla NS: Ca2+ – and Mg2+ – dependent ATPase activities in the deoxycholate-treated rat heart sarcolemma. Int J Biochem 14: 387–397, 1982
32. Post JA, Langer GA, Op den Kamp JAF, Verkleij AJ: Phospholipid asymmetry in cardiac sarcolemma. Analysis of intact cells and 'gas-dissected' membranes. Biochim Biophys Acta 943: 256–266, 1988
33. Dhalla NS, Pierce GN: Isolation and characterization of the sarcolemmal membrane from the heart. In: NS Dhalla (ed) Methods for Studying Cardiac Membranes, vol 1. CRC Press, Boca Raton, Florida, 1984, pp 3–18
34. Sanderman H: Regulation of membrane enzymes by lipids. Biochim Biophys Acta 515: 209–237, 1978
35. Bouchard A, Watters TA, Wu S, Parmley WW, Stone RD, Botvinick E, Sievers R, Jasmin G, Wikman-Coffelt J: Effect of altered coronary perfusion pressure on function and metabolism of normal and cardiomyopathic hamster hearts. J Mol Cell Cardiol 19: 1011–1023, 1987

Molecular and Cellular Biochemistry **116**: 95–101, 1992.

Annexins in cardiac tissue: cellular localization and effect on phospholipase activity

Marc van Bilsen,[1] Chris P.M. Reutelingsperger,[2] Peter H.M. Willemsen,[1] Robert S. Reneman,[1] and Ger J. van der Vusse[1]
Department of [1] Physiology and [2] Biochemistry Cardiovascular Research Instistute Maastricht, University of Limburg, Maastricht, The Netherlands

Abstract

Stimulation of cardiac phospholipid metabolism has diverse biological effects, ranging from subtle changes in cellular function to severe cellular damage. Accordingly, knowledge of the factors governing the activity of cardiac phospholipases is of great biological importance. A possible role of annexins, intracellular proteins that bind to membranes in a calcium dependent manner, as modulators of phospholipase activity has been proposed. In this study we investigated the cell type specific distribution of Annexin V and VIII in the heart. Recombinant Annexin V was used to examine the effect of this type of Annexin on cardiac phospholipase activity. Western blot analysis shows that annexin V is abundantly present in the heart. Using isolated myocytes and cultured cardiac endothelial and fibroblast-like cells, it is demonstrated that the localization of Annexin V is confined to non-myocytes. No positive bands matching the Mw of recombinant Annexin VIII are found in any of the cell types examined. *In vitro* studies demonstrate that recombinant Annexin V potently inhibits the activity of cardiac membrane-bound phospholipases, acting on their natural surrounding substrate, in a calcium dependent manner. Interestingly, annexin V also inhibits triacylglycerol hydrolysis. In conclusion, the expression of annexins is cell-type specific and suggests a cell-type specific function of these proteins in the heart. The absence of Annexin V in cardiac myocytes dismisses involvement of this annexin in cardiomyocyte phospholipid metabolism. The presence of Annexin V in cardiac endothelial and fibroblasts suggests a regulating role in the phospholipid homeostasis of non-myocyte cell types in the heart. (Mol Cell Biochem **116:** 95–101, 1992)

Key words: cardiac annexins; cellular damage; intracellular proteins; endothelial cells; fibroblast

Introduction

In the myocardium the turnover of membrane lipids is a finely tuned process. Enhanced hydrolysis of phospholipids (PLs) as a result of receptor stimulation leads to the production of biologically active products, like diacylglycerol and arachidonic acid. In addition, degradation of phospholipids is considered to play a pivotal role in the sequence of events leading to irreversible cell damage during myocardial ischemia and reperfusion [1, 2]. Accordingly, insight into the factors governing cardiac PL hydrolysis, and more specifically phospholipase activity, is of great biological importance. In this respect, the role of annexins has recently received widespread attention.

In vitro studies, using pancreatic phospholipase A_2 as

Address for offprints: G.J. van der Vusse, Department of Physiology, University of Limburg, PO Box 616, 6200 MD Maastricht, The Netherlands a model enzyme, have shown that annexins inhibit PL hydrolysis. A role of annexins in the regulation of cardiac phospholipid metabolism has been proposed [3, 4]. Annexins are found in a variety of tissues, including heart tissue [5–8]. The annexins represent a family of structurally related proteins that interact with phospholipid bilayers in a calcium dependent manner [6]. Recently Pula *et al.* [8] demonstrated that Annexin V is by far the most prominent annexin in the heart. In this organ the content of other annexins, such as Annexin VI, seems to be low [5, 8]. To be able to assess the function of cardiac annexins it is not only important to be informed of the type of annexins present in the heart, but also to know their cellular distribution. Besides, it is of interest to know whether cardiac annexins affect the activity of endogenous phospholipases in a manner comparable to the action of annexins towards pancreatic phospholipases.

In the present study, we investigated the prevalence and cellular distribution of Annexin V and Annexin VIII in rat cardiac tissue. Human recombinant Annexin V and VIII (formerly referred to as Vascular Anti-Coagulant-α and -β, respectively) have been cloned and expressed in E. coli [9–11]. Western blot analysis was used to examine the presence of these annexins in heart tissue and in various cell types derived from the heart. The availability of recombinant Annexin V enabled us to study the effect of this protein on cardiac phospholipid metabolism. These results have been previously reported in preliminary form [12].

Methods

Tissue handling

Hearts of ether anaesthetized male Lewis rats (250–300 g) were removed from the body and rapidly flushed with ice-cold phosphate buffered saline (PBS). For Western blot analysis ventricular tissue was homogenized (15% homogenate, w/v) in SET buffer (250 mM sucrose, 2 mM EDTA, 50 mM Tris-HCl, pH 7.2). After centrifugation (1,000 G, 10 min) the pellet was homogenized and centrifuged once again. Both supernatant fractions were combined ('post-nuclear supernatant' or PNS fraction). To assess the effects of annexins on lipid hydrolysis in crude homogenates, ventricular tissue was homogenized (15% w/v) in ST buffer (SET buffer minus EDTA). The membrane fraction was obtained by centrifugation of the PNS fraction (105,000 G, 60 min). The pellet was washed with SET buffer and subsequently with ST buffer, and thereafter resuspended in ST buffer.

Myocytes, endothelial cells and fibroblasts were isolated separately from rat hearts. The isolation of myocytes and the isolation, culturing, and characterization of cardiac endothelial cells and fibroblast-like cells have been described in detail before [13].

Western blot analysis

After addition of Ca^{2+} (final concentration 5 mM), the PNS fraction was incubated at 4° C for 15 min. After centrifugation (105,000 G, 60 min) the annexin containing pellet was resuspended in a large volume of SET buffer. After 15 min of incubation the suspension was centrifuged again. The annexin containing supernatant was concentrated by ultrafiltration. The concentrated samples were diluted 1 : 1 in 4% sodium dodecylsulfate (SDS) and 5 mM β-mercaptoethanol and subjected to SDS-gel electrophoresis. Isolated myocytes, endothelial and fibroblast-like cells were lysed in buffer (pH 6.8) containing 1% Triton X-100, 1% SDS, 1 mM EDTA, and protease inhibitors (0.1 mM leupeptin, 1 mM phenyl methyl sulfonyl fluoride (PMSF). The destruction of DNA was accomplished by ultra-sonication. Thereafter, the lysed cell preparations were adjusted to 2% SDS and 2.5 mM β-mercaptoethanol, and 20 μg of cellular protein per lane was subjected to SDS-gel electrophoresis. The separated proteins were subsequently transferred to nitrocellulose sheats using electroblotting [14]. The annexins were stained by an indirect immunoperoxidase method using immunoaffinity purified antibodies against human recombinant Annexin V (Annexin V_r) and Annexin VIII (Annexin VIII_r). The apparent Mw of positive bands was estimated by comparison with Mw markers after the gels had been stained with Coomassie Blue.

Lipid hydrolysis

Aliquots of the cardiac homogenate (250 μl) and membrane fraction (250 μl) were transferred to glass tubes. Subsequently, $CaCl_2$ and recombinant Annexin V (Annexin V_r) were added in the concentrations indicated. The final concentrations in the assay mixture were: sucrose 214 mM, Tris-HCl 46 mM, NaCl 14 mM, EGTA 0.1 mM (total volume 350 μl, pH 7.2). After flushing with nitrogen gas (to avoid oxidation of fatty acids) the tubes were capped and incubated at 37° C. Incubations were stopped by quick freezing in liquid nitrogen. The deeply frozen samples were thawed in 7.5 ml of ice-cold chloroform/methanol (2: 1 by volume, containing 0.01% BHT) and extracted as described in detail before [15]. The lipid extract was subjected to thin layer chromatography (TLC) [15]. After development the fatty acid (FA), triacylglycerol (TG) and phospholipid (PL) spots weer scraped from the TLC plate. To determine the relationship between FA and lysophospholipid production, in a subset of experiments the lipid extract was applied to silicagel column chromatography [16]. The PL and apolar lipid fraction were separated by stepwise elution from the column. Apolar lipids were separated by TLC as described above. The PL fraction was further separated by TLC as described previously [17]. After elution from the silicagel powder the lipid fractions were (trans)methylated using borontrifluoride. The fatty acid methyl esters were quantitated by gas chromatography [16].

Results

Western blot analysis

Incubation of Western blots of cardiac homogenates with anti-Annexin V_r antibody resulted in a single band with an apparent molecular weight (Mw) of 32 kD (not shown). The 32 kD antigen detected with anti-Annexin V_r antibodies is very likely the rat counterpart of human annexin V since conservation between human and rat is about 89% and the antibody used does not cross-react with other annexins [11]. Antibodies against Annexin VIII_r (apparent Mw 36 kD) gave rise to several faint bands with apparent Mw's that differed from Annexin VIII_r (not shown).

To examine the cellular distribution of these annexins, preparations of cardiac myocytes, endothelial cells and fibroblast-like cells were subjected to Western Blot analysis. Figure 1 shows that Annexin V is not present in myocytes, but is confined to endothelial and fibroblast-like cells. When anti-Annexin VIII_r antibodies were used, the myocyte extract displayed some cross-reactivity with a band of apparent Mw of 67 kD. No obvious cross reactive bands were observed in the extracts of the other two cell types (Fig. 1).

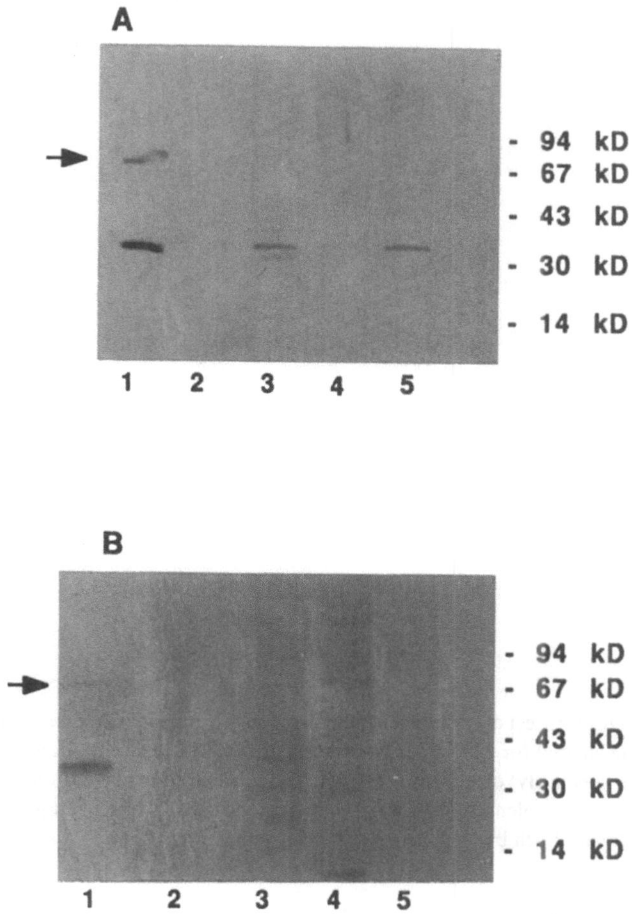

Fig. 1. Western blot analysis of different cardiac cell types with anti-annexin V and anti-annexin VIII antibodies. Extracts of isolated cardiomyocytes, cardiac fibroblasts and endothelial cells were prepared as described in the Method section. 2-μg of protein extract was submitted to SDS-gel electrophoresis and Western blot analysis using anti-annexin V-IgG (panel A) and anti-annexin VIII-IgG (panel B). Lanes 3: fibroblasts, lanes 4: myocytes and lanes 5: endothelial cells. Either 200 ng of purified recombinant annexin V (panel A, lane 1) and annexin VIII (panel A, lane 2) or 20 ng of annexin V (panel B, lane 2) and annexin VIII (panel B, lane 1) were run as references. The arrows indicate dimers of the annexins, that constitute 1–4% of the protein.

Effects on endogenous phospholipase activity

To investigate whether Annexin V_r was able to inhibit cardiac phospholipase, in general, and membrane-bound cardiac phospholipases, in particular, crude homogenates and membrane fractions from rat hearts were incubated in the presence of different concentrations of Ca^{2+} and Annexin V_r.

To characterize this *in vitro* system, homogenates were incubated for a period of up to 3 hr and the time course of the formation of fatty acids (FAs), lysophosphatidylcholine (LPC) and lysophosphatidylethanola-

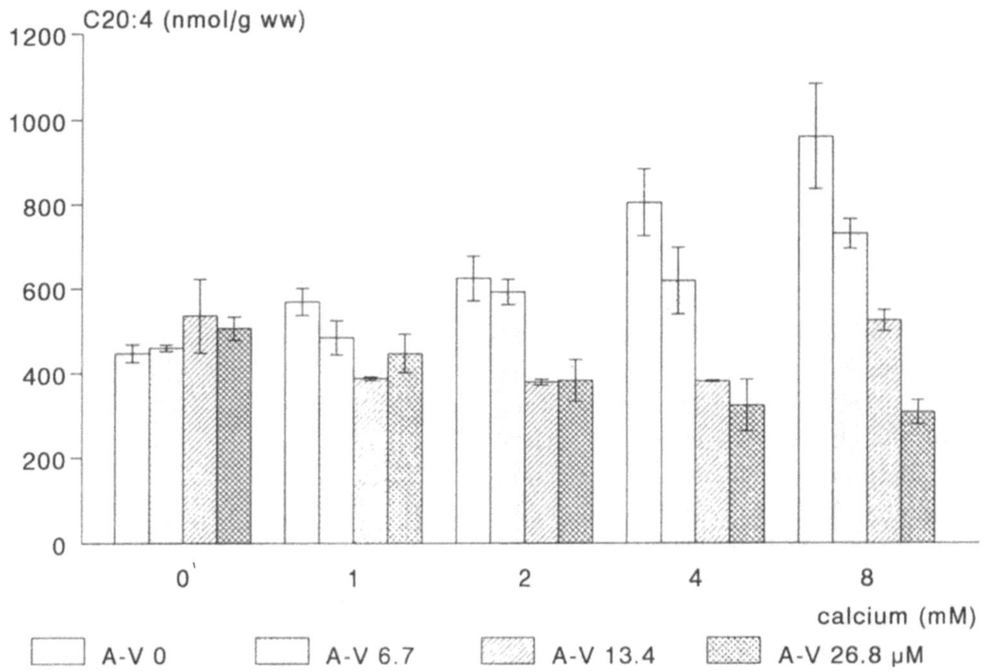

Fig. 2. Effect of calcium and human recombinant Annexin V on arachidonic acid (C20 : 4) accumulation in heart homogenates. Homogenates were incubated for 3 hr under N_2-atmosphere at 37° C concentrations instead of amounts in the absence (0 μM) or presence (6.7, 13.4 or 26.8 μM, respectively) of annexin V (A–V). The concentration of endogenous phospholipids in the incubation mixture averaged 6.5 mM (measured as fatty acid equivalents). The C20 : 4 content at the end of 3 h incubation is shown. Initial C20 : 4 levels were 33 ± 6 per nmol per gram wet weight (ww) of tissue. Each bar represents the mean value of n = 2–4.

mine (LPE), and the degradation of triacylglycerols (TG's) was followed (conditions: 0.1 mM EGTA, 1 mM Ca^{2+}, pH 7.2, 37° C, N_2-atmosphere). During these 3 hr the rates of FA, LPC and LPE formation were linear. The FA, LPC, and LPE content increased on the average from 0.3, 0.1 and 0.2 to 5.2, 0.4 and 0.4 μmol/g wet weight of tissue, respectively, indicating that less than 3% of the PL's were degraded. FA accumulation exceeded lysophospholipid accumulation by about 10 fold, suggesting that the majority of the lysophospholipids formed was further degraded.

The formation of arachidonic acid (AA) was linear in time. Based upon the disappearance of AA from the TG pool during incubation, it is estimated that less than 6% of the accumulating AA could be attributed to the liberation of AA from TG's. Hence, the rise of AA was largely due to PL degradation and was considered to be a reliable index for PL degradation.

The effect of increasing concentrations of Ca^{2+} and Annexin V_r on AA accumulation in the cardiac homogenate is shown in Fig. 2. In the absence of free Ca^{2+} the AA content rose on the average from 33 to 447 nmol/g wet wt during 3 hr of incubation. Increasing the Ca^{2+} concentration to 8 mM enhanced the production of AA by 114%. The presence of Annexin V_r in increasing amounts resulted in the inhibition of AA formation in a Ca^{2+}-dependent manner. At a concentration of 8 mM Ca^{2+} the accumulation of AA was reduced by almost 70% when Annexin V_r was present at a concentration of 27 μM. At this concentration of Annexin V_r the ratio of PLs over annexin approximated 120 : 1. AA accumulation was not reduced when Annexin V_r was denatured prior to addition to the assay mixture (data not shown).

In the experiments with cardiac membrane fractions, the effects of Ca^{2+} and Annexin V_r were comparable to those observed in homogenates (Fig. 3). Incrreceived Ca^{2+} concentrations stimulated the accumulation of AA. This effect was completely reversed when Annexin V_r was administered in increasing amounts. Interestingly, Ca^{2+} and Annexin V_r also affected TG degradation in the crude homogenates (Fig. 4). Calcium was able to potentiate the hydrolysis of TG's, whereas Annexin V_r counteracted this effect.

Discussion

The findings in the present study demonstrate that An-

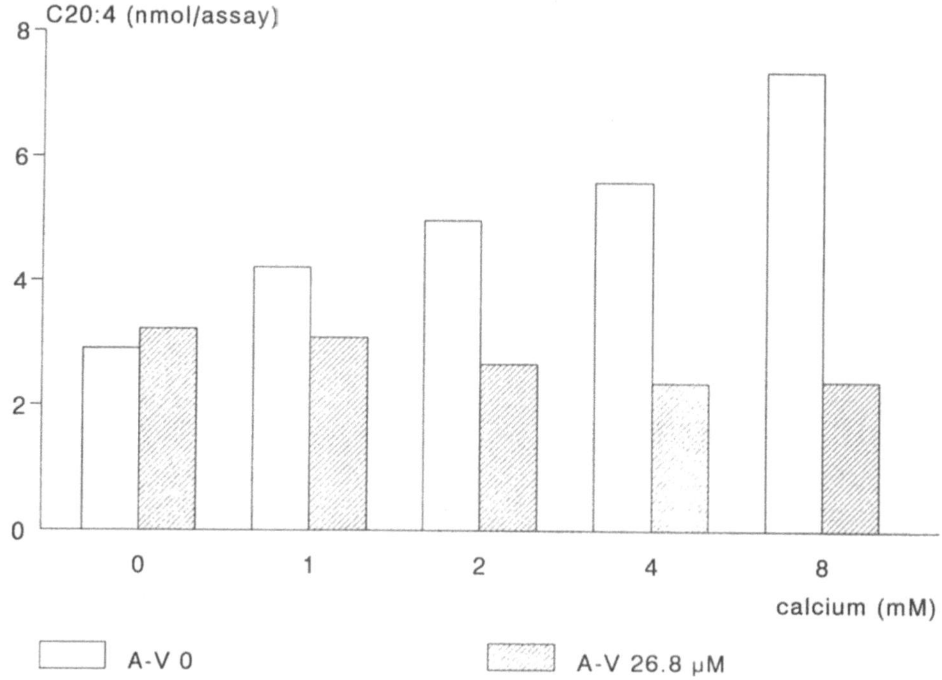

Fig. 3. Effect of calcium and human recombinant Annexin V (A–V) on arachidonic acid (C20 : 4) accumulation using a cardiac membrane fraction. The concentration of membrane phospholipids in the incubation mixture averaged 1.5 mM (measured as fatty acid equivalents). The bars represents the C20 : 4 content/assay after 3 h incubation of aliquots of cardiac membrane fractions in a typical experiment. Initial C20 : 4 levels were 0.85 ± 0.19 nmol/assay. Assay conditions were similar as described in Fig. 2.

nexin V, the quantitatively most important annexin in the heart [8], inhibits the activity of endogenous membrane bound phospholipases in a Ca^{2+}-dependent manner. Annexin V is exclusively localized in non-myocyte cells. At present, the significance of the protein cross-reacting with anti-Annexin VIII_r, is unknown. It is noteworthy that several annexins with apparent Mw of arround 67 kD have been described. Recently, Annexin VIII has been found in other rat tissues, like the lung (CPHM Reutelingsperger, unpublished observations). These results indicate that the expression of annexins is cell-type specific which suggests a cell-type specific function of these proteins in the heart. Until now the physiological significance of the annexins is largely unknown. There is increasing evidence that they are involved in various membrane-associated processes, such as membrane fusion, coagulation and PL homeostasis.

Beside anticoagulant activity, Annexin V_r and Annexin VIII_r were found to inhibit the activity of soluble pancreatic phospholipase A_2, when (^{3}H)-oleic acid labeled E. coli membranes are used as substrate [10]. Using comparable *in vitro* systems other members of the annexin family were found to inhibit the activity of soluble or solubilized phospholipases [13–20]. In the isolated lung human recombinant lipocortin I (Annexin I) was able to suppress thromboxane synthesis, when added to the perfusion medium [21]. The latter finding suggests that annexins are also able to affect phospholipase activity in intact cell systems.

Apparently, the inhibition of PL degradation is mediated by interaction with the PL substrate, rather than interaction with the phospholipase enzyme itself [18, 19]. Through their binding to the membrane surface, which is Ca^{2+} dependent, the annexins are thought to diminish the accessibility of the lipid substrate. This mechanism of action is referred to as the 'substrate depletion model' [18]. This model has been found useful for studies employing soluble or solubilized phospholipases. As stated above, conclusions concerning the phospholipase inhibitory effect and mechanism of action of annexins are primarily based on studies using soluble pancreatic phospholipase A_2 as model enzyme and artificial membranes as substrate [10, 18–20]. The availability of large amounts of recombinant Annexin V enabled us to study the effect of annexins on PL hydrolysis in a more natural setting, i.e. endogenous phospholipases acting on their natural surrounding substrate. The experiments with cardiac homogenates indicate that Annexin V_r inhibits the Ca^{2+}-stimulated overall (soluble plus membrane-bound) cardiac phos-

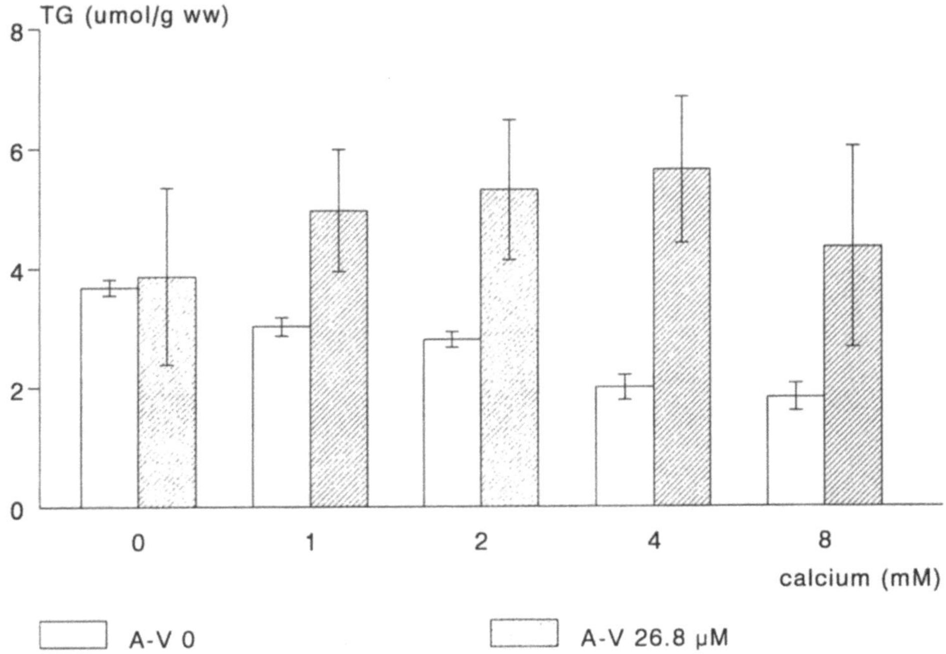

Fig. 4. Effect of calcium and human recombinant Annexin V (A–V) on the degradation of triacylglycerols (TG) in cardiac homogenates (n = 3). Initial TG levels were 4.8 ± 1.0 μmol fatty acid equivalents/g wet weight (ww) of tissue. Bars represent the content of TG (expressed as μmol/g ww) at the end of 3 h incubation. Assay conditions and procedures are described in the legends to Fig. 2.

pholipase activity when endogenous PLs are used as substrate. Moreover, using a cardiac membrane preparation, it is shown that the added Annexin V_r potently inhibits the Ca^{2+}-dependent activity of membrane-bound cardiac phospholipases. It should be noted that relatively high molar ratios of annexins over PL's are required to observe marked effects on PL hydrolysis. The latter finding supports the notion that Annexin V does not affect PL hydrolysis through direct interaction with phospholipases, but rather through interaction with the PL substrate. The present observation that cardiac TG degradation is also affected extends the already known effects of annexins on lipid hydrolysis and poses new questions concerning their intracellular function and their mechanisms of action involved.

The fact that annexins act in an aspecific manner does not dismiss a potential role in PL homeostasis. A variety of factors may influence the activity of annexins *in situ*. The binding of annexins to PL bilayers is highly PL species dependent. Pula and coworkers [8] showed that annexin V has some binding preference for sarcolemmal membranes. Moreover, the affinity for membranes can be potentiated by other divalent cations than Ca^{2+} [22], and can be modulated through phosphorylation [6].

The absence of annexin V in cardiac myocytes excludes a role of this annexin in cardiomyocyte PL homeostasis. The relevance of the 67 kD protein band cross reacting with anti-Annexin VIII_r in this cell type remains to be determined. Annexin V has been shown to be a potent inhibitor of cardiac membrane-bound phospholipases and is abundantly present in cardiac non-myocytes, especially endothelial cells. The latter observations suggest that in these cells Annexin V might play an important role in the regulation of membrane-associated processes involving phospholipases.

Acknowledgements

Marc van Bilsen is supported by a grant of the Netherlands Organization for Scientific Research (NWO) and Chris P.M. Reutelingsperger is a fellow of the Royal Netherlands Academy of Sciences (KNAW).

References

1. Corr PB, Gross RW, Sobel BE: Amphipathic metabolites and membrane dysfunction in ischemic myocardium. Circ Res 55: 135–154, 1984
2. Van der Vusse GJ, Van Bilsen M, Sonderkamp T, Reneman RS: Hydrolysis of phospholipids and cellular integrity. In Piper HM

(ed): Pathophysiology of severe ischemic myocardial injury, Dordrecht, Kluwer Acad Publ, pp 167–193, 1990

3. Dennis EA: Regulation of eicosanoid production: role of phospholipases and inhibitors. Bio/technology 5: 1294–1300, 1987
4. Hatch HM, O K, Choy PC: Regulation of phosphatidylcholine metabolism in mammalian hearts. Biochem Cell Biol 67: 67–77, 1989
5. Wallner BP, Mattaliano RJ, Hession C, Cate RL, Tizard R, Sinclair LK, Foeller C, Chow EP, Browning JL, Ramachandran KL, Pepinsky RB: Cloning and expression of human lipocortin, a phospholipase A_2 inhibitor with potential anti-inflammatory activity. Nature 320: 77–81, 1986
6. Klee CB: Ca^{2+}-dependent phospholipid- (and membrane-) binding proteins. Biochemistry 27: 6645–6653, 1988
7. Kaetzel MA, Hazarika P, Dedman JR: Differential tissue expression of three 35-kDa annexin calcium-dependent phospholipid binding proteins. J Biol Chem 364: 14463–14470, 1989
8. Pula G, Bianchi R, Ceccarelli P, Giambanco I, Donato R: Characterization of mammalian heart annexins with special reference to CaBP 33 (annexin V). FEBS Lett 277: 53–58, 1990
9. Reutelingsperger CPM, Hornstra G, Hemker HC: Isolation and partial purification of a novel anticoagulant from arteries of human umbilical cord. Eur J Biochem 151: 625–629, 1985
10. Hauptmann R, Maurer-Fogy I, Krystek E, Boco G, Andree H, Reutelingsperger CPM: Vascular anticoagulant beta (VAC-beta: a novel human Ca^{++}/phospholipid binding protein that inhibits coagulation and phospholipase A_2 activity. Eur J Biochem 185: 63–71, 1989
11. Maurer-Fogy I, Reutelingsperger CPM, Pieters J, Bodo G, Stratowa C, Hauptmann R: Cloning and expression fo cDNA for human vascular anticoagulant, a Ca^{2+}-dependent phospholipid-binding protein. Eur J Biochem 174: 585–592, 1988
12. Van Bilsen M, Reutelingsperger CPM, Van der Vusse GJ: Are annexins involved in cardiac phospholipid metabolism? J Mol Cell Cardiol 23 (suppl V): S.51, 1991
13. Linssen MCJG, Vork MM, de Jong YF, Glatz JFC, Van der Vusse GJ: Fatty acid oxidation capacity and fatty acid-binding protein content of different cell types isolated from rat heart. Mol Cell Biochem 98: 19–25, 1990
14. Towbin H, Staehelin T, Gordon J: Electrophoretic transfer of proteins from polyacrylamide gels to nitrocellulose sheats: Procedure and some applications. Proc Natl Acad Sci USA 76: 4350–4354, 1979
15. Van der Vusse GJ, Roemen THM, Prinzen FW, Coumans WA, Reneman RS: Uptake and tissue content of fatty acids in dog myocardium under normoxic and ischemic conditions. Circ Res 50: 538–546, 1982
16. Roemen THM, Keizer H, Van der Vusse GJ: Quantitative assessment of gamma-linolenic acid in human blood and plasma with capillary gas chromatography. J Chromatogr (Biomed Appl) 528: 447–452, 1990
17. Van Bilsen M, Van der Vusse GJ, Willemsen PHM, Coumans WA, Roemen THM, Reneman RS: Lipid alterations in isolated, working hearts during ischemia and reperfusion: Its relation to myocardial damage. Circ Res 64: 304–314, 1989
18. Davidson FF, Dennis EA, Powell M, Genney JR: Inhibition of phospholipase A_2 by 'lipocortins' and calpactins: an effect of binding to substrate phospholipids. J Biol Chem 262: 1698–1705, 1987
19. Aarsman AJ, Mynbeek G, Van den Bosch H, Rothhut B, Prieur B, Comera C, Jordan L, Russo-Marie F: Lipocortin inhibition of extracellular and intracellular phospholipase A_2 is substrate concentration dependent. FEBS Lett 219: 176–180, 1987
20. Fauvel J, Vicendo P, Roques V, Ragab-Thomas J, Garnier C, Vilgrain I, Chambaz E, Rochat H, Chap H, Douste-Blazy L: Isolation of two 67 kDa calcium-binding proteins from pig lung differing in affinity for phospholipids and in anti-phospholipase A_2 activity. FEBS Lett 221: 397–402, 1987
21. Cirino G, Flower RJ, Browning JL, Sinclair LK, Pepinski RB: Recombinant human lipocortin 1 inhibits thromboxane release from guinea-pig isolated perfused lung. Nature 328: 270–272, 1987
22. Andree HSM, Reutelingsperger CPM, Hauptman R, Hemker HC, Hermens WT, Willems GM: Binding of vascular anticoagulant-alpha (VAC-α) to planar phospholipid bilayers. J Biol Chem 265: 4923–4928, 1990

Molecular and Cellular Biochemistry **116**: 103–109, 1992.

Myocardial fatty acid oxidation during ischemia and reperfusion

René Lerch, Christian Tamm, Irene Papageorgiou and Richard H. Benzi
Cardiology Center, University Hospital, Geneva, Switzerland

Abstract

Inhibition of fatty acid oxidation is an early event in myocardial ischemia that most likely contributes to tissue injury by the accumulation of potentially toxic intermediates such as acylCoA and acylcarnitine. After reperfusion both myocardial oxygen consumption and fatty acid oxidation may rapidly recover to preischemic levels, even when contractile function remains depressed. The mechanisms underlying the apparent dissociation between contractile function and oxidative metabolism early during reperfusion are still controversial. In isolated rat hearts subjected to 60 min of no-flow ischemia myocardial oxygen consumption and oxidation of palmitate were lowered during reperfusion by 3 mM of $NiCl_2$ and by 6 μM of ruthenium red. The results provide indirect evidence for the hypothesis that intracellular calcium transport may be involved in the mechanisms responsible for the high oxidative metabolic rate early after reperfusion. (Mol Cell Biochem **116:** 103–109, 1992)

Key words: metabolism; fatty acid, glucose, reperfusion

Introduction

Ischemic heart disease is a main cause of human morbidity and mortality. During acute ischemia, inhibition of oxidative metabolic pathways and impairment of the removal of metabolic intermediates initiate a cascade of events that causes myocardial dysfunction and ultimately lead to cell death. Lipid metabolism of ischemic myocardium has been extensively characterized and results have been summarized in several review articles [1–5]. The introduction of thrombolytic therapy in the treatment of myocardial infarction has recently stimulated interest in the metabolism of reperfused myocardium. In animal experiments, sudden reperfusion of ischemic myocardium is often associated with loss of ionic homeostasis of the myocytes [6], production of free radicals [7], enhancement of myocardial contracture [8], acceleration of the loss of enzymes [9] and delayed recovery of the contractile function [10]. Characterization of alterations of fatty acid metabolism after reperfusion is of clinical interest for at least two reasons. First, alterations of oxidative fatty acid metabolism after reperfusion may be involved in the mechanisms underlying contractile dysfunction ('stunning'), reperfusion arrhythmias and progression to irreversible injury after reperfusion ('reperfusion injury'). Second, fatty acid utilization after reperfusion may serve as an index of tissue viability, amenable to noninvasive assessment by new imaging modalities such as positron emission tomography [11]. This review will first provide a short survey of metabolic changes during acute ischemia, and then focus on oxidative metabolism in postischemic myocardium.

Address for offprints: R. Lerch, Cardiology Center, University Hospital, Geneva, Switzerland

Fatty acid metabolism during acute ischemia

Oxygen deprivation of the mitochondria during ischemia impairs electron flux in the respiratory chain and hence removal of the reducing equivalents originating from substrate oxidation. As a consequence, the ratios $NADH/NAD^+$ and $FADH_2/FAD$ increase within the first minute after the onset of ischemia [2], resulting in rapid inhibition of β-oxidation [13, 14]. In contrast to β-oxidation, activation of fatty acids with Coenzyme A (CoA) may continue for some time because the Km for ATP of the acylCoA synthetase reaction is comparatively low at 0.8 mM [15]. Based on a cytosolic ATP concentration in normal myocardium of approximately 7 mM [16] and on reported rates of ATP depletion during no-flow ischemia [17, 18], one would predict that fatty acid activation is not limited by lack of ATP for at least 30–60 min. Indeed, myocardial accumulation of appreciable amounts of nonesterified fatty acids is a rather late phenomenon during ischemia starting after 20 to 30 min [15, 19]. Van Bilsen *et al.* [15] have concluded, based on simultaneous measurement of fatty acid accumulation and high energy phosphate content, that in isolated rat hearts subjected to no-flow ischemia accumulation of AMP, rather than lack of ATP, initiates inhibition of fatty acid activation.

In contrast to free fatty acids, accumulation of fatty acid esters of Coenzyme A and of carnitine is detectable as early as 2 min after the onset of ischemia [13]. The extent of accumulation of acylCoA and acylcarnitine during ischemia depends on several factors including the residual perfusion and the arterial fatty acid concentration [13, 20–23].

Since long-chain fatty acids, acylCoA and acylcarnitine are capable of inhibiting a number of enzyme systems, and may compromise membrane function by other mechanisms, accumulation of these amphiphiles during ischemia has been implicated in the mechanisms underlying arrhythmias and irreversible myocyte injury [For a review see 3, 5, 24].

Increased cytosolic content of acylCoA combined with increased anaerobic production of glycerol-3-phosphate in the glycolytic pathway promote synthesis of triacylglycerols in ischemic myocardium [25]. Indeed, net accumulation of triacylglycerols during ischemia has been observed in several studies [13, 26, 27], but is not a constant finding [28]. There is evidence that not only synthesis, but also hydrolysis of triacylglycerols is accelerated during ischemia [15, 29, 30]. Because thioesterification of fatty acids with Coenzyme A requires ATP, hydrolysis and subsequent resynthesis of triacylglycerols results in an energy-consuming metabolic cycle [29, 30]. On the basis of observed rates of glycerol release [29, 30] in isolated rat hearts subjected to low-flow ischemia, the rate of ATP utilization in this metabolic cycle can be estimated to be of the order of 0.10–0.25 μmol ATP·min^{-1}·g wet weight^{-1} [29, 30]. Expressed in metabolic equivalents these values correspond to oxygen consumption of 0.015–0.040 μmol·min^{-1}·g wet weight^{-1} or lactate production from exogenous glucose of 0.10–0.25 μmol·min^{-1}·g wet weight^{-1}. Thus, energy 'wasted' in the metabolic cycle by hydrolysis and resynthesis of triacylglycerols is small, but may contribute to loss of high energy phosphate stores in ischemic myocardium.

Fatty acid metabolism during reperfusion

Overall oxidative metabolic rate in reperfused myocardium

Recovery of normal myocyte function after reperfusion requires restoration of mitochondrial function and resumption of oxidative substrate metabolism. We have studied recovery of oxidative substrate metabolism in isovolumically-beating isolated rat hearts perfused with erythrocyte-enriched medium (hematocrit 30%) containing 11 mM of glucose and 0.4 mM of palmitate bound to 0.4 mM of albumin [31]. In hearts reversibly injured by 60 min of low-flow ischemia (5% of control flow) myocardial oxygen consumption had returned to preischemic levels as early as 5 min after the onset of reperfusion when left ventricular pressure development was still significantly reduced to 60% of control. Thus, recovery of myocardial oxygen consumption, in this model, is disproportionately high after reperfusion when compared with recovery of contractile function. Similar trends have been observed in dogs [32, 33] and pigs [34] with transient coronary occlusion. In these studies with regional ischemia, myocardial oxygen consumption in mechanically 'stunned' postischemic myocardium was similar to that in a normally contracting control region of the same heart.

It has been proposed that recovery of oxidative metabolism after reperfusion may presage later recovery of contractile function [35]. To determine the effect of more advanced ischemic injury on recovery of oxidative metabolic rate we subjected isolated rat hearts to 60 min of complete cessation of myocardial perfusion. Surprisingly, also in this group of hearts, myocardial oxygen consumption had recovered almost completely after 5 min of reperfusion, to 94% of the value measured in control hearts, despite marked diastolic contracture and persistent depression of left ventricular pressure development to less than 30% of control. Electron microscopy examination of the hearts, which were perfusion-fixed after 10 min of reperfusion, revealed evidence consistent with irreversible myocyte injury. Thus, in this model, appreciable rates of oxidative metabolism may also be observed immediately after reperfusion in myocardium that exhibits evidence of irreversible injury after reperfusion. Because in a subsequent series of experiments ruthenium red administered during the reperfusion period exerted a protective effect on these hearts [36], it is likely that part of the irreversible injury occurred during the early reperfusion period rather than during ischemia (reperfusion injury). It may be speculated that oxidative metabolism recovers immediately after reperfusion, not only in reversibly injured myocardium, but also in myocardium with more advanced injury which becomes irreversible during the early reperfusion period.

When no-flow ischemia was extended to 120 min in isolated rat hearts there was no recovery of oxidative metabolism after reperfusion.

Substrate pattern of oxidative metabolism after reperfusion

In most instances in normal myocardium, fatty acids represent the predominant fuel for oxidative metabolism [2, 37]. To determine the contribution of glucose and of palmitate to oxidative metabolism after reperfusion we supplemented the perfusate in isovolumically-beating rat hearts with trace amounts of either (U-^{14}C)-glucose or (1-^{14}C)-palmitate and monitored myocardial release of $^{14}CO_2$ [31]. In control hearts, oxidation of exogenously supplied palmitate and glucose accounted for 52% and 32% of total oxygen consumption respectively. Fifteen minutes after reperfusion following 60 min of low-flow ischemia, oxidation of palmitate had completely recovered and accounted for 48% of total myocardial oxygen consumption. During the initial 15 min of reperfusion myocardial release of $^{14}CO_2$ from (U-^{14}C)-glucose was significantly increased above baseline levels, suggesting a transient increase of glucose oxidation, which, however, resolved within 30 min [31].

The observation of complete recovery of fatty acid oxidation to preischemic levels within 15 min after reperfusion is consistent with the reports by Lopaschuk *et al.* [38] in isolated working rat hearts subjected to 25 min of no-flow ischemia, and by Liedtke *et al.* [39] in working swine hearts subjected to 45 min of low-flow ischemia. Compared to our experiments, in these latter studies [38, 40] the contribution of glucose oxidation was lower both during control conditions and after reperfusion, most likely because the arterial fatty acid concentration was higher. In the study by Lopaschuk *et al.* [38], the rate of glucose oxidation was markedly higher in reperfused isolated rat hearts when palmitate was absent in the perfusate and glucose was the only substrate. Thus, oxidation of glucose is inhibited by the oxidation of fatty acids after reperfusion, as is the case in normal myocardium [38].

Possible causes of the dissociation between oxidative metabolism and contractile function after reperfusion

The experiments cited above indicate that in postischemic myocardium both myocardial oxygen consumption and oxidation of palmitate may rapidly recover to preischemic values despite persistent depression of contractile function [31]. Several mechanisms may contribute to the apparent dissociation between contractile function and oxidative metabolic rate after reperfusion. First, since wall tension is a major determinant of myocardial oxygen consumption [41], energy requirements of contractile elements may still be high despite poor pressure development or segment shortening during reperfusion. However, in isovolumically beating isolated rat hearts reperfused after 60 min of no-flow ischemia, unloading of the myocardium by deflation of the left ventricular ballon resulted in only a small, 15% reduction of oxygen consumption (unpublished observation). Thus, energy needs for maintainance of wall tension are most likely not responsible for the bulk of myocardial oxygen consumption after reperfusion. Second, metabolic cycling of fatty acids initiated during ischemia and involving triacylglycerol synthesis and hydrolysis [29, 30] may continue during early reperfusion. In support of increased triacylglycerol synthesis we have observed in isolated rat hearts increased incorporation of exogenously-supplied (1-^{14}C)-palmitate into triacylglycerols after reperfusion [31]. A similar obser-

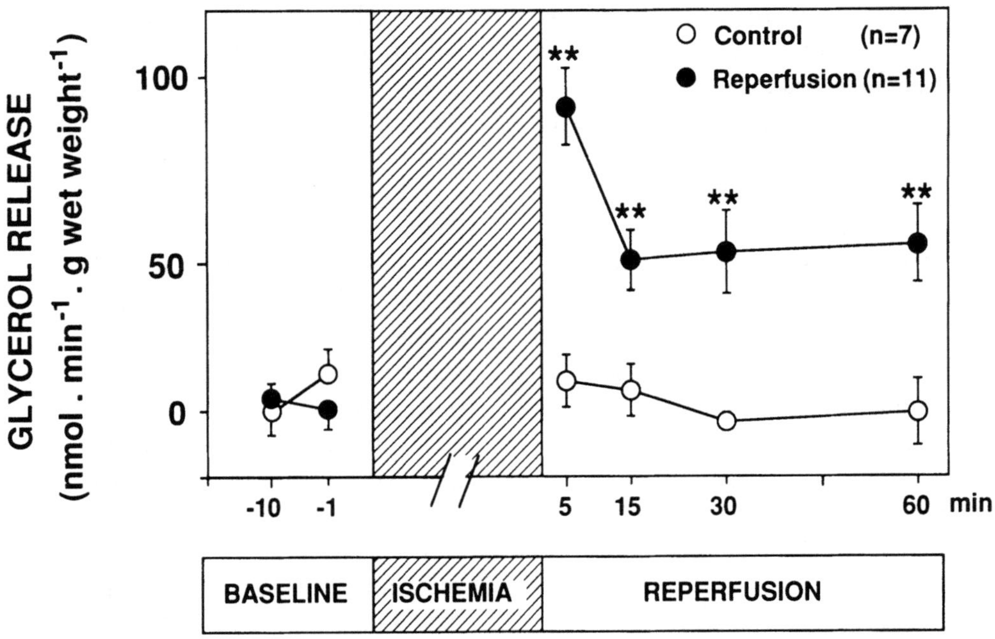

Fig. 1. Myocardial release of glycerol in isovolumically-beating isolated rat hearts perfused for 140 min at control flow (2 ml·min^{-1}·g wet weight^{-1}; open circles) or subjected to 60 min of no-flow ischemia followed by 60 min of reperfusion (filled circles). The perfusate consisted of erythrocyte-enriched (hematocrit 30%) Krebs-Henseleit buffer containing 0.4 mM of albumin, 0.4 mM of palmitate and 11 mM of glucose. Release of glycerol was increased during the entire reperfusion period. Values are means ± SEM. ** $p < 0.01$ vs control hearts.

vation has been reported by Lopaschuk *et al.* [38]. Furthermore, in isolated rat hearts subjected to 60 min of no-flow ischemia, we found that during postischemic reperfusion myocardial release of glycerol, after the initial washout-peak, remained markedly elevated during reperfusion, at approximately 45 nmol·min^{-1}·g wet weight^{-1} (Fig. 1). Assuming resynthesis of triacylglycerol at a rate corresponding to release of glycerol, ATP consumed by metabolic cycling would be equivalent to oxygen consumption of approximately 0.05 μmol·min^{-1}·g wet weight^{-1}. This accounts for less than 5% of the myocardial oxygen consumption observed in these hearts after reperfusion. Thus, although metabolic cycling of fatty acids may continue after reperfusion, energy consumed by this process is likely to account for only a minor fraction of the oxygen consumption. Third, mitochondrial damage may lead to uncoupling between electron transport in the respiratory chain and oxidative phosphorylation [42]. Fourth, respiratory energy may be required by cellular processes other than contractile element work after reperfusion. Since cytosolic calcium accumulation occurs during ischemia [43], and often further increases after reperfusion [44], activation of calcium transport processes at the sarcoplasmic reticulum and at the inner mitochondrial membrane may contribute to enhanced energy expenditure after reperfusion [45–47].

We sought to investigate the potential role of intracellular calcium transport in the high metabolic rate after reperfusion in isovolumically-beating isolated rat hearts perfused retrogradely with erythrocyte-enriched buffer containing 0.4 mM of palmitate and 11 mM of glucose and reperfused after 60 min of no-flow ischemia. Without intervention, these hearts exhibited, as discussed above, complete recovery of oxygen consumption after reperfusion, despite severe depression of contractile function [31]. In a first series of experiments the medium was supplemented with 3 mM of $NiCl_2$ during the initial 30 min of reperfusion. In the presence of Ni^{2+}, which inhibits transsarcolemmal calcium influx after reperfusion [48], myocardial oxygen consumption, oxidation palmitate and oxidation of glucose were reduced by 71%, 71%, and 86%, respectively, compared with the values measured in untreated postischemic hearts (Fig. 2). The observation suggested that the high oxidative metabolic rate after reperfusion is related to cytosolic calcium overload. In a second series of experiments, 6 μM of ruthenium red was administered during the first 40 min of reperfusion [36]. Ruthenium red inhibits calcium transport at the inner mitochondrial membrane both directly [49, 50] and indirectly by reduction of calcium release from the sarcoplasmic reticulum [51]. Fifteen minutes after the onset of reperfusion ruthenium red-treated hearts exhibited a significant reduction of myocardial oxygen consumption, palmitate oxidation and glucose oxidation compared to untreated reperfused hearts by 35%, 37% and 42% respectively [36]. Furthermore, recovery of contractile function was markedly improved during ruthenium red infusion, despite the lower oxidative metabolic rate. The initially severely abnormal relationship between contractile function and myocardial oxygen consumption, indicated by a marked reduction of the ratio of developed pressure to myocardial oxygen consumption, returned to preischemic values during ruthenium red infusion. The results provide circumstantial evidence for the hypothesis that energy expenditure for intracellular calcium transport may be involved in the mechanisms underlying the dissociation between the oxidative metabolic rate and contractile function after reperfusion.

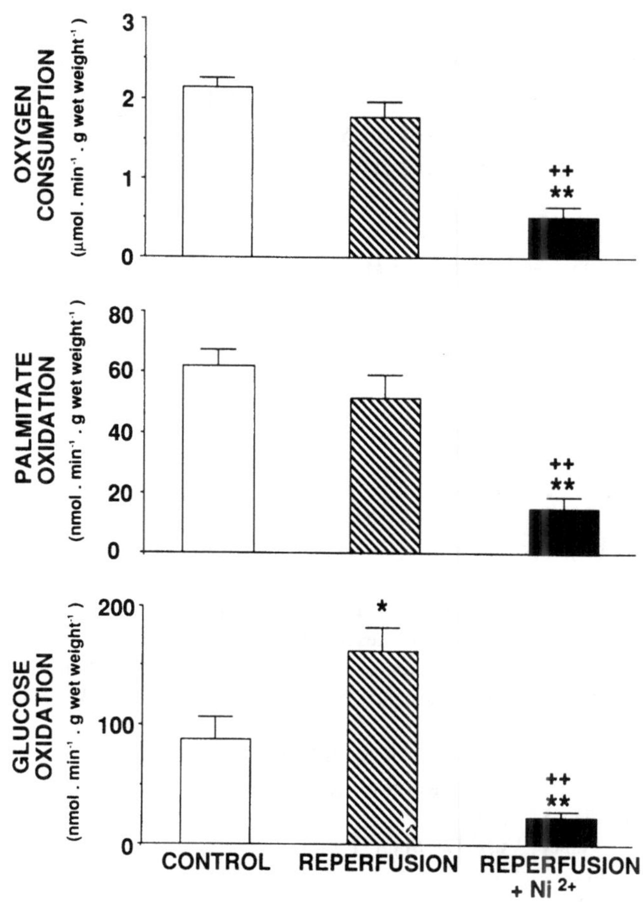

Fig. 2. Effect of addition of 3 mM of Ni^{2+} on myocardial oxygen consumption, palmitate oxidation and glucose oxidation during reperfusion in isolated rat hearts. Hearts were perfused retrogradely with erythrocyte-enriched (hematocrit 30%) Krebs-Henseleit buffer containing 0.4 mM of albumin, 0.4 mM of palmitate and 11 mM of glucose. Control hearts were perfused without intervention. Two groups of hearts were subjected to 60 min of no-flow ischemia. One group of hearts was reperfused with standard medium (REPERFUSION), the other group with medium containing 3 mM of Ni^{2+} (REPERFUSION + Ni^{2+}). Oxidation of palmitate and glucose were determined based on measurement of $^{14}CO_2$ release from (1-^{14}C)-palmitate or (1-^{14}C)-glucose respectively [31, 36]. Measurements were made 15 min after the onset of reperfusion. Myocardial oxygen consumption, oxidation of palmitate and oxidation of glucose were markedly reduced in the presence of Ni^{2+}. Values are means ± SEM of at least 5 hearts. * $p < 0.05$, ** $p < 0.01$ vs control group. ++ $p < 0.01$ vs reperfusion group.

Imaging of myocardial fatty acid metabolism during reperfusion

The observation of rapid recovery of fatty acid oxidation after reperfusion is at first glance at variance with studies employing positron emission tomography and (1-^{11}C)-palmitate. In dogs subjected to transient coronary occlusion, Schwaiger *et al.* [52, 53] have shown that clearance of myocardial radioactivity was considerably delayed after reperfusion for hours [52] or days [53] depending on the severity of the preceding ischemic insult. The clearance pattern was similar to that observed during ischemia and hypoxia [54] and therefore was interpreted as persistent impairment of fatty acid oxidation after reperfusion. Two recent observations may reconcile some of the apparently conflicting findings between external monitoring of ^{11}C-clearance from (1-^{11}C)-palmitate [52, 53] and direct analysis of the coronary effluent for $^{14}CO_2$ release from (1-^{14}C)-palmitate [31, 38, 39]. First, results of Lopaschuk *et al.* [38] and from our laboratory [31] suggested that, although oxidation of (1-^{14}C)-palmitate in isolated perfused rat hearts rapidly returns to preischemic levels, incorporation of the tracer into triacylglycerols is increased early during reperfusion. Therefore it is conceivable that in the postischemic myocardium, as compared to normal myocardium, a larger fraction of extracted palmitate is incorporated into triacylglycerols prior to oxidation. Thus, reduced clearance of radioactivity from (1-^{11}C)-palmitate after reperfusion may reflect passage of an increased fraction of the tracer in a slow turnover lipid pool rather than reduction of overall fatty acid oxidation. Second, in contrast to studies with positron emission tomography, studies with direct measurement of fatty acid oxidation have essentially concentrated on the first hour after reperfusion [31, 38, 39].

A preliminary report by Liedtke *et al.* [55] recently suggested that in isolated working swine hearts fatty acid oxidation is reduced four days after 60 min of low-flow ischemia. Thus, fatty acid oxidation may decline again later during reperfusion despite initial recovery.

Acknowledgements

Supported by the Swiss National Science Foundation grant 32-26373.89.

References

1. Opie LH: Effects of regional ischemia on metabolism of glucose and fatty acids. Relative rates of aerobic and anaerobic energy production during myocardial infarction and comparison with effects of anoxia. Circ Res 38 (suppl I): I52–I86, 1976
2. Liedtke JA: Alterations of carbohydrate and lipid metabolism in the acutely ischemic heart. Progress Cardiovasc Res 23: 321–336, 1981
3. Corr PB, Gross RW, Sobel BE: Amphopathic metabolites and membrane dysfunction in ischemic myocardium. Circ Res 55: 135–154, 1984
4. Van der Vusse GJ, Stam H: Lipid and carbohydrate metabolism in the ischaemic heart. Basic Res Cardiol 82 (suppl 1): 149–154, 1987
5. Liedtke JA, Shrago E: Detrimental effects of fatty acids and their derivatives in ischemic and reperfused myocardium. In: HM Piper (ed) Pathophysiology of severe myocardial injury. Kluwer Academic Publishers, Dordrecht, 1990, pp 149–166
6. Bourdillon PD, Poole-Wilson PA: The effects of verapamil, quiescence, and cardioplegia on calcium exchange and mechanical function in ischemic rabbit myocardium. Circ Res 50: 360–368, 1982
7. Garlick PB, Davies MJ, Hearse DJ, Slater TF: Direct determination of free radicals in the reperfused rat heart using electron spin resonance spectroscopy. Circ Res 61: 757–760, 1987
8. Siegmund B, Klietz T, Schwartz P, Piper HM: Temporary contractile blockade prevents hypercontracture in anoxic-reoxygenated cardiomyocytes. Am J Physiol 260: H426–H435, 1991
9. Hearse DJ, Humphrey SM, Chain EB: Abrupt reoxygenation of the anoxic potassium-arrested perfused rat heart: A study of myocardial enzyme release. J Mol Cell Cardiol 5: 395–407, 1973
10. Heyndrickx GR, Millard RW, McRitchie RJ, Maroko PR, Vatner SF: Regional myocardial function and electrophysiological alterations after brief coronary artery occlusion in conscious dogs. J Clin Invest 56: 978–985, 1975
11. Bergmann SR, Lerch RA, Fox KAA, Ludbrook PA, Welch MJ, Ter-Pogossian MM, Sobel BE: Temporal dependence of beneficial effects of coronary thrombolysis characterized by positron tomography. Am J Med 73: 573–581, 1982
12. Kanaide H, Taira Y, Nakamura M: Transmural anoxic wave front and regional dysfunction during early ischemia. Am J Physiol 253: H240–H247, 1987
13. Whitmer JT, Idell-Wenger JA, Rovetto MJ, Neely JR: Control of fatty acid metabolism in ischemic and hypoxic hearts. J Biol Chem 253: 4305–4309, 1978
14. Healy-Moore K, Radloff JF, Hull FE, Sweeley CC: Incomplete fatty acid oxidation by ischemic heart: β-hydroxy fatty acid production. Am J Physiol 239: H257–H265, 1980
15. Van Bilsen M, Van der Vusse GJ, Willemsen PHM, Coumans WA, Roemen THM, Reneman RS: Lipid alterations in isolated, working rat hearts during ischemia and reperfusion: Its relation to myocardial damage. Circ Res 64: 304–314, 1989
16. Allen DG, Orchard CH: Myocardial contractile function during ischemia and hypoxia. Circ Res 60: 153–168, 1987
17. Reimer KA, Jennings RB, Hill ML: Total ischemia in dog hearts, *in vitro*. 2. High energy phosphate depletion and associated defects in energy metabolism, cell volume regulation, and sarcolemmal integrity. Circ Res 49: 901–911, 1981
18. Nishioka K, Jarmakani JM: Effect of ischemia on mechanical function and high-energy phosphates in rabbit myocardium. Am J Physiol 242: H1077–H1083, 1982
19. Chien KR, Han A, Sen A, Buja M, Willerson JT: Accumulation of unesterified arachidonic acid in ischemic canine myocardium. Relationship to a phosphatidylcholine deacylation-reacylation cycle and the depletion of membrane phospholipids. Circ Res 54: 313–322, 1984
20. Paulson DJ, Schmidt MJ, Romens J, Shug AL: Metabolic and physiological differences between zero-flow and low-flow myocardial ischemia: effects of L-acetylcarnitine. Basic Res Cardiol 79: 551–561, 1984
21. Ichihara K, Neely JR: Recovery of ventricular function in reperfused ischemic rat hearts exposed to fatty acids. Am J Physiol 249: 492–497, 1985
22. Neely JR, Feuvray D: Metabolic products and myocardial ischemia. Am J Pathol 102: 282–291, 1981
23. Moore KH, Bonema JD, Solomon FJ: Long chain acyl-CoA and acylcarnitine hydrolase activities in normal and ischemic rabbit hearts. J Mol Cell Cardiol 16: 905–913, 1984
24. Katz AM, Messineo FC: Lipid-membrane interactions and the pathogenesis of ischemic damage in the myocardium. Circ Res 48: 1–16, 1981
25. Schoonderwoerd K, van der Kraij T, Hülsmann WC, Stam H: Hormones and triacylglycerol metabolism under normoxic and ischemic conditions. Mol Cell Biochem 88: 129–137, 1989
26. Bilheimer D, Buja LM, Parkey RW, Bonte FJ, Willerson JT: Fatty acid accumulation and abnormal lipid deposition in peripheral border zones of experimental myocardial infarcts. J Nucl Med 19: 276–283, 1978
27. Crass MF, Sterrett PR: Distribution of glycogen and lipids in the ischemic canine left ventricle: Biochemical and light and electron microscopic correlates. In: PE Roy and G Rona (eds) Recent advances in studies on cardiac structure and metabolism. University Park Press, Baltimore, 1975, pp 251–263
28. Van der Vusse GJ, Roemen THM, Prinzen FW, Coumans WA, Reneman RS: Uptake and tissue content of fatty acids in dog myocardium under normoxic and ischemic conditions. Circ Res 50: 538–546, 1982
29. Trach V, Buschmans-Denkel E, Schaper W: Relation between lipolysis and glycolysis during ischemia in the isolated rat heart. Basic Res Cardiol 81: 454–464, 1986
30. Schoonderwoerd K, Broekhoven-Schokker S, Hülsmann WC, Stam H: Enhanced lipolysis of myocardial triglycerides during

low-flow ischemia and anoxia in the isolated rat heart. Basic Res Cardiol 84: 165–173, 1989
31. Görge G, Chatelain P, Schaper J, Lerch R: Effect of increasing degrees of ischemic injury on myocardial oxidative metabolism early after reperfusion in isolated rat hearts. Circ Res 68: 1681–1692, 1991
32. Stahl LD, Weiss HR, Becker LC: Myocardial oxygen consumption, oxygen supply/demand heterogeneity, and microvascular patency in regionally stunned myocardium. Circulation 77: 865–872, 1988
33. Laxson DD, Homans DC, Dai X-Z, Sublett E, Bache RJ: Oxygen consumption and coronary reactivity in postischemic myocardium. Circ Res 64: 9–20, 1989
34. Schott RJ, Rohmann S, Braun ER, Schaper W: Ischemic preconditioning reduces infarct size in swine myocardium. Circ Res 66: 1133–1142, 1990
35. Brown MA, Nohara R, Vered Z, Perez JE, Bergmann SR: The dependence of recovery of stunned myocardium on restoration of oxidative metabolism (abstract). Circulation 78 (suppl II): II–467, 1988
36. Lerch R, Papageorgiou I, Benzi R: Role of mitochondrial calcium transport in myocardial hypermetabolism after reperfusion (abstract). Circulation 82 (suppl III): III–756, 1990
37. Neely JR, Rovetto MJ, Oram JF: Myocardial utilization of carbohydrate and lipids. Progress Cardiovasc Dis 15: 289–329, 1972
38. Lopaschuk GD, Spafford MA, Davies NJ, Wall SR: Glucose and palmitate oxidation in isolated working rat hearts reperfused after a period of transient global ischemia. Circ Res 66: 546–553, 1990
39. Liedtke AJ, Demaison L, Eggleston AM, Cohen LM, Nellis SH: Changes in substrate metabolism and effects of excess fatty acids in reperfused myocardium. Circ Res 62: 535–542, 1988
40. Renstrom B, Nellis SH, Liedtke AJ: Metabolic oxidation of glucose during early myocardial reperfusion. Circ Res 65: 1094–1101, 1989
41. Sarnoff SJ, Braunwald E, Welch GH Jr, Case RB, Stainsby WN, Macruz R: Hemodynamic determinants of oxygen consumption of heart with special reference to tension-time index. Am J Physiol 192: 148–156, 1958
42. Huang XQ, Liedtke AJ: Alterations in fatty acid oxidation in ischemic and reperfused myocardium. Molec Cell Biochem 88: 145–153, 1989
43. Marban E, Kitakaze M, Koretsune Y, Yue DT, Chacko VP, Pike MM: Quantification of $(Ca^{2+})i$ in perfused hearts. Critical evaluation of the 5F-BABTA and nuclear magnetic resonance method as applied to the study of ischemia and reperfusion. Circ Res 66: 1255–1267, 1990
44. Nayler WG, Ferrari R, Williams A: Protective effect of pretreatment with verapamil, nifedipine and propranolol on mitochondrial function in the ischemic reperfused myocardium. Amer J Cardiol 46: 242–248, 1980
45. Kusuoka H, Koretsune Y, Chacko VP, Weisfeldt ML, Marban E: Excitation-contraction coupling in postischemic myocardium. Does failure of activator Ca^{2+} transients underlie stunning? Circ Res 66: 1268–1276, 1990
46. Carafoli E: The homeostasis of calcium in heart cells. J Mol Cell Cardiol 17: 203–212, 1985
47. Vercesi A, Reynafarje B, Lehninger A: Stoichiometry of H^+ ejection and Ca^{2+} uptake coupled to electron transport in rat heart mitochondria. J Biol Chem 253: 6379–6385, 1978
48. Poole-Wilson PA, Harding DP, Boudillon PDV, Tones MA: Calcium out of control. J Mol Cell Cardiol 16: 175–187, 1984
49. Peng CF, Kane JJ, Straub KD, Murphy ML: Improvement of mitochondrial energy production in ischemic myocardium by *in vivo* infusion of ruthenium red. J Cardiovasc Pharmacol 2: 45–54, 1980
50. Henry PD, Shuchleib R, Davis J, Weiss ES, Sobel BE: Myocardial contracture and accumulation of mitochondrial calcium in ischemic rabbit heart. Am J Physiol 233: H677–H684, 1977
51. Chamberlain BK, Volpe P, Fleischer S: Inhibition of calcium-induced calcium release from purified cardiac sarcoplasmic reticulum vesicles. J Biol Chem 259: 7547–7553, 1984
52. Schwaiger M, Schelbert HR, Keen R, Vinten-Johansen J, Hansen H, Selin C, Barrio J, Huang SC, Phelps ME: Retention and clearance of ^{11}C-palmitic acid in ischemic and reperfused canine myocardium. J Am Coll Cardiol 6: 311–320, 1985
53. Schwaiger M, Schelbert HR, Ellison D, Hansen H, Yeatman L, Vinten-Johansen J, Selin C, Barrio J, Phelps ME: Sustained regional abnormalities in cardiac metabolism after transient ischemia in the chronic dog model. J Am Coll Cardiol 6: 336–347, 1985
54. Lerch RA, Bergmann SR, Ambos HD, Welch MJ, Ter-Pogossian MM, Sobel BE: Effect of flow-independent reduction of metabolism on regional myocardial clearance of ^{11}C-palmitate. Circulation 65: 731–738, 1982
55. Liedtke AJ, Renstrom B, Nellis SH, Whitesell LF: Compromised metabolism in chronically reperfused pig hearts (abstract). J Am Coll Cardiol 17: 37A, 1991

Molecular and Cellular Biochemistry **116**: 111–116, 1992.

The relative contribution of glucose and fatty acids to ATP production in hearts reperfused following ischemia

Gary D. Lopaschuk and Maruf Saddik
Cardiovascular Disease Research Group and Lipid and Lipoprotein Research Group, Departments of Pediatrics and Pharmacology, Faculty of Medicine, The University of Alberta, Edmonton, Alberta, Canada

Abstract

High levels of fatty acids decrease the extent of mechanical recovery of hearts reperfused following a transient period of severe ischemia. Glucose oxidation rates during reperfusion are low under these conditions, which can result in a decreased recovery of mechanical function. Stimulation of glucose oxidation with the carnitine palmitoyl transferase I inhibitor, Etomoxir, or by directly stimulating pyruvate dehydrogenase activity with dichloroacetate (DCA) results in an improvement in mechanical function during reperfusion of previously ischemic hearts. Addition of DCA (1 mM) to hearts perfused with 11 mM glucose and 1.2 mM palmitate results in an increase in contribution of glucose oxidation to overall ATP production from 6 to 23%, with a parallel decrease in that of fatty acid oxidation from 90 to 69%. In aerobic hearts, endogenous myocardial triglycerides are an important source of fatty acids for β-oxidation. Using hearts in which the myocardial triglycerides were pre-labeled, the contribution of both endogenous and exogenous fatty acid oxidation to myocardial ATP production was determined in hearts perfused with 11 mM glucose, 1.2 mM palmitate and 500 μU/ml insulin. In hearts reperfused following a 30 min period of global no flow ischemia, 91.9% of ATP production was derived from endogenous and exogenous fatty acid oxidation, compared to 87.7% in aerobic hearts. This demonstrates that fatty acid oxidation quickly recovers following a transient period of severe ischemia. Furthermore, therapy aimed at overcoming fatty acid inhibition of glucose oxidation during reperfusion of ischemic hearts appears to be beneficial to recovery of mechanical function. (Mol Cell Biochem **116:** 111–116, 1992)

Key words: pyruvate dehydrogenase, glucose oxidation, glycolysis, β-oxidation

Introduction

Considerable interest has focussed on the contribution of fatty acids to myocardial ischemic injury. High levels of fatty acids are thought to contribute to infarct size and mortality in humans [see 1 and 2 for reviews], and have been shown to potentiate ischemic injury in several experimental models, including pig, dog, rabbit, and rat hearts [3–7]. Fatty acid levels in the blood are markedly elevated in patients suffering myocardial infarction or undergoing cardiac bypass surgery [8, 9]. As a result, reperfusion of previously ischemic myocardium exposes the heart muscle to very high levels of fatty acids. The exact mechanisms by which fatty acids exert their deleterious effects have not been completely delineated, but increasing evidence suggests that inhibition of myocar-

Address for offprints: G.D. Lopaschuk, 423 Heritage Medical Research Bldg., The University of Alberta, Edmonton, Canada, T6G 2S2

dial glucose utilization is involved [1, 2, 6, 7, 10, 11]. This appears to be related not only to a decrease in glycolysis during ischemia [10], but also to a decrease in glucose oxidation during reperfusion following ischemia [6, 7, 11].

Despite the presence of high levels of fatty acids during reperfusion of previously ischemic myocardium there is no uniform concensus as to the relative contribution of glucose and fatty acids to overall myocardial ATP production during aerobic reperfusion. Studies have suggested that fatty acid oxidation is impaired [12, 13], unchanged [7, 14], or accelerated [15] compared to non-ischemic hearts. If indeed fatty acid oxidation is impaired, an alternative source of ATP must be provided in order to maintain heart function. However, our previous studies have suggested that glucose oxidation, which is the other major source of myocardial ATP production, is very low in hearts reperfused in the presence of high concentrations of fatty acids [7, 11].

To better understand the contribution of glucose and fatty acids as sources of ATP production, we measured glycolysis, glucose oxidation, and both exogenous and endogenous fatty acid oxidation in isolated working rat hearts reperfused following a transient period of global ischemia.

Materials and methods

Heart perfusions

Hearts from Na^+ pentobarbital anesthetized male Sprague-Dawley rats (200–250 g) were excised, the aorta cannulated, and a retrograde perfusion with Krebs-Henseleit buffer, (pH 7.4, gassed with 95% O_2-5% CO_2,) initiated, as described previously [6]. Spontaneously beating hearts were used throughout the studies, with heart rate, peak systolic pressure (PSP), and developed pressure ($\triangle P$) being measured by a Gould P21 pressure transducer in the aortic outflow line.

Energy substrate utilization in spontaneously beating hearts, perfused at an 11.5 mm Hg left atrial preload and 80 mm Hg aortic afterload, was measured under aerobic conditions, or in hearts reperfused under aerobic conditions following ischemia. Global no-flow ischemia was induced by clamping off both the left atrial and aortic flow. Hearts were maintained at 37° C throughout the ischemic period. Following 30 min of no-flow ischemia, left atrial and aortic flow was restored, and recovery of mechanical function was monitored for a further 60 min period.

Measurement of glycolysis, glucose oxidation and fatty acid oxidation

In hearts used for glycolysis and glucose oxidation measurements, perfusate containing 11 mM [2-^{3}H/U-^{14}C]-glucose, 1.2 mM palmitate, 3% albumin, ± 500 μU/ml insulin. Steady state glycolysis and glucose oxidation rates were determined as described previously [16]. Quantitative ^3H_2O production was used to measure steady state glycolytic rates (^3H_2O is liberated at the phosphoglucoisomerase step of glycolysis), while quantitative $^{14}CO_2$ production was used to measure glucose oxidation ($^{14}CO_2$ is liberated at the level of pyruvate dehydrogenase (PDH) and in the tricarboxylic acid (TCA) cycle. Hearts used for palmitate oxidation were perfused with 11 mM glucose and 1.2 mM [1-^{14}C]-palmitate, and 3% albumin. Measurement of $^{14}CO_2$ production was determined as described previously [6].

Endogenous fatty acid oxidation was measured as described previously [16]. After a 10 min equilibration period, hearts were switched to the working mode, and perfused with Krebs Henseleit buffer containing 11 mM glucose, 1.2 mM [1-^{14}C] palmitate, and 500 μU/ml in-

Table 1. Effect of dichloroacetate on glycolysis, glucose oxidation, and palmitate oxidation in hearts perfused with 11 mM glucose and 1.2 mM palmitate.

Perfusion condition	Glycolysis (μmol/g dry wt·min)	Glucose oxidation (μmol/g dry wt·min)	Palmitate oxidation (μmol/g dry wt·min)
– no addition	2.18 ± 0.48	0.181 ± 0.026	0.757 ± 0.048
– 1 mM DCA	3.77 ± 0.44*	0.641 ± 0.137*	0.538 ± 0.083*

– Values are the mean ± S.E.M. of at least 5 hearts in each group. Hearts were perfused under aerobic conditions with 11 mM glucose, and 1.2 mM palmitate. Glycolysis and glucose oxidation were determined in hearts perfused with [U-^{14}C/2-^{3}H]-glucose, while palmitate oxidation was determined in hearts perfused with [1-^{14}C]-palmitate. *, significantly different from hearts perfused in the absence of dichloroacetate (DCA) sulin. Following this initial one hour 'pulse' period, hearts were subjected to either a 10 min Langendorff drip-out perfusion, or a 30 min period of global no-flow ischemia. During this period the buffer used during 'pulse' was drained, the perfusion system was washed thoroughly with distilled water, and the 'chase' buffer was added to the system. This 'chase' buffer contained 11 mM glucose, 1.2 mM [9,10-^{3}H] palmitate and 500 μU/ml insulin. During the 60 min reperfusion period, $^{14}CO_2$ production was used as a measure of endogenous fatty acid oxidation, while ^3H_2O production was a measure of exogenous fatty acid oxidation rates. Perfusate and gaseous samples were collected at 10 min intervals during the 'chase', and $^{14}CO_2$ production measured as described above.

Results and discussion

The contribution of glucose to overall ATP production in the heart depends to a large extent on the concentration of fatty acids in the blood [17, 18]. In intact hearts, high levels of fatty acids inhibit both glycolysis and glucose oxidation [17], with the most dramatic effect being inhibition of glucose oxidation [16]. For example, in the presence of 1.2 mM palmitate, glucose oxidation rates are 11 to 15 times lower than rates seen in hearts perfused in the absence of fatty acids, and provides only 5 to 10% of overall ATP production [7, 16]. The carnitine palmitoyltransferase I inhibitor, Etomoxir, is able to partially overcome this inhibition, resulting in a 2 fold increase in glucose oxidation in fatty acid perfused hearts [6, 7]. However, the increase in glucose oxidation is not accompanied by a parallel decrease in fatty acid oxidation. This contrasts the marked inhibition of fatty acid oxidation observed in isolated mitochondria treated with CPT I inhibitors [19, 20]. These apparently contradictory results are probably explained by the fact that any decrease in fatty acid oxidation in the intact heart would need to be accompanied by a parallel increase in ATP production from other sources. If fatty acid oxidation is providing over 90% of ATP production in hearts, any significant decrease in fatty acid oxidation would have to be accompanied by a very large increase in glucose oxidation. Although Etomoxir stimulates glucose oxidation, its effect is not dramatic, and at best increases the contribution of glucose oxidation to ATP production from 5 to 10% [7]. As a result, demand for intramitochondrial acyl CoA for β-oxidation in the intact heart probably overcomes the direct effects of Etomoxir on CPT I.

Table 2. Effect of dichloroacetate on steady state ATP production in hearts perfused with 11 mM glucose and 1.2 mM palmitate

ATP source	Steady state ATP production (μmol/g dry wt·min)	
	Control	DCA (1 mM)
From Glycolysis	4.3 ± 1.0 (4%)	7.5 ± 0.9 (8%)
From glucose oxidation	6.5 ± 4.9 (6%)	23.1 ± 4.9 (23%)
From palmitate oxidation	97.7 ± 6.2 (90%)	69.4 ± 10.7 (69%)
Total ATP production	108.5	100.2

– ATP production was calculated from values obtained from Table 1.

To further investigate the relationship between glucose and fatty acid use in the intact heart, both glucose utilization and fatty acid oxidation was measured in a series of hearts perfused with dichloroacetate (DCA), since it is a potent stimulator of glucose oxidation [11, 21, 21]. As shown in Table 1, in hearts perfused in the presence of 1.2 mM palmitate, glycolytic rates were substantially greater than glucose oxidation rates. Addition of 1 mM DCA to the perfusate resulted in an increase in both glycolysis and glucose oxidation, with the most dramatic effect being stimulation of glucose oxidation. Paralleling the increase in glucose oxidation was a decrease in palmitate oxidation. Steady state ATP production calculated from these rates is shown in Table 2. A value of 2 ATP produced from glucose during glycolysis, 36 ATP produced during oxidation of glucose, and 129 ATP produced during oxidation of palmitate was used in these calculations. Overall ATP production from exogenous substrates was essentially similar in both groups. However, DCA caused a 21% increase in ATP production from glycolysis and glucose oxidation, with a parallel decrease in palmitate oxidation. This demonstrates the tightly regulated coupling of fatty acid and carbohydrate metabolism in the heart.

The key site at which fatty acids inhibit glucose oxidation is at the level of pyruvate dehydrogenase [17, 18]. This is regulated by a phosphorylation-dephosphorylation cycle, with a PDH kinase inhibiting PDH and a PDH phosphatase activating PDH [23]. The PDH kinase is activated by an increase in intramitochondrial acetyl CoA/CoA and NADH/NAD ratios. As a result, high levels of fatty acids decrease the activity of PDH, accounting for the dramatic decrease in glucose oxidation observed in intact hearts. DCA can overcome this inhibition by inhibiting PDH kinase, with a resultant stimulation of PDH and glucose oxidation [11]. The data in Table 2 demonstrates the close coupling of this process. As inhibition of PDH is overcome, intramitochondrial acetyl CoA production from glucose oxidation increases, with a parallel decrease in acetyl CoA production from β-oxidation.

Despite the fact that glucose oxidation rates are low during reperfusion of ischemic hearts, there is controversy as to the relative contribution of glucose and fatty acids to overall ATP production in these hearts. Positron emission tomography studies suggest a shift from predominant fatty acid utilization to predominant glucose utilization in reperfused/ischemic hearts [11]. [^{14}C]-palmitate infusion studies in human have also suggested that palmitate utilization decreases during reperfusion [13]. Measurement of glucose and palmitate extraction during reperfusion of ischemic dog hearts, however, suggests that fatty acid utilization does not decrease during reperfusion [14]. Recent direct measurements of fatty acid oxidation during reperfusion of ischemic hearts has also shown that fatty acid oxidation either quickly recovers to pre-ischemic rates [7], or is even higher than pre-ischemic rates [15].

To further address this issue we measured the relative contribution of ATP production from the major energy substrates in both aerobic and reperfused/ischemic hearts. Endogenous myocardial triglycerides, through lipolysis, are also a potential source of fatty acids for oxidative metabolism [16]. In recent studies, we have characterized an experimental model in which triglyceride turnover could be directly measured in the intact heart [16]. The protocol allows us to directly measure endogenous triglyceride fatty acid oxidation while simultaneously measuring exogenous fatty acid oxidation. This is accomplished by initially prelabeling the myocardial lipid pools during a 'pulse' perfusion with 1.2 mM [^{14}C] palmitate. We used this methodology to measure both endogenous and exogenous fatty acid oxidation during an aerobic 'chase' perfusion following a transient period of global ischemia.

Overall ATP production derived from glycolysis, glucose oxidation and both endogenous and exogenous fatty acid oxidation during reperfusion is shown in Table 3. Glycolysis and glucose oxidation were simultaneously measured in one series of hearts, while endogenous and exogenous fatty acid oxidation were measured in a second series of hearts. In aerobic hearts, endogenous and exogenous fatty acid oxidation provided 87.7% of total ATP production. In reperfused/ischemic hearts, fatty acid oxidation quickly recovered, and provided 91.9% of total ATP production. This data demonstrates that in reversibly injured ischemic hearts, fatty acid oxidation remains the primary source of ATP production. In fact, following ischemia the contribution of fatty acids to ATP production is probably greater than under aerobic conditions.

Table 3 demonstrates that contribution of glucose oxidation to ATP production is very low in hearts reperfused following ischemia in the presence of high levels of fatty acids. If glucose oxidation rates are increased during reperfusion, an enhanced degree of functional recovery will occur [7, 11]. The actual mechanism by which this occurs has yet to be determined. One possible explanation may relate to the initial rate of return of oxidative metabolism during reperfusion. Stimulating pyruvate dehydrogenase activity shifts the mitochondrial redox state towards reduction and increases the thermodynamic driving force, resulting in a stimulation of respiration [23–25]. The net effect of this is to increase the initial rate of energization of the heart during reperfusion. As a result, flux through PDH in early reperfusion could potentially facilitate the rate of recovery of oxidative metabolism. Although the rate of

Table 3. Myocardial ATP production during aerobic reperfusion following ischemia in hearts perfused with 11 mM glucose, 1.2 mM palmitate, and 500 μU/ml insulin

ATP source	Steady state ATP production (μmol/g dry wt·min)	
	Aerobic perfusion	Reperfusion following ischemia
Glucose oxidation	9.61 ± 0.84 (7.4%)	6.81 ± 1.01 (4.5%)
Glycolysis	5.06 ± 0.64 (3.9%)	5.37 ± 0.76 (3.6%)
Endogenous fatty acid oxidation	9.41 ± 1.87 (7.3%)	7.85 ± 2.08 (5.2%)
Exogenous fatty acid oxidation	105.3 ± 17.5 (81.4%)	130 ± 17.2 (86.7%)
Total ATP production	129.4	150.1

Values are the mean ± SEM of at least six hearts in each group. Glycolysis, glucose oxidation and endogenous and exogenous palmitate oxidation are as described in 'Methods'. Percentage contribution to overall ATP utilization is shown in brackets.

return of oxidative metabolism is important in the functional recovery of post-ischemic hearts, it appears unlikely that overall oxidative metabolism remains depressed in hearts that recover, since fatty acid oxidation will return to pre-ischemic levels by 10 min post-ischemia [7]. A second possibility why low glucose oxidation rates are detrimental to recovery of function may be related to the rate of recovery from the acidosis that occurs during ischemia. In the presence of high concentration of fatty acids, glycolytic rates are substantially greater than glucose oxidation rates [16]. If glucose passes through glycolysis to lactate, the overall H^+ ion balance is the production of 2 H^+ ions. In contrast, the H^+ ion balance of glucose to CO_2 (glucose oxidation) is zero. As a result, fatty acid inhibition of glucose oxidation in control hearts may result in a greater H^+ ion load on the heart during the critical period in which the heart is attempting to recover from the acidosis that occurred during ischemia. This could potentially contribute to Ca^{2+} overload during reperfusion. To date, however, it has not been determined if high concentrations of fatty acids actually accelerate Ca^{2+} overload during reperfusion. These experiments are presently being performed.

In summary, we demonstrate that fatty acid oxidation quickly recovers during aerobic reperfusion of previously ischemic hearts, and provides over 90% of overall ATP production. Furthermore, in the presence of high concentrations of fatty acids, glucose oxidation rates are low in both aerobic and reperfused/ischemic hearts. Earlier studies have demonstrated that stimulation of glucose oxidation during reperfusion will significantly improve the degree of functional recovery. These data suggest that a potential therapeutic goal in myocardial ischemia is to overcome fatty acid inhibition of glucose oxidation, as opposed to directly inhibiting fatty acid oxidation.

Acknowledgements

Supported by a grant from the Medical Research Council of Canada. GDL is a scholar of the Alberta Heritage Foundation for Medical Research. MS is a post-doctoral fellow of the Medical Research Council, and a student trainee of the Alberta Heritage Foundation for Medical Research.

References

1. Opie LH: Metabolism of free fatty acids, glucose, and catecholamines in acute myocardial infarction. Am J Cardiol 36: 938–953, 1975
2. Vik-Mo H, Mjos OD: Influence of free fatty acids on myocardial oxygen consumption and ischemic injury. Am J Cardiol 48: 361–365, 1981
3. Liedtke AJ, Nellis SH, Neely JR: Effects of excess free fatty acids on mechanical and metabolic function in normal and ischemic myocardium in swine. Circ Res 43: 652–661, 1978
4. Mjos OD: Effect of free fatty acids on myocardial function and oxygen consumption in intact dogs. J Clin Invest 50: 1386–1389, 1978
5. Johnson DL, Lewandowski ED: Fatty acid metabolism and contractile function in the reperfused myocardium. Multinuclear NMR studies of isolated rabbit hearts. Circ Res 68: 714–725, 1991
6. Lopaschuk GD, Wall SR, Olley PM, Davies NJ: Etomoxir, a carnitine palmitoyltransferase I inhibitor, protects hearts from fatty acid-induced ischemic injury independent of changes in long chain acylcarnitine. Circ Res 63: 1036–1043, 1988
7. Lopaschuk GD, Spafford MA, Davies NJ, Wall SR: Glucose and palmitate oxidation in isolated working rat hearts reperfused after a period of transient global ischemia. Circ Res 66: 546–553, 1990
8. Oliver MF, Kurien VA, Greenwood TW: Relation between serum-free-fatty acids and arrythmia and death after myocardial infarction. Lancet 1: 71–715, 1968
9. Svensson S, Svedjeholm R, Ekroth R, Milocco I, Nilsson F, Sabel KG, William-Olsson G: Trauma metabolism and the heart. Uptake of substrates and effects of insulin early after cardiac operations. J Thorac Cardiovas Surg 99: 1063–1073, 1990
10. Owen P, Dennis S, Opie LH: Glucose flux rate regulates onset of ischemic contracture in globally underperfused rat hearts. Circ Res 66: 344–354, 1990
11. McVeigh JJ, Lopaschuk GD: Dichloroacetate stimulation of glucose oxidation improves recovery of ischemic rat hearts. Am J Physiol 259: H1070–H1085, 1990
12. Schwaiger M, Schelbert HR, Ellison D, Hansen H, Yeatman L, Vinten-Johansen J, Selin C, Barrio J, Phelps ME: Sustained regional abnormalities in cardiac metabolism after transient ischemia in the chronic dog model. J Am Coll Cardiol 6: 311–320, 1985
13. Teoh KH, Mickle DAG, Weisel RD, Fremes SE, Christakis GT, Romaschin AD, Harding RS, Madonik MM, Ivanov J: Decreased postoperative myocardial fatty acid oxidation. J Surg Res 44: 36–44, 1988
14. Myears DW, Sobel BE, Bergmann SR: Substrate use in ischemic and reperfused canine myocardium: Quantitative considerations. Am J Physiol 253: H107–H114, 1987
15. Liedtke AJ, Demaison L, Eggleston AM, Cohen LM, Nellis SH: Changes in substrate metabolism and effects of excess fatty acids in reperfused myocardium. Circ Res 62: 535–542, 1988
16. Saddik M, Lopaschuk GD: Myocardial triglyceride turnover and contribution to energy substrate utilization in isolated working rat hearts. J Biol Chem 266: 8162–8170, 1991

Kerbey AL, Randle PJ, Cooper RH, Whitehouse S, Pask HT, Denton RM: Regulation of pyruvate dehydrogenase in rat heart. Biochem J 154: 327–348, 1976

18. Newsholme EA, Randle PJ, Manchester KL: Inhibition of the phosphofructokinase reaction in perfused rat heart by respiration of ketone bodies, fatty acids and pyruvate. Nature 193: 270–271, 1962
19. Turnbull DM, Bartlett K, Younan SIM, Sherratt SA: The effects of 2(5(4-chlorophenyl)pentyl)oxirane-2-carbonyl-CoA on mitochondrial oxidations. Biochem Pharmacol 33: 475–481, 1984
20. Tutwiler GF, Mohrbacher R, Ho W: Methyl 2-tetradecyl-glycidate, an orally effective hypoglycaemic agent that inhibits long-chain fatty acid oxidation selectively. Diabetes 28: 511–519, 1979
21. McAllister A, Allison SP, Randle PJ: Effects of dichloroacetate on the metabolism of glucose, pyruvate, acetate, 3-hydroxybutyrate and palmitate in rat diaphragm and heart muscle *in vitro* and on extraction of glucose, lactate, pyruvate and free fatty acids by dog hearts *in vivo*. Biochem J 134: 1067–1081, 1973
22. Stacpoole PW: The pharmacology of dichloroacetate. Metabolism 38: 1124–1144, 1989
23. Patel MS, Roche TE: Molecular biology and biochemistry of pyruvate dehydrogenase complexes. FASEB J 4: 3224–3233, 1990
24. Bunger R, Mallet RT, Hartman DA: Pyruvate-enhanced phosphorylation potential and inotropism in normoxic and postischemic isolated working heart. Eur J Biochem 180: 221–233, 1989
25. Mallet RT, Hartman DA, Bunger R: Glucose requirement for postischemic recovery of perfused working heart. Eur J Biochem 188: 481–493, 1990
26. Zimmer SD, Ugurbil K, Michurski SP, Mohanakrishnan P, Ulstad VK, Foker JE, From AHL: Alterations in oxidative function and respiratory regulation in the post-ischemic myocardium. J Biol Chem 264: 12402–12411, 1989

Molecular and Cellular Biochemistry **116**: 117–123, 1992.

Effects of palmitoyl CoA and palmitoyl carnitine on the membrane potential and Mg^{2+} content of rat heart mitochondria

D. Siliprandi, C. Biban, S. Testa, A. Toninello and N. Siliprandi
Department of Biological Chemistry, University of Padova and 'Centro Fisiologia dei Mitocondri, CNR', Via Trieste 75, Padova, Italy

Abstract

Palmitoyl CoA and palmitoyl carnitine added to rat heart mitochondria in amounts above 20 and 50 nmoles/mg protein, respectively, induced a fall in transmembrane potential and loss of endogenous Mg^{2+}. The dissipation of membrane potential by low concentrations of palmitoyl CoA in the presence of Ca^{2+}, but not that of high concentrations of palmitoyl CoA alone, was prevented by either ruthenium red, Cyclosporin A or Mg^{2+}, but reversed only by Mg^{2+}. The fall of membrane potential induced by palmitoyl carnitine was not prevented by any of these factors. It is suggested that the action of both palmitoyl CoA and palmitoyl carnitine at high concentrations is due to a non specific disruption of membrane architecture, while that of low concentrations of palmitoyl CoA in the presence of Ca^{2+} is associated specifically with energy dissipation due to Ca^{2+} cycling. (Mol Cell Biochem **116:** 117–123, 1992)

Key words: cardiac mitochondria, carnitine, palmitoyl carnitine, palmitoyl CoA, Cyclosporin A

Abbreviations: LCACoA – Long-Chain Acyl CoAs; LCAcar – Long-Chain Acyl carnitines; Pcar – Palmitoyl carnitine; PCoA – Palmitoyl CoA; $\Delta\Psi$ – Transmembrane Potential

Introduction

Although long-chain acyl CoA (LCACoA) and long-chain acyl carnitines (LCAcar) are intermediates in fatty acid metabolism, they are potentially toxic, especially when they accumulate as a result of hypoxia [1]. Thus both may alter the properties of cellular membrane and the activity of membrane bound enzymes and carriers [2, 3]. In the case of LCACoA, one consequence of this effect is a fall in the efficiency of the oxidative phosphorylation [5]. Both in heart [4] and liver [5] mitochondria LCACoA inhibit the adenylate translocase activity. The high selectivity of this inhibition may be the consequence of the concurrent binding of LCACoA to both polar and non polar domains of the translocator [6]. Furthermore, LCACoA also inhibit, though with lower affinity the carriers for di- and tricarboxylates and for phosphate [5]. Inhibition of these carriers will deeply restrict the entry of succinate, hence may inhibit indirectly the mitochondrial processes normally energized by this substrate [7]. This could be one explanation for the reported action of LCACoA on

Address for offprints: D. Siliprandi, Department of Biological Chemistry, University of Padova and 'Centro Fisiologia dei Mitocondri, CNR', Via Trieste 75, Padova, Italy

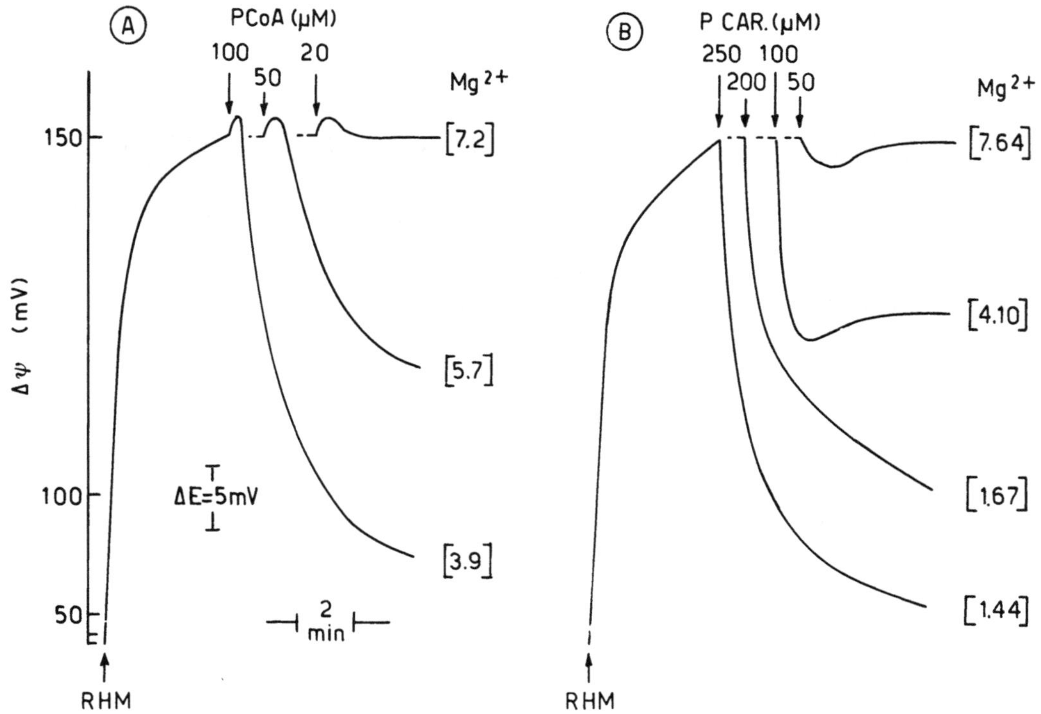

Fig. 1. Effects of palmitoyl CoA (PCoA) and palmitoyl carnitine (Pcar) on $\Delta\Psi$ and Mg^{2+} content of rat heart mitochondria. RHM were incubated as described in Materials and methods in the standard medium in the presence of 2 μM TPP+ and 1 mM EGTA. As indicated PCoA (A) and Pcar (B) were added at different concentrations. Numbers in brackets alongside each curve indicate nanomoles Mg^{2+}/mg mit. prot. determined in the pellet at the end of incubation.

transmembrane potential [8] and Ca^{2+} release [9] in rat liver mitochondria energized with succinate.

In contrast to LCACoA, the action of LCAcar on mitochondria, as on other cellular organelles and membranes [2, 10], is considered to be due mainly if not exclusively, to its non-specific detergent properties; but the two esters have not previously been systematically compared.

We now report on the influence of both LCACoA and LCAcar on some energy linked functions (transmembrane potential, Ca^{2+} transport) and Mg^{2+} content in rat heart mitochondria. We have found that especially in the presence of Ca^{2+}, palmitoyl CoA (PCoA) is a more potent inhibitor than palmitoyl carnitine (Pcar). We shall show that the two esters differ significantly in their properties. In particular the action of PCoA, but not that of Pcar, is prevented by ruthenium red, Cyclosporin A, carnitine and either prevented or reversed by Mg^{2+}.

Materials and methods

Rat heart mitochondria were isolated according to Sordhal *et al.* [11] and their protein content was determined by the biuret method with serum albumin as a standard. Incubation of mitochondria was carried out at 20° C with 1 mg mitochondrial protein/ml in the following standard medium: 200 mM sucrose, 10 mM Na-Hepes, 5 mM Na-glutamate, 5 mM Na-malate, 1 mM Na-phosphate, pH 7.4. Other additions in the medium are indicated in the description of the specific experiments. Membrane potential ($\Delta\Psi$) was measured by monitoring the distribution of the lipophylic cation tetraphenylphosphonium (TPP+) across the mitochondrial membrane with a selective electrode prepared according to published procedure [12, 13] and an Ag/AgCl reference electrode. The $\Delta\Psi$ measured with the TPP+ electrode was calibrated as proposed by Jensen *et al.* [14]. Oxygen uptake was measured with a Clark oxygen electrode. TPP+ and O2 concentrations were registered simultaneously in the same measuring cell. Mg^{2+} content in the mitochondrial pellet was determined by atomic spectroscopy [15]. Calcium movements were measured with a Ca^{2+} selective electrode (Radiometer F2112) with a calomel electrode (Radiometer K401) as the response electrode.

Cyclosporin A was a gift of Sandoz Pharmaceuticals.

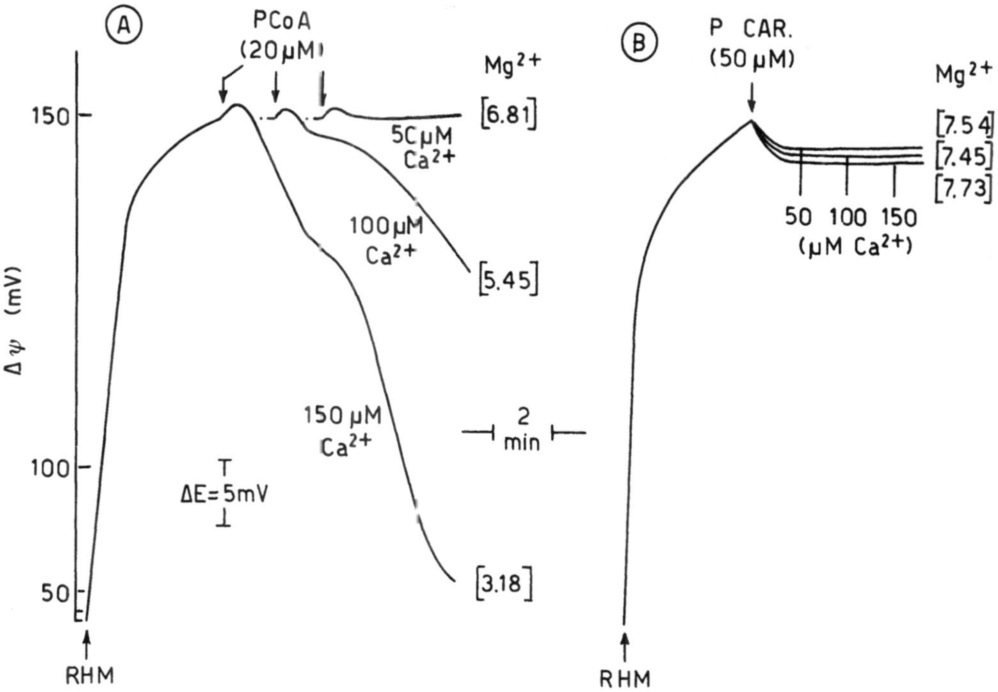

Fig. 2. Influence of Ca^{2+} on the action of PCoA and Pcar on $\Delta\Psi$ and Mg^{2+} content of cardiac mitochondria. RHM were added to the standard medium in the presence of 2 μM TPP+ and different Ca^{2+} concentrations as indicated. Numbers in brackets alongside each curve indicate nanomoles Mg^{2+}/mg mit. prot. determined in the pellet at the end of incubation. In the absence of PCoA or Pcar, Ca^{2+} induce a decrease of $\Delta\Psi$ only at concentrations above 500 μM.

Palmitoylcarnitine was generously supplied by Sigma Tau Co (Pomezia, Rome, Italy) and palmitoyl CoA was purchased from Sigma (Munich, Germany).

Results

PCoA and Pcar added to suspensions of rat heart mitochondria at 20° C could each induce falls in the transmembrane potential ($\Delta\Psi$). The concentrations of Pcar required to achieve the same fall as PCoA were, however, much greater (Figs 1A and 1B). Thus the minimum concentrations for any effects (threshold values) were, for PCoA, 20 nmoles/mg protein and for Pcar 50 nmoles, while for the complete collapse of $\Delta\Psi$ they were, for PCoA, 100 nmoles/mg protein and for Pcar, 250 nmoles. When, after a fall in $\Delta\Psi$, the washed mitochondrial sediment was subsequentially assayed for Mg^{2+}, it was found that in each case, the bigger the fall in $\Delta\Psi$, the lower was the residual Mg^{2+} content (Figs 1 and 2).

An even more striking difference between the two esters was shown also adding Ca^{2+} to the suspension medium. Ca^{2+} had no effect of its own on $\Delta\Psi$ at concentrations below 500 nmoles/mg protein, but its concurrent addition (above 50 nmoles/mg protein) to a threshold amount of PCoA induced a fall in $\Delta\Psi$ which increased in magnitude as the Ca^{2+} increased. In contrast, in the presence of the threshold amount of Pcar, Ca^{2+}, within the same concentration range, had no significant effect on $\Delta\Psi$ (Figs 2A and 2B).

The fall in $\Delta\Psi$ induced by Ca^{2+} and PCoA under these conditions was inhibited by ruthenium red, a known inhibitor of Ca^{2+} transport (Fig. 3A). Ruthenium red, however, had no effect on the falls in $\Delta\Psi$ obtained either with high concentrations of PCoA alone (results not reported) or Pcar (Fig. 3B).

Cyclosporin A, a lipophilic immunosuppressor which inhibits the respiration uncoupled by PCoA + Ca^{2+} (Fig. 4A), similarly prevented the fall in $\Delta\Psi$ in the presence of threshold PCoA and Ca^{2+} (Fig. 4A), but had no effect on the fall of $\Delta\Psi$ induced either by high PCoA alone, or by Pcar (results not reported). In this system Cyclosporin A also prevented the efflux of previously accumulated Ca^{2+} (Fig. 4B).

The fall in $\Delta\Psi$ induced by PCoA and Ca^{2+} was completely prevented by either carnitine or Mg^{2+} (both at 1 mM). The former may act by depleting PCoA to form Pcar (as catalyzed by carnitine palmitoyl transferase) while the latter may do so by maintaining the endoge-

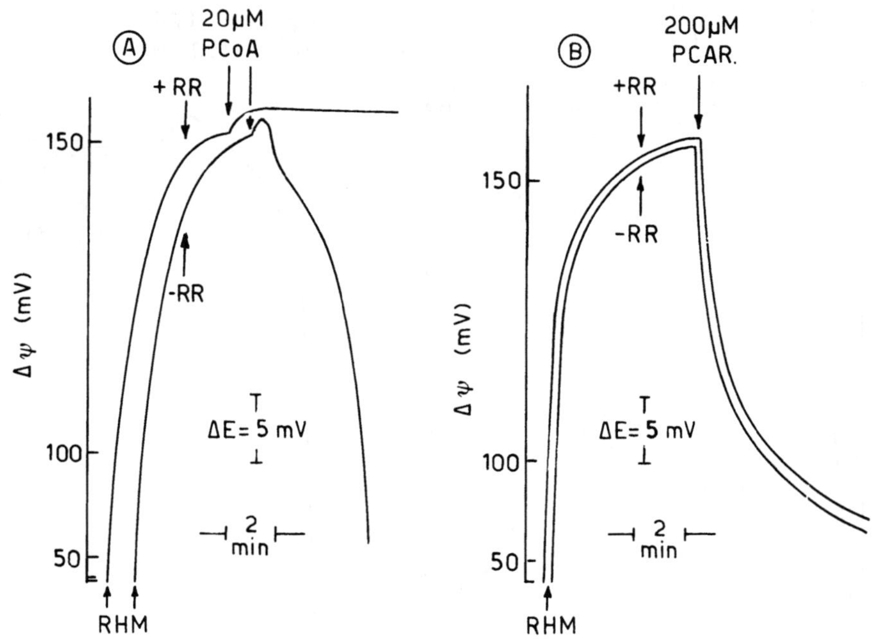

Fig. 3. Effect of ruthenium red on ΔΨ of heart mitochondria incubated in the presence of PCoA and Pcar. RHM were incubated in the standard medium in the presence of 2 μM TPP+ and 150 μM Ca^{2+}. When added RR was 0.16 μM. 20 μM PCoA and 200 μM Pcar were added at the times indicated.

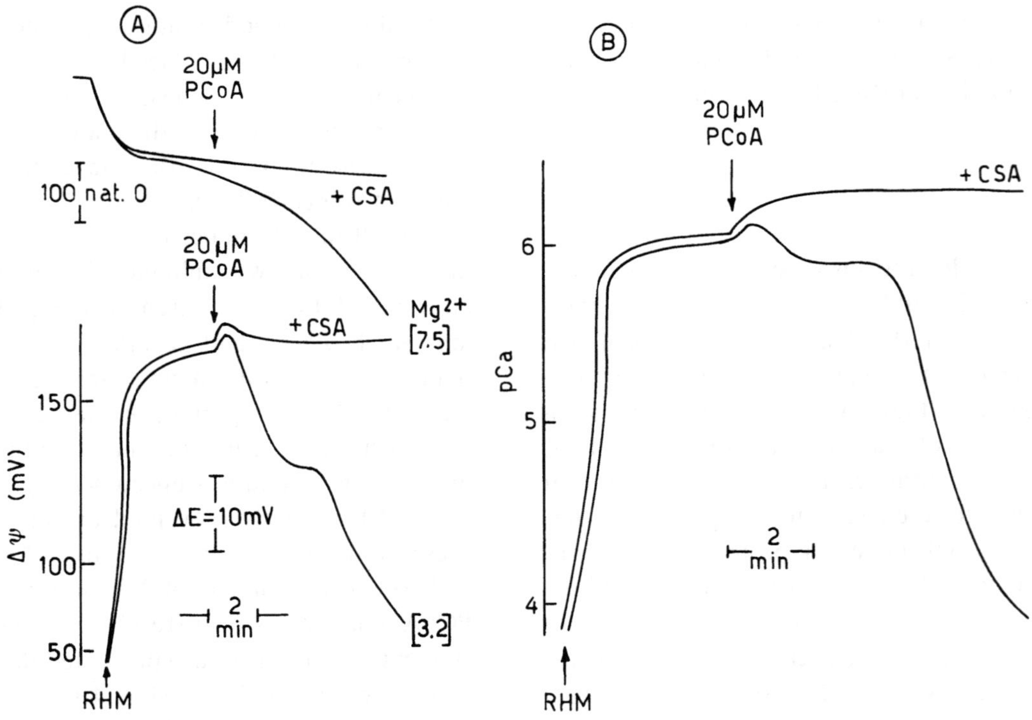

Fig. 4. Effect of Cyclosporin A on ΔΨ, state 4 respiration and Ca^{2+} retention of cardiac mitochondria incubated in the presence of PCoA. RHM were incubated in the standard medium in the presence of 150 μM Ca^{2+}. In A 2 μM TPP+ was present. 20 μM PCoA was added at the time indicated. When present 0.5 μM cyclosporin A (CSA).

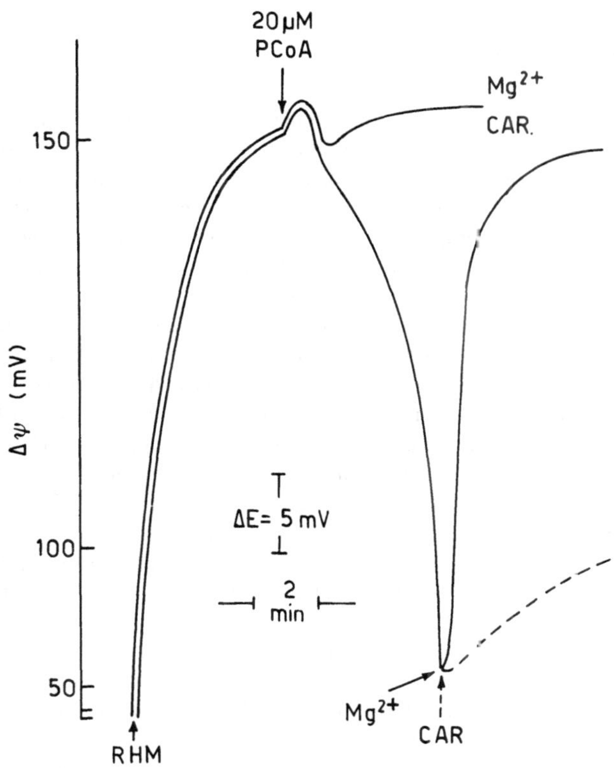

Fig. 5. Protection either by L-Carnitine or Mg^{2+} of the collapse of ΔΨ produced by PCoA. RHM were incubated in the standard medium in the presence of 2 μM TPP+ and 150 μM Ca^{2+}. 20 μM PCoA was added at the time indicated. When present in the incubation medium or added at times indicated 1 mM Mg^{2+} and 1 mM L-carnitine.

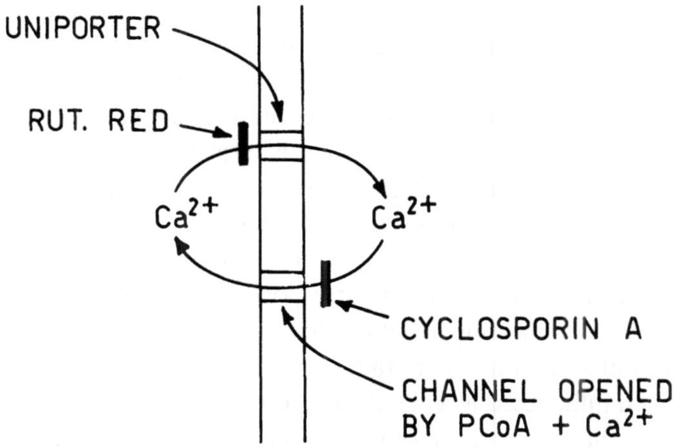

Fig. 6. Suggested mechanism for the futile Ca^{2+} cycling in and out the inner mitochondrial membrane promoted by PCoA.

nous Mg^{2+} content. Thus ΔΨ, collapsed by Ca^{2+} and PCoA, could be restored to its initial value by subsequent addition of Mg^{2+}, whereas subsequent addition of carnitine had little effect (Fig. 5).

It has to be emphasized that both L-carnitine and Mg^{2+} did not protect mitochondria against the damaging action of elevated concentrations of Pcar (results not reported).

Discussion

Both PCoA and Pcar impair transmembrane potential of cardiac mitochondria. At high concentrations either ester does so, perhaps due to their amphiphilic properties [2]. However, ΔΨ is much more sensitive to PCoA than to Pcar. A plausible explanation is that, though both may act by deforming the membrane, PCoA may bind to it more tenaciously.

It is unlikely that these high ester concentrations can ever be reached in the intact myocardium, even under pathological conditions. This is especially the case for LCAcar because these esters can bind with high affinity to fatty acid binding protein [17], and may also be rapidly eliminated because of their high diffusibility.

Of greater relevance to conditions '*in vivo*' is the increased sensitivity to Ca^{2+} in the presence of low concentrations of PCoA but not Pcar. *In vitro* Ca^{2+} alone can collapse ΔΨ of heart mitochondria, but only at very high concentrations. The dramatic fall in this resistance to Ca^{2+} induced by small amounts of PCoA may be relevant to the pathogenesis of cardiac ischemia, where a concomitant increase of the two species does occur [1].

The fall in ΔΨ induced by PCoA + Ca^{2+} may be the result of their stimulating a futile 'Ca^{2+} cycling' process [16]. A possible mechanism is illustrated by the scheme of Fig. 6. This process consists in the energy dependent uptake of Ca^{2+}, mediated by the Ca^{2+} uniporter and its concurrent release, which in our conditions might occur through putative pores opened by the associated action of PCoA and Ca^{2+}. In support of this proposed mechanism is the observed inhibition by ruthenium red, and inhibitor of the Ca^{2+} uniporter and by Cyclosporin A. The latter does not inhibit Ca^{2+} uptake, but is said [18] to prevent the opening of non specific pores which is triggered by Ca^{2+} under unfavorable conditions. In this way it inhibits the non specific outward diffusion of ions (Ca^{2+} as well as Mg^{2+}) and non ionic compounds from their intramitochondrial compartments.

Added Mg^{2+} may prevent to fall in ΔΨ due to PCoA + Ca^{2+} by preventing the depletion of endogenous Mg^{2+}. Such depletion is itself known to deenergize mitochondria [15]. Since exogenous Mg^{2+} fully restores ΔΨ after its collapse, it is clear that the structural prerequisites for energy conservation were not irreversibly damaged by these two agents. Interestingly Mg^{2+} also counteracts the action of PCoA alone at high concentrations and this may be due to the cationic barrier shielding the membrane bilayer from the attack of PCoA.

Pcar seems to be less specific in its effect on mitochondrial membranes than PCoA (see Introduction) and this view is consistent with the observation that, unlike the latter ester, it is unresponsive to the presence of Mg^{2+}, indicating that it produces irreversible damage to mitochondrial membranes.

In intact cardiac cells PCoA accumulates mainly inside the matrix space of mitochondria and Pcar mainly outside [19], whereas under our experimental conditions both were initially present outside. However, when PCoA was generated within the mitochondria from added Pcar it still did not depress $\Delta\Psi$. It is deduced that either internally generated PCoA did not reach the critical concentration requested to affect $\Delta\Psi$, or that endogenous Mg^{2+} prevents this action.

The effect of PCoA + Ca^{2+} on $\Delta\Psi$ of heart mitochondria resembles that of Ca^{2+} + phosphate on liver mitochondria [20, 21]. The latter combination also dissipates $\Delta\Psi$ with a concomitant efflux of mitochondrial Mg^{2+} (together with adenine nucleotides) and is either prevented or restored by added Mg^{2+}, although in this case ADP is also necessary. The effects of Ca^{2+} and phosphate have been observed in rat heart mitochondria too, but a higher temperature (35° C) and a lower pH (6.8) were necessary [22].

Our previously expressed view [20] that mitochondrial Mg^{2+} and adenine nucleotides are components of a permeability barrier necessary for maintaining mitochondrial $\Delta\Psi$ accords with the present results. The described protection by Mg^{2+} and Cyclosporin A is consistent with the suggestion by Panov *et al.* [23] that adenine nucleotide translocase centers may operate as gate pores regulated by ADP and Mg^{2+}. Whether these 'gate pores' may be identified with the non specific pores opened by Ca^{2+} and blocked by Cyclosporin A [18, 24] is a problem deserving further consideration.

Acknowledgements

We thank Prof. Peter Jocelyn (Edinbourgh) for critical reading of the manuscript. This study was supported in part by a grant from Sigma Tau (Rome, Italy).

References

1. Neely JR, Garber D, McDonough K, Idell-Wenger J: Relationship between ventricular function and intermediates of fatty acid metabolism during myocardial ischemia: effects of carnitine. In: MM Winbury and Y Abisko (eds) Ischemic myocardium and antianginal drugs. Raven Press, New York, 1979, pp 225–234
2. Brecher P: The interaction of long-chain acyl CoA with membranes. Mol Cell Biochem 57: 3–15, 1983
3. Lamers JMJ, Stinis HT, Montfoort AD, Hulsmann WC: The effect of lipid intermediates on Ca^{2+} and Na^+ permeability and (Na^+ + K^+) ATPase of cardiac sarcolemma. A possible role in myocardial ischemia. Biochim Biophys Acta 774: 127–137, 1984
4. Pande SV, Blanchaer MC: Reversible inhibition of mitochondrial adenosine diphosphate phosphorylation by long acyl coenzyme A esters. J Biol Chem 246: 402–411, 1971
5. Morel F, Laquin C, Lunardi J, Duszynski J, Vignais PV: An appraisal of the functional significance of the inhibitory effect of long chain acyl-CoAs on mitochondrial transports. FEBS Lett 39: 133–138, 1974
6. Devaux PF, Bienvenue A, Lauquin C, Brisson AD, Vignais M, Vignais PV: Interaction between spin-labeled acyl-coenzyme A and the mitochondrial adenosine diphosphate carrier. Biochemistry 14: 1272–1280, 1975
7. Beatrice MC, Pfeiffer DR: The mechanism of palmitoyl-CoA inhibition of Ca^{2+} uptake in liver and heart mitochondria. Biochem J 194: 71–77, 1981
8. Di Lisa F, Menabò R, Miotto G, Bobyleva-Guarriero V, Siliprandi N: Ca^{2+}-mediated action of long-chain acyl-CoA on liver mitochondria energy-linked processes. Biochim Biophys Acta 973: 185–188, 1989
9. Le Quoc K, Le Quoc D: Involvement of the ADP/ATP carrier in calcium-induced perturbations of the mitochondrial inner membrane permeability: Importance of the orientation of the nucleotide binding site. Arch Biochem Biophys 265: 249–257, 1988
10. Piper MH, Sezer O, Schwartz P, Hutter JF, Schweikhardt C, Spieckermann PG: Acylcarnitine effects on isolated cardiac mitochondria and erythrocites. Basic Res Cardiol 79: 186–198, 1984
11. Sordhal LA, Johnson C, Blailock ZR, Schwartz A: The mitochondrion. Methods in Pharmacol 1: 247–283, 1971
12. Kamo N, Muratsugu M, Hongoh R, Kobatake Y: Membrane potential of mitochondria measured with an electrode sensitive to tetraphenylphosphonium and relationship between proton electrochemical potential and phosphorylation potential in steady state. Membr Biol 49: 105–121, 1979
13. Affolter H, Siegel E: A simple system for the measurement of ion activities with solvent polymeric membrane electrodes. Anal Biochem 97: 315–319, 1979
14. Jensen SD, Gunter KK, Gunter TE: The efficiencies of the component steps of oxidative phosphorylation ii. Experimental determination in mitochondria and examination of the equivalence of membrane potential and pH gradient in phosphorylation. Arch Biochem Biophys 248: 305–323, 1986
15. Toninello A, Siliprandi D, Siliprandi N: Mg^{2+} restores membrane potential in rat liver mitochondria deenergized by Ca^{2+} and phosphate movements. FEBS Lett 142: 63–66, 1982
16. Stucki JW, Ineichen EA: Energy dissipation by calcium recycling

and the efficiency of calcium transport in rat liver mitochondria. Eur J Biochem 48: 365–375, 1974

17. Fournier NC, Richard MA: Fatty acid-binding protein, a potential regulator of energy production in the heart. J Biol Chem 263: 14471–14479, 1988
18. Crompton M, Ellinger H, Costi A: Inhibition by cyclosporin A of a Ca^{2+}-dependent pore in heart mitochondria activated by inorganic phosphate and oxidative stress. Biochem J 255: 357–360, 1988
19. Idell-Wenger JA, Grotyohann LW, Neely JR: Coenzyme A and carnitine distribution in normal and ischemic hearts. J Biol Chem 253: 4310–4318, 1978
20. Zoccarato F, Rugolo M, Siliprandi D, Siliprandi N: Correlated effluxes of adenine nucleotides, Mg^{2+} and Ca^{2+} induced in rat liver mitochondria by external Ca^{2+} and phosphate. Eur J Biochem 114: 195–199, 1981
21. Toninello A, Siliprandi D, Siliprandi N: On the mechanism by which Mg^{2+} and adenine nucleotides restore membrane potential in rat liver mitochondria deenergized by Ca^{2+} and phosphate. Biochim Biophys Res Comm 111: 792–797, 1983
22. Siliprandi D, Di Lisa F, Toninello A: On the role of Mg^{2+} in heart mitochondria. In: MC Berman, W Gevers, LH Opie (eds) Membranes and Muscle. Published for the ICSU Press by IRL Press, 1985, pp 378–379
23. Panov A, Filippova S, Lyakhovich V: Adenine nucleotide translocase as a site of regulation by ADP of the rat liver mitochondria permeability to H^+ and K^+ ions. Arch Biochem Biophys 199: 420–426, 1980
24. Griffiths EJ, Halestrap AP: Further evidence that cyclosporin A protects mitochondria from calcium overload by inhibiting a matrix peptidil-prolyl cis-trans isomerase. Biochem J 274: 611–614, 1991

Molecular and Cellular Biochemistry **116**: 125–129, 1992.

Carnitine requirement of vascular endothelial and smooth muscle cells in imminent ischemia

W.C. Hülsmann[1,2] and M.L. Dubelaar[2]
[1] Thorax Centre, Erasmus University, Rotterdam and [2] Department of Physiology, Cardiovascular Research Institute, University of Limburg, Maastricht

Abstract

Vascular endothelial and -smooth muscle cells have been shown to use fatty acids as substrates for oxidative phosphorylation. Endothelial cells are more vulnerable to oxidative stress than muscle cells and are prone to loose carnitine early during hypoperfusion. This has been suggested by two observations. The first is that incubation of isolated endothelial cells in a low carnitine medium leads to oleate oxidation, dependent upon carnitine addition, whereas smooth muscle cells do not depend on carnitine addition during *in vitro* incubation, although aminocarnitine, a specific inner-membrane carnitine palmitoyltransferase inhibitor, inhibits fatty acid oxidation. The second observation is that rat hearts labeled *in vivo* with ^{14}C-carnitine loose, as paced Langendorff heart, only 4% of their carnitine in 20 min perfusion, following 60 min global ischemia. The carnitine released had a much higher specific radioactivity than the carnitine that was not released. It indicates compartmentation of carnitine in heart. As earlier and presently discussed work shows endothelial vulnerability, it is to be expected that this cell type may become carnitine deficient during pacing and ischemia. Endothelial incompetence in flow regulation could be delayed by the presence of carnitine and fatty acids in pre-ischemia. It is speculated how activated fatty acids could protect endothelium. (Mol Cell Biochem **116:** 125–129, 1992)

Key words: fatty acid, carnitine, smooth muscle, endothelium, ischemia, heart, skeletal muscle

Introduction

We have observed acute stimulation of function of paced muscle by intravenous administration of L-carnitine [1, 2]. This was accompanied by the increase of flow, as determined by microsphere injection [2]. The effect of carnitine is probably due to stimulation of fatty acid metabolism [1, 2]. Earlier we observed [3] that during hypoperfusion of heart endothelial cells are most vulnerable, judged by the early release of xanthine oxidase, an endothelial marker enzyme. This enzyme was probably involved in the production of oxygen free radicals in the interstitium, as the rapid loss of cardiac function due to hypoperfusion, could acutely be diminished by superoxide dismutase (SOD) administration. As we have observed that fatty acids are vasodilators [4] and delay acidosis-induced loss of vascular flow [5], we decided to investigate fatty acid metabolism in vascular cells.

Fatty acid oxidation in vascular endothelium, has recently been demonstrated by us [6]. Fatty acid oxidation in vascular smooth muscle cells will be reported in the present communication. We will also discuss the particular sensitivity of vascular endothelial cells to ischemia and the possible mechanisms by which fatty

Address for offprints: W.C. Hülsmann, Thorax Centre, P.O. Box 1738, 3000 DR Rotterdam, The Netherlands acids and (acyl)carnitines delay the loss of flow in preischemia.

Materials and methods

Cell culture

Human umbilical artery smooth muscle cells were provided by Mr. A. Verkerk, Dept. Cell Biol. Erasmus University, Rotterdam. They were isolated from pieces of arterial wall and cultured in a fibronectin coated 24 well plate (Costar) in medium M199 supplemented with 10% fetal calf serum, 10% human serum, 100 IU/ml penicillin, 0.1 mg/ml streptomycin. The muscle cells were positively identified with a monoclonal antibody against smooth muscle alpha-actin (Sigma clone no 1 A4) and subcultivated by 1:3 splits when the culture reached confluency. A 7 h incubation was carried out in (carnitine-free) Waymouth MB 721/1 medium supplemented with 2% fatty acid free bovine serum albumin and 0.9 mM (9,1)-^{3}H)oleate. After incubation the ^3H_2O formed was determined as described [6]. The carnitine and aminocarnitine used were gifts from Dr. F. Maccari (Sigma Tau, Pomezia, Italy).

Rat heart perfusion

Rat heart perfusions were carried out according to the Langendorff technique with a modified Tyrode solution, containing 11 mM glucose as substrate, as described before [4]. Ischemia was produced by completely interrupting flow and keeping the heart at 37° C in medium saturated with 95% N_2 and 5% CO_2. After 60 min of ischemia, reperfusion was started and effluents were collected over 20 min. The radioactivity in the effluents and of the residual activity in the heart were determined. The percentual release of carnitine was calculated, based on radioactivity measurements, to include loss of acylcarnitines. The specific activities of carnitine in hearts and effluents were determined. Carnitine was measured by the method of Cederblad [7].

Table 1. Oleate oxidation in cultured smooth muscle cells

Addition	nmol oleate/mg prot/h
None	13.4 ± 1.6
L-carnitine 60 μM	16.0 ± 4.9
Aminocarnitine 30 μM	0.5 ± 0.6

7 h incubation of cultured arterial cells in Waymouth medium enriched with 2% BSA and 0.9 mM ^{3}H-oleate; n = 4. Values are mean ± SD.

Isolation of mitochondria; assay of mitochondrial components

Rat hearts were finely minced with scissors and homogenized in a Potter-Elvehjem, teflon-glass homogenizer. Debris and nuclei were removed at 5 min 600 × g and the supernatant spun for 10 min at 6000 × g to sediment the mitochondria. The isolation medium contained 0.1 M KCl, 0.05 M Tris-HCl (pH 7.4), 1 mM EDTA, 5 mM $MgCl_2$, 1 mM ATP, 0.8 mg/ml BSA. After isolation, the mitochondria were frozen in batches for analysis. Protein was determined by the biuret method. Carnitine-acetyltransferase [8] and PropionylCoA-carboxylase [9] were measured as described.

Results

Fatty acid oxidation in vascular smooth muscle

From Table 1 it can be seen that in cultured smooth muscle cells the addition of L-carnitine to carnitine-free incubation medium slightly stimulates oxidation of ^{3}H-labeled oleate. The carnitine content of the culture medium (present in serum) was 6 μM and apparently enough to obtain an optimal intracellular carnitine level that remained high during the subsequent 7 h incubation. The strong inhibition by aminocarnitine, a specific inhibitor of CPT-2 [10], indicated the requirement of carnitine-dependent transport of (activated) fatty acid

Table 2. Oleate oxidation in adventitia- and endothelium-free rat aorta

Addition	nmol/h/mg DNA
None	11.3 ± 0.1
L-carnitine 80 μM	14.7 ± 1.4
Aminocarnitine 40 μM	7.8 ± 0.4

Three rat aortas, cleaned in saline at room temp., cut in small pieces, were incubated in SPM with 11 mM glucose + 2% BSA + 1.7 mM ^{3}H-oleate (1 μmol = 0.835 μCi) for 4 h at 37° C in 9 siliconized scintillation vials under 95% O_2 + 5% CO_2. For ^3H_2O and DNA determinations see ref. 6. Values are mean ± SD.

into the mitochondrial matrix space. Aminocarnitine also inhibited oleate oxidation in adventitia- and endothelium-free rat aorta (Table 2). This effect of carnitine addition was also small. The low dependence of fatty acid oxidation in smooth muscle on carnitine addition (Tables 1 and 2), compared to that in endothelial cells [6] during *in vitro* incubation, suggests a higher rate of carnitine loss from the latter cell type during *in vitro* incubation. More experiments are required to settle this point.

Relative resistance of muscle cells against ischemia

During hypoperfusion of Langendorff rat hearts, we observed that xanthine oxidase leaked through endothelial plasmalemma, while smooth muscle cells were functionally still intact as judged by prompt responses to vasodilator (nitroprusside) and vasoconstrictor (vasoxin). As there was not loss of myoglobin, we concluded that vascular endothelium is most vulnerable to ischemia [3].

After overnight *in vivo* labeling of rats with ^{14}C-L-carnitine, Langendorff perfusion of the electrically stimulated heart led to no measurable loss of label after the removal of blood (not shown). After one hour of global ischemia at 37° C the loss of carnitine was still very low and the specific activity of the released carnitine much higher than that of the residual cardiac carnitine (Table 3). It suggests leakage of a carnitine containing cell compartment.

From the early leakage of enzymes from endothelium [3] it is most likely that leakage of carnitine after brief ischemia of muscle, is also predominantly from endothelial origin. Also the mitochondrial inner membrane of cardiomyocytes is relatively resistant to ischemia.

From Table 4 we can conclude that 30 min global ischemia of rat heart at 37° C, followed by the isolation of mitochondria, at 0–5° C, resulted in preservation of matrix components. That carnitine acetyl transferase (CAT) is not lost, might be due to its firm binding to the innermembrane, but propionyl-CoA carboxylase is a loosely-bound matrix protein [11] and could be lost easily if mitochondrial leakage would occur. This also holds for low molecular compounds, like carnitine (Table 4). The latter compound increased in concentration in the mitochondria, a phenomenon observed already by Idell-Wenger *et al.* [12], who found that carnitine shifts from the cytosol to the mitochondria due to ischemia. They also found retention of coenzyme A in the matrix mitochondrialis [12].

The observed loss of carnitine from heart- or skeletal muscle during pacing or strenuous exercise [13–16] may be expected to be at least partly due to its loss from endothelial cells. This leads to our hypothesis that observed acute beneficial effects of intravascular carnitine administration, during pacing or exercise [1, 2, 17], may be due to restoration of regulatory functions of endothelium on muscle flow.

Table 3. Specific activity and loss of carnitine from *in vivo* ^{14}C-carnitine labeled rat hearts (n = 4)

S.A. in effluent (pCi/nmol)	S.A. in heart (pCi/nmol)	% label removed after isch. (%)
2710 ± 730	22.36 ± 2.29	4.36 ± 1.58

15 h after *in vivo* labeling with 5 μCi ^{14}C-carnitine by i.p. injection of each rat of about 250 g, perfusions were carried out as described under Methods. Values are mean ± SD.

Discussion

Endothelial cells possess a large capacity for anaerobic glycolytic energy production [18, 19]. In aerobiosis they also oxidize fatty acids [6]. Vascular smooth muscle cells also catalyse fatty acid oxidation (Tables 1 and 2), as may have been expected [20]. Vascular fatty acid oxidation might still proceed in imminent ischemia, as vascular mitochondria may still have access to some oxygen when mitochondria within the muscle tissue are already fully anaerobic, due to the presence of an oxygen concentration gradient. Perhaps this situation makes the endothelium most suitable for oxygen free radical formation in imminent ischemia. That endothelial cells can be damaged by oxygen radicals has been known for some time [21]. Indeed, an early indication

Table 4. Mitochondrial matrix components in control and ischemic hearts

Mitochondrial source	Carn. acetyl-transferase mU/mg pr.	Prop. CoA carboxylase mU/mg pr.	L-carnitine nmol/mg pr.
Control hearts	290 ± 15	21 ± 2	1.3 ± 0.2
Ischemic hearts	308 ± 32	20 ± 2	2.6 ± 0.3

Control and ischemic (30 min at 37° C) hearts were rapidly cooled in ice-cold isolation medium and mitochondria isolated as shown under Methods. Values are mean ± SD of 4 experiments.

of endothelial alteration is its failure to regulate flow [22, 23]. We have been able to show [3], that the loss of cardiac function during hypoperfusion of rat Langendorff hearts, could partly be overcome by the addition of SOD to the perfusion medium. This indicates that at least part of the damaging free radicals are present extracellularly, probably in the interstitium. Intracellularly, the radicals may be formed by various subcellular fractions [24], many cell types are equipped with SOD. For endothelium, this may be illustrated by the observation of Mügge *et al.* [25], that endothelium derived relaxing factor (EDRF) can not be secreted from rabbit aorta *after* inhibition of intracellular SOD. In our study [3], the beneficial effect of added, extracellular, SOD indicates neutralization of the negative effect of $O_2^{\cdot-}$ resulting in increased capillary flow. The addition of a vasoconstrictor instead of SOD had a similar effect on heart function, which may be explained by improved perfusion pressure during flow limitation. The targets of the interstitial free radicals under our conditions of hypoperfusion have not been studied. It could be the endothelial cell itself, affecting its vasoregulatory messengers (endothelins, EDRF, prostacyclin or other autacoids), or their receptors on smooth muscle cells. It is probably not the smooth muscle alpha-adrenergic receptor or guanylate cyclase of smooth muscle cells, as judged by prompt reactions upon vasoxin or nitroprusside addition [3]. The observed effect of fatty acids on delaying imminent ischemia, as induced by acidosis [5], is probably located in endothelial cells, the most likely source of $O_2^{\cdot-}$ in preischemia [3, 23, 26]. Our experience that fatty acid and carnitine are both needed to improve force of the paced latissimus dorsi muscle [1], suggests that fatty acids must be activated in order to be effective. Activated fatty acid can protect being a substrate of oxidative phosphorylation. A high rate of ATP generation limits glycolysis and prevents lactic acidosis. More efficient ATP generation may be expected to lower free radical formation, as low cytosolic AMP limits the extracellular generation of substrates for xanthine oxidase, as discussed above. It might help the regulation of intracellular Ca^{2+} and the reacylation of membrane lysophospholipids, formed by endogenous phospholipase action, and perhaps by membrane stabilization of long-chain acylcarnitine [27, 28].

Acute effects of carnitine on stimulated muscle have been discussed above and may well be based on improved flow [28]. Carnitine derivatives, such as acetyl- and propionyl-carnitine have also been shown to be effective in the prevention of disorders related to hypoperfusion [29, 30]. Sassen *et al.* [31] observed that propionyl-carnitine improved transmural cardiac flow in pigs subjected to ischemic conditions. Acetyl- and propionyl-carnitine might be efficient carnitine donors [29]. Propionyl-carnitine could have an additional advantage over carnitine and acetylcarnitine as after carboxylation, it could generate citric acid cycle intermediates [32]. However, the beneficial effect of acetyl-carnitine [30], would then require a different explanation as it is not an anaplerotic agent and could lead to further increase the acetyl-CoA/CoA ratio in the mitochondrial matrix. Another possibility is the protection of mitochondria from depletion of CoASH by its conversion to short-chain acyl-CoA, before CoASH is oxidized to CoASSCoA (inactive, oxidized-coenzyme A) by free radicals, known to affect the sulfhydryl status of mitochondria [24]. The fate of added carnitine and their short-chain derivatives in imminent ischemia should be studied in endothelial cells, as muscle cell mitochondria are relatively resistant towards ischemia (Table 4 and ref. 33). Both propionyl- and acetyl-carnitine have been shown to decrease cytoplasmic Ca^{2+} levels in endothelial cells, subjected to oxidative stress [34].

Recent experiments indicate that carnitine addition to fatty acid and glucose containing media increases NADPH in endothelial cells. This is required for EDRF and GSH generation and maybe largely responsible for the observed improvement of flow during oxidative stress.

Acknowledgements

The authors wish to thank Mr. A. Verkerk, dept. of Cell Biology I Erasmus University Rotterdam, for the culture of smooth muscle cells. The Netherlands Heart Foundation, The Hague and Sigma Tau Pharmaceutical of Rome, Italy, are thanked for financial support.

References

1. Dubelaar M-L, Lucas CMHB, Hülsmann WC: Acute effect of 1-carnitine upon skeletal muscle force tests in the dog. Am J Physiol 260: E189–E193, 1991
2. Dubelaar M-L, Lucas CMHB, Hülsmann WC: The effect of 1-carnitine on force development of m. latissimus dorsi in dogs. J Cardiac Surgery 6, suppl 1: 270–275, 1991
3. Hülsmann WC, Dubelaar M-L: Early damage of vascular endothelium during cardiac ischemia. Cardiovasc Res 21: 674–677, 1987

4. Hülsmann WC: Coronary vasodilation by fatty acids. Basic Res Cardiol 71: 179–191, 1976
5. Hülsmann WC, De Wit LEA, Schneydenberg C, Verkleij AJ: Loss of cardiac contractility and severe morphologic changes by acutely lowering the pH of the perfusion medium, protection by fatty acids. Biochim Biophys Acta 1033: 214–218, 1990
6. Hülsmann WC, Dubelaar M-L: Aspects of fatty acid metabolism in vascular endothelial cells. Biochimie 70: 681–686, 1988
7. Cederblad G, Lindstedt CD: A method for the determination of carnitine in the picomolar range. Clin Chim Acta 37: 235–243, 1972
8. Pearson DJ, Chase JFA, Tubbs PK: The assay of carnitine and its O-acylderivatives. In: SP Colowick and NO Kaplan (eds) Methods of Enzymol 14: 612–622, Academic Press Inc, 1969
9. Scholte HR, Luyt-Houwen IEM, Dubelaar M-L, Hülsmann WC: The source of malonyl-CoA in rat heart. The calcium paradox releases acetyl-CoA carboxylase and not propionyl-CoA carboxylase. FEBS Lett 198: 47–50, 1986
10. Hülsmann WC, Schneydenberg C, Verkleij AJ: Accumulation and excretion of long-chain acylcarnitines by rat heart, studies with aminocarnitine. Biochim Biophys Acta, in press, 1991
11. Wit-Peeters EM, Scholte HR, Van De Akker F, DeNie I: Intramitochondrial localization of palmitoyl-CoA-dehydrogenase, β-hydroxy-acyl-CoA-dehydrogenase and enoyl-CoA-hydrase in Guinea pig heart. Biochim Biophys Acta 231: 23–31, 1971
12. Idell-Wenger JA, Grotyohann LW, Neely JR: Coenzyme A and carnitine distribution in normal and ischemic hearts. J Biol Chem 253: 4310–4318, 1978
13. Ferrannini E, Buzzigoli G, Bevilaqua S, Boni C, DelChiaro D, Oleggini M, Brandi L, Maccari F: Interaction of carnitine with insulin-stimulated glucose metabolism in man. Am J Physiol 255: E964–E952, 1988
14. Siliprandi N, Di Lisa F, Pieralsi G, Ripari P, Maccari F, Menabo R, Giamberardino MA, Vecchiet L: Metabolic changes induced by maximal exercise in human subjects following L-carnitine administration. Biochim Biophys Acta 1034: 17–21, 1990
15. Spagnoli LG, Corsi M, Villaschi S, Palmieri G, Maccari F: Myocardial carnitine deficiency in acute myocardial infarction. Lancet i: 1419–1420, 1982
16. Hiatt WR, Regensteiner JG, Wolfel EE, Ruff L, Brass EP: Carnitine and acylcarnitine metabolism during exercise in humans. -dependence on skeletal muscle metabolic state. J Clin Invest 84: 1167–1173, 1989
17. Ferrari R, Cucchini F, Visioli O: The metabolic effects of L-carnitine in angina pectoris. Int J Cardiol 5: 213–216, 1984
18. Dobrina A, Rossi F: Metabolic properties of freshly isolated bovine endothelial cells. Biochim Biophys Acta 762: 295–301, 1983
19. Mertens S, Noll T, Spahr R, Krützfeldt A, Piper HM: Energetic response of coronary endothelial cells to hypoxia. Am J Physiol 258: H689–H694, 1990
20. Odessey YY, Chace KV: Utilization of endogenous lipid, glycogen, and protein by rabbit aorta. Am J Physiol 243: H128–H132, 1982
21. Sacks T, Moldow CF, Craddock PR, Bowers TK: Oxygen radicals mediate endothelial cell damage by complement-stimulated granulocytes. An *in vitro* model of immune vascular damage. J Clin Invest 61: 1161–1167, 1978
22. Dauber JM, Van Benthuysen KM, McMurthy JF, Wheeler GS, Lesnefsky EJ, Horwitz LD, Neil JV: Functional coronary microvascular injury evident as increased permeability due to brief ischemia and reperfusion. Circ Res 66: 986–998, 1990
23. Tsao PS, Lefer AM: Time course and mechanism of endothelial dysfunction in isolated ischemic- and hypoxic-perfused rat hearts. Am J Physiol 259: H1660–H1666, 1990
24. Chance B, Sies H, Boveris H: Hydroperoxide metabolism in mammalian organs. Physiol Rev 59: 527–605, 1979
25. Mügge A, Elwell JH, Peterson TE, Harrison DG: Release of intact endothelium derived relaxing factor depends on endothelial superoxide dismutase activity. Am J Physiol 260: C219–C225, 1991
26. Lefer AM, Tsao PS, Lefer DJ, Ma XL: Role of endothelial dysfunction in the pathogenesis of reperfusion injury after myocardial ischemia. FASEB J 5: 2029–2034, 1991
27. Hülsmann WC, Dubelaar M-L, Lamers JMJ, Macari F: Protection by acyl-carnitines and phenylmethylsulfonylfluoride of rat heart subjected to ischemia and reperfusion. Biochim Biophys Acta 847: 62–66, 1985
28. Hülsmann WC, Dubelaar M-L: Carnitine in metabolism of paced cardiac and skeletal muscles; prevention of acidosis and improvement of vascular flow. In: L. Carnitine and its role in medicine, from function to therapy. R. Ferrari, S. Di Mauro, G. Sherwood, eds. Academic Press Ltd (1992) pp. 345–358.
29. Paulson DJ, Traxler J, Schmidt M, Nooman J, Shug AL: Protection of the ischemic myocardium by L-propionylcarnitine: effects on the recovery of cardiac output after ischemia and reperfusion, carnitine transport and fatty acid oxidation. Cardiovasc Res 20: 536–541, 1986
30. Liedtke AJ, DeMaison L, Nellis SH: Effect of L-propionylcarnitine on mechanical recovery during reflow in intact hearts. Am J Physiol 255: H169–H176, 1988
31. Sassen LMA, Bezstarosti K, Koning MMG, Van Der Giessen WJ, Lamers JMJ: Effects of administration of L-propionylcarnitine during ischemia on the recovery of myocardial function in the anaesthetized pig. Cardioscience 1: 155–161, 1990
32. Di Lisa F, Menabo R, Siliprandi N: L-propionylcarnitine protection of mitochondria in ischemic rat hearts. Mol Cell Biochem 88: 169–173, 1989
33. Maxwell L, Gavin JB: The contribution of ischaemia to the development of microvascular incompetence in the myocardium. Cardiovasc Res 25: 491–495, 1991
34. Van Hinsberg VWM, Scheffer MA: Effect of L-propionylcarnitine on human endothelial cells. Cardiovasc Res Ther 5: 97–106, 1991

Molecular and Cellular Biochemistry **116**: 131–137, 1992.

Protection of the ischemic diabetic heart by L-propionylcarnitine therapy

Dennis J. Paulson, Austin L. Shug[1] and Jiansheng Zhao
Department of Physiology, Chicago College of Osteopathic Medicine, 555 31st Street, Downers Grove, Illinois 60515, USA ([1] Present address: Department of Neurology, University of Wisconsin, Madison, WI 33705, USA)

Abstract

Diabetics suffer from an increased incidence of myocardial infarction and are less likely to survive an ischemic insult. Since L-propionylcarnitine (LPC) has been shown to protect against ischemic/reperfusion injury, we hypothesized that LPC may be of even greater benefit to the diabetic heart. Diabetes was induced by i.v. streptozotocin, 60 mg/kg; duration: 12 wks. The chronic effect of LPC was determined by daily i.p. injections (100 mg/kg) for 8 wks. The acute effects of LPC were determined by adding it to the perfusion medium (5 mM) of control and diabetic hearts. Initial cardiac contractile performance of isolated perfused working hearts was assessed by varying left atrial filling pressure. Hearts were then subjected to 90 min of low flow global ischemia followed by 30 min reperfusion. Chronic LPC treatment had no effect on initial cardiac performance in either control or diabetic hearts. Acute addition of LPC to the perfusion medium enhanced pump performance of control hearts, but had no effect in diabetic hearts. Both acute and chronic LPC significantly improved the ability of control and diabetic hearts to recover cardiac contractile performance after ischemia and reperfusion, however, chronic treatment was more effective in diabetic hearts. (Mol Cell Biochem **116:** 131–137, 1992)

Key words: diabetes, ischemia, L-propionylcarnitine, diabetic cardiomyopathy, reperfusion injury

Introduction

Diabetic patients are known to have a high frequency of atherosclerosis [1]. In particular, coronary artery disease is the ultimate cause of death in more than half of all diabetic patients [2–4]. Diabetic patients are less likely to survive an acute ischemic episode than non-diabetics [3, 4]. In addition, diabetic patients exhibit an increased incidence of heart failure which in some cases is not attributable to coronary artery disease. In such cases the heart failure that occurred has been attributed to a condition termed diabetic cardiomyopathy [5]. Some studies [6–11], but not all [12, 13], using experimental animal models of diabetes have also provided evidence that the diabetic heart exhibits an increased vulnerability to ischemia/reperfusion injury as well as depressed cardiac contractile performance.

There is a growing body of evidence that acute administration of L-propionylcarnitine (LPC) improves the functional recovery of normal hearts subjected to ischemia/reperfusion injury [14–18] and reduces infarct size [18]. In addition, LPC reportedly limits ischemia and lipid amphiphile-induced electrical alterations [19–22]. The mechanism for the beneficial effects of LPC remains unproven; however, it has been suggested that LPC stimulates substrate utilization by the ischemic/

Address for offprints: D.J. Paulson, Department of Physiology, Chicago College of Osteopathic Medicine, 555 31st Street, Downers Grove, IL, 60515, USA reperfused heart [14, 17, 23, 24]. Since, substrate utilization is altered in the diabetic heart [25], we hypothesized that LPC therapy may be of even greater benefit to the diabetic heart. The present study tests this hypothesis by investigating whether chronic intraperitoneal (i.p.) LPC treatment administered to control and diabetic rats would: a) improve cardiac contractile performance of isolated perfused working hearts subjected to varying preloads and b) enhance the recovery of cardiac pump performance of hearts subjected to 90 min period of low-flow ischemia followed by reperfusion. In addition the acute effects of LPC were determined by adding it directly to the perfusion medium of untreated control and diabetic hearts.

Methods

Introduction of diabetes

Male Sprague-Dawley rats weighing between 225–275 g were anesthetized with ether and made diabetic by intravenous injection of streptozotocin (60 mg/kg) via the tail vein. Over the next 4 weeks, the induction of diabetes was confirmed by glucose testing of the urine using Ames-Keto-Diastix and measuring body weight. The total duration of diabetes was 12 wks.

Treatment protocols

The acute effects of LPC were determined by adding it to the perfusion medium of isolated perfused control and diabetic hearts at a concentration of 5 mM. The pH of the LPC solution was adjusted to 7.4 with NaOH prior to addition. To maintain the Na^+ concentration of the perfusion buffer at 143 mM, the NaCl concentration of the perfusion buffer was adjusted. LPC was added at the beginning of the working heart perfusion. For the chronic studies, LPC solution was filter-sterilized and injected i.p. (100 mg/kg/day). Chronic treatment with LPC was initiated after the rats were diabetic for 4 weeks and continued for the next 7 to 9 weeks.

Perfusion of hearts

Each rat was anesthetized with ketamine (80 mg/kg) and xylazine (8 mg/kg). The chest was opened, the heart excised, placed in cold perfusion buffer and mounted on the perfusion apparatus within 60 sec. Each heart was initially perfused in a Langendorff manner for 10 min. The perfusion medium consisted of a Krebs-Henseleit bicarbonate buffer containing 11 mM glucose gassed with 95% O_2 and 5% CO_2 and maintained at 37° C. The heart was then converted to a working mode by perfusing the buffered medium into the left atrium at 10 cm H_2O left atrial filling pressure (LAFP) with an aortic resistance of 100 cm H_2O. This perfusion medium contained 3% albumin bound to 1.2 mM palmitate and 22 mM glucose. A 3F Millar pressure transducer was inserted through the wall of the left ventricle in order to measure pressure development. After 5 min, the hearts were paced at 300 beats/min. Aortic output and coronary flow were measured by timed collections. After a 10 min equilibration period, the LAFP was decreased to 5 cm H_2O and then sequentially increased at 5 min intervals to 10, 15, 20 and 25 cm H_2O. Cardiac pump performance and pressure development were measured at each LAFP. The LAFP was then lowered to 10 cm H_2O and the heart was perfused for an additional 10 min prior to initiating the ischemic episode.

The heart was then subjected to 90 min of low-flow global ischemia which was produced by clamping both the left atrial and aortic lines and by infusing buffer through the aorta into the coronary arteries. Ischemic coronary flow was to adjusted to 4.4 ml/min/g dry wt using a Harvard Apparatus syringe 22 infusion pump. Because of differences in heart weights between control and diabetic rats, the ischemic coronary flow for control hearts was set at a rate of 1.25 ml/min. For the smaller diabetic hearts an ischemic coronary flow rate of 0.85 ml/min was used. Electrical pacing of the heart was maintained during this interval. After ischemia, the heart was reperfused for 30 min. The hearts were not paced for the first 5 min of reperfusion. Cardiac performance measurements were made thereafter at 5 min intervals during this reperfusion period. The percent recovery of cardiac hemodynamic parameters was determined by dividing the final reperfusion value by the value obtained just prior to ischemia.

Statistical analysis

Data obtained for each group were compared to determine statistical significant differences using a one-way analysis of variance test and a least significant difference test for post hoc comparisons among treatment groups. A $p < 0.05$ was accepted as significant.

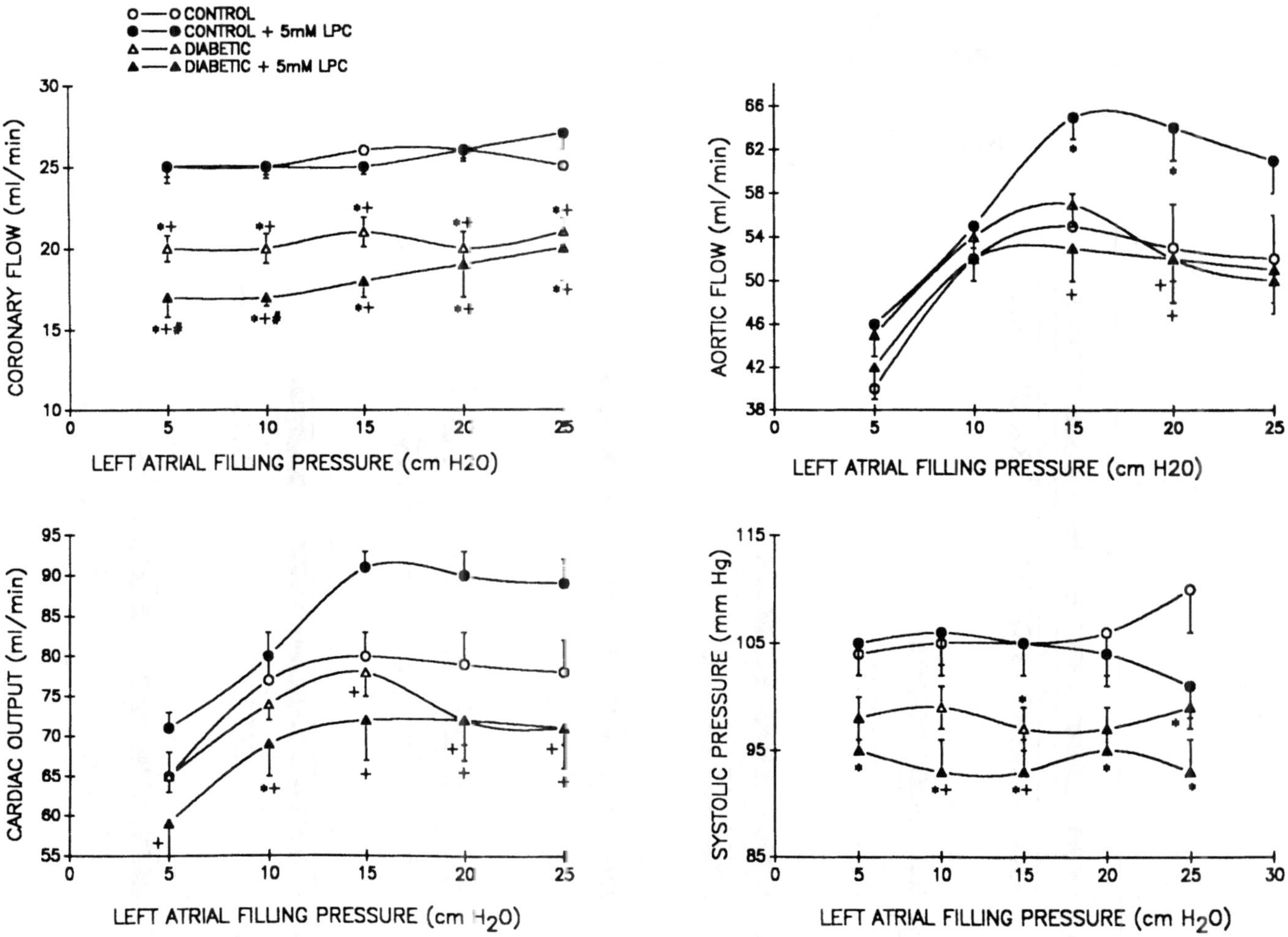

Fig. 1. Effects of the addition of 5 mM LPC to the perfusion medium on coronary flow, aortic flow, cardiac output and left ventricular systolic pressure in response to varying left atrial filling pressures. All values are mean ± S.E.M. for n = 8 to 17. * Significantly different from control (P < 0.05). + Significantly different from control + 5 mM LPC (P < 0.05). # Significantly different from diabetic (p < 0.05).

Results

The effects of acute administration of LPC on cardiac contractile performance of control and diabetic hearts subjected to varying LAFP is shown in Fig. 1. The smaller diabetic hearts had lower values for coronary flow. Aortic flow values were similar between untreated control and diabetic hearts, but systolic pressure development was decreased in diabetic hearts. No other differences were noted for other hemodynamic parameters with the exception of − dP/dt which was depressed in untreated diabetic versus untreated control hearts (data not shown). The addition of LPC to the perfusion medium produced a significant increase in aortic flow of control hearts, but had no effect in diabetic hearts. Other hemodynamic parameters were not affected by LPC.

The acute effects of LPC on the percent recoveries of the various hemodynamic parameters after ischemia and reperfusion are shown in Fig. 2. In control hearts, the addition of LPC to the perfusion medium significantly improved the recovery of aortic flow and cardiac output, but had no effect on the recovery of pressure development which was already 90% in untreated control hearts. The recovery of untreated diabetic hearts was identical to untreated control hearts. The addition of LPC to the perfusion medium of diabetic hearts tended to improve the recovery of aortic flow, but this increase was not significant. However, the recovery of + dP/dt was significantly increased in this group.

Chronic i.p. treatment with LPC to control had no effect on the various contractile performance parameters in perfused hearts subjected to varying LAFP (data not shown). In diabetic hearts, chronic LPC treatment increased coronary flow, but had no significant

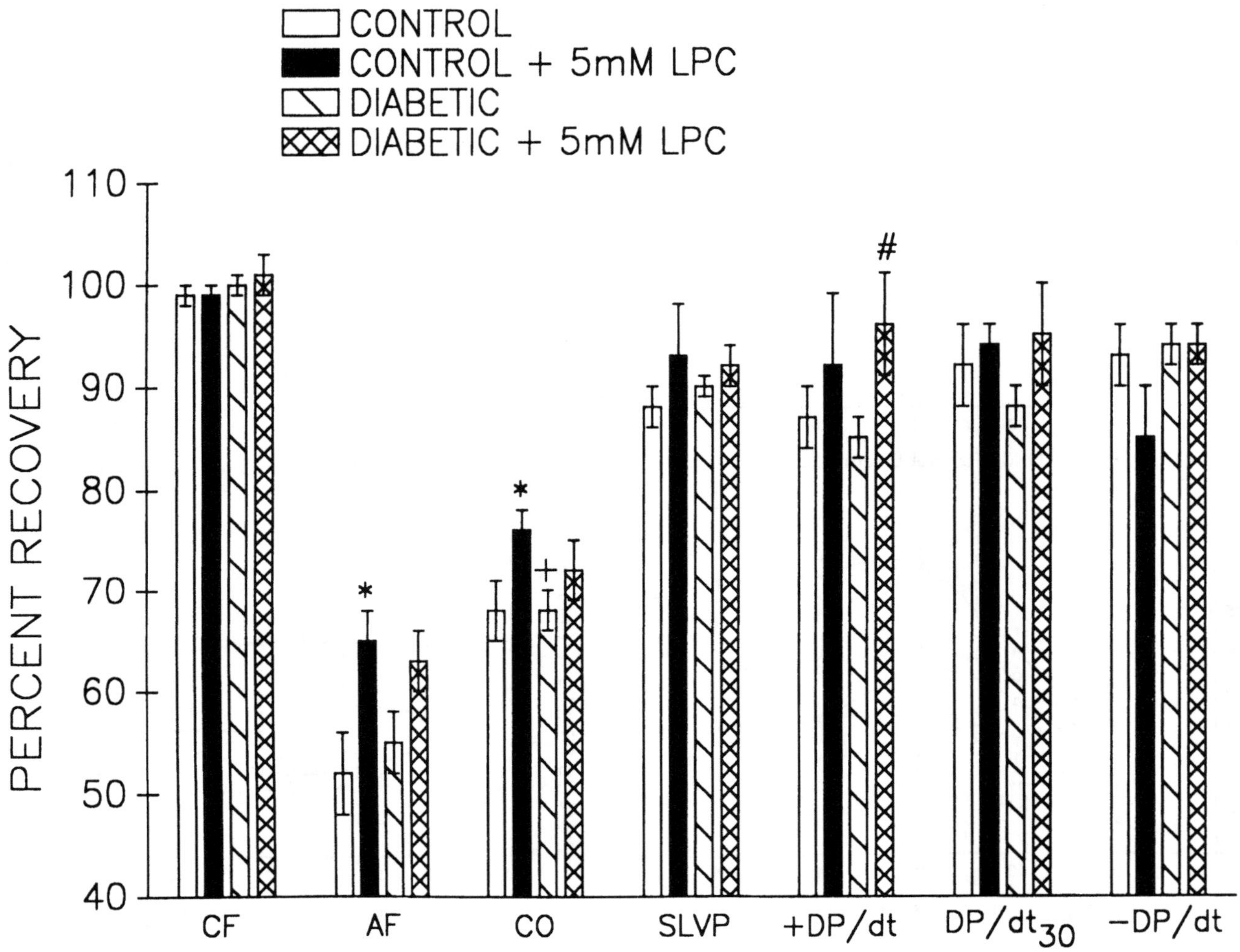

Fig. 2. Effects of addition of 5 mM LPC to the perfusion medium on the percent recovery of various hemodynamic parameters after 90 min of low flow ischemia and 30 min of reperfusion. All values are mean ± S.E.M. for n = 8 to 17. * Significantly different from control ($P < 0.05$). + Significantly different from control + 5 mM LPC ($P < 0.05$). # Significantly different from diabetic ($p < 0.05$).

effect on other parameters. In response to ischemia and reperfusion, chronic treatment of diabetic rats with LPC produced a dramatic improvement in the recovery of aortic flow, cardiac output, left ventricular systolic pressure, peak + dP/dt, and + dP/dt measured at 30 mmHg developed pressure (Fig. 3). Chronic LPC treatment to control rats was still effective in improving the recovery of aortic flow and cardiac output, but the magnitude of this effect was less than that found in diabetic hearts.

Discussion

The results of the present study support previous findings [14–22] indicating that the addition of LPC to the perfusion medium of isolated perfused normal hearts is effective in improving the recovery of cardiac contractile performance after ischemia and reperfusion. In addition, the data on the acute effects of LPC in elevating aortic flow of control hearts in response to varying LAFP provides further evidence that LPC does elicit a positive inotropic effect [15, 16]. In contrast to control hearts, acute administration of LPC to diabetic hearts had no effect on cardiac contractile performance in response to changing LAFP and produced only a modest protection after ischemia and reperfusion. These differing effects of acute LPC treatment on control and diabetic hearts may be explained by the studies that have demonstrated that carnitine transport is impaired in the diabetic heart [26]. Since LPC may be transported into the cardiac cell using this carrier, impaired transport of LPC into the diabetic heart may have limited any beneficial effects that LPC may have had in this system.

Chronic administration of LPC to control or diabetic rats had no effect on cardiac contractile performance

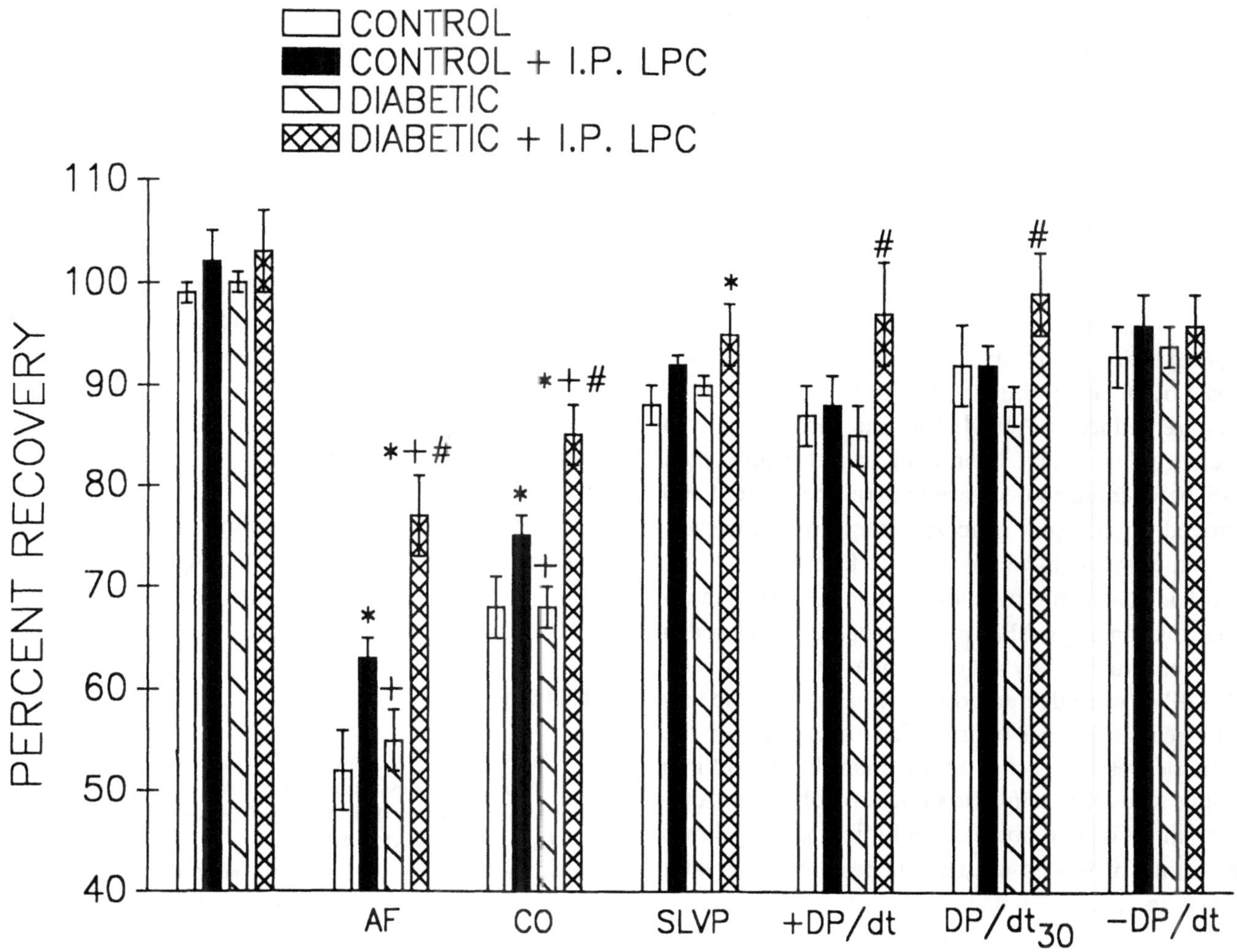

Fig. 3. Effects of chronic i.p. LPC treatment on the percent recovery of various hemodynamic parameters after 90 min of low flow ischemia and 30 min of reperfusion. All values are mean ± S.E.M. for n = 9 to 17. * Significantly different from control ($P < 0.05$). # Significantly different from diabetic ($p < 0.05$).

under normoxic conditions, thus indicating that the positive inotropic effect of LPC described above is an acute effect. This treatment, however, did result in a significant improvement in the ability of both control and diabetic hearts to recover from a period of low flow ischemia and reperfusion. Interestingly, this beneficial effect of chronic LPC treatment was much greater in the diabetic compared to the control hearts. The LPC treated diabetic rats recovered aortic flow to a level that was significantly greater than that of the untreated control and diabetic groups as well as the LPC-treated control group. The explanation for these differing effects of chronic LPC treatment on control and diabetic hearts remains to be determined.

The mechanisms for the beneficial effect of acute and chronic LPC are unknown and may not necessarily be the same. Recently, it has been proposed that LPC may prevent free radical damage induced when the ischemic heart is reperfused [16, 27, 28]. While these findings are noteworthy, the metabolic effects of LPC on glucose and fatty acid fluxes of the diabetic heart as well as the ischemic and reperfused myocardium may also be an important mechanism. LPC can be converted to propionyl coenzyme A by carnitine acyl transferase and then be metabolized to succinyl CoA by the enzymes: propionyl CoA carboxylase, methylmalonyl-CoA racemase and methylmalonyl-CoA mutase [23, 24]. It is likely that this conversion of LPC may produce a stimulation of the citric acid cycle flux [17]. In addition, the L-carnitine released by this reaction is available to buffer the acetyl CoA/free ratio which could conceivably increase pyruvate dehydrogenase activity [23]. This conversion of pyruvate to acetyl CoA may limit the degree of lactic acidosis and stimulate glucose oxidation which would provide more ATP to the ischemic/reperfused heart. The L-carnitine released may also enhance fatty acid oxidation thus preventing the accumulation of lipid intermediates which may be harmful to the ischemic myocardium. This overall stimulation of substrate utilization by LPC may be one of the mechanisms for its protective effect against myocardial ischemic/reperfusion injury. This metabolic action of LPC may be of even more importance in a situation where glucose utilization is impaired, such as the myocardium of the diabetic [25, 29]. The diabetic heart relies on fatty acids almost exclusively for its production of energy. This increased reliance on fatty acid oxidation by the diabetic heart occurs even though myocardial total carnitine levels are reduced [11, 30]. Chronic LPC therapy to diabetic rats may have been beneficial if it prevented the diabetes-induced decrease in carnitine content and improved myocardial glucose utilization, however, these experiments have not been performed.

Cardiac contractile function of diabetic hearts in response to changing LAFP was only slightly depressed in the present study. Only left ventricular systolic pressure and − dP/dt were decreased in untreated diabetic versus untreated control hearts. The magnitude of the cardiac depression is slightly less than that found in a previous study [31] using the same dose of streptozotocin and duration of diabetes. In addition, we had shown that 90 min of low flow ischemia resulted in greater damage in untreated diabetic hearts versus control hearts [9]. However, in the present study, the recovery of cardiac contractile performance between untreated control and diabetic hearts was similar. One possible explanation for this discrepancy between the present and previous study [9] was the ischemic coronary flow rates. In the previous study ischemic coronary flow rate was 1.0 ml/min for both control and diabetic rats. However, because of differences in heart weights between control and diabetic rats the ischemic flow rates for control and diabetic hearts was actually on a per gram dry wt basis 4.0 and 4.9 ml/min/g dry wt, respectively. In the present study coronary flow rates were adjusted for heart weight differences between control and diabetic rats, 1.18 ml/min for control and 0.85 ml/min for diabetic or 4.4 ml/min/g dry wt for both control and diabetic hearts. Tani and Neely [13] found that diabetic hearts exposed to zero-flow ischemia exhibited enhanced recovery of cardiac function with reperfusion. This decrease vulnerability of the diabetic heart to zero flow ischemia may be due to a depressed activity of the sarcolemmal Na^+/H^+ exchanger in the diabetic heart [12, 32]. Decreased activity of this exchanger is conceivably beneficial because it would limit the accumulation of calcium during reperfusion. These findings suggest that the diabetic heart may have decreased vulnerability to zero flow or very low flow ischemia, but when ischemic flow rate is increased this protection is lost.

The results of the present study provide further evidence that LPC is effective in protecting the ischemic/reperfused heart. Chronic treatment of LPC was particularly beneficial to the diabetic heart in improving the ability of the hearts to recover from a period of low flow ischemia and reperfusion, but had no effect on cardiac contractile performance under normoxic conditions.

Acknowledgements

This work was supported by grants from the National Institutes of Health DK 39200 and Sigma Tau Pharmaceuticals.

References

1. Barrett-Connor EL, Cohn BA, Wingard DL, Edelstein SL: Why is diabetes mellitus a stronger risk factor for fatal ischemic heart disease in women than in men? The Rancho Bernardo Study. JAMA 265: 627–631, 1991
2. Assmann G, Schulte H: Diabetes mellitus and hypertension in the elderly: Concomitant hyperlipidemia and coronary heart disease risk. Am J Cardiol 63: 33H–37H, 1989
3. Henning R, Lundman T: Swedish cooperative CCV study: a study of 2009 patients with acute myocardial infarction from twelve Swedish hospitals with coronary care units. I: A description of the early stage. ACTA Med Scand 198: 1–64, 1975
4. Kotler TS, Diamond GA: Myocardial ischemia in diabetic patients. Ann Intern Med 109: 678–678, 1988
5. Kannel WB: Role of diabetes in cardiac disease: conclusions from population studies. In: S Zonaraich (ed.) Diabetes and the Heart. Thomas, Springfield, IL, 1978, pp. 97–112
6. Hearse DJ, Steward DA, Green DG: Diabetes and the survival and recovery of the anoxic myocardium. J Mol Cell Cardiol 7: 397–415, 1975
7. Hearse DJ, Steward DA, Green DG: Myocardial susceptibility to ischemic damage: a comparative study of disease models of rats. Eur J Cardiol 75: 437–450, 1978
8. Ingebretsen CG, Moreau P, Hawelu-Johnson C, Ingebretsen WR: Performance of diabetic rat hearts: effects of anoxia and increased work. Am J Physiol 239: H614–H620, 1980
9. Paulson DJ, Kopp SJ, Peace DG, Tow JP: Improved postischemic recovery of cardiac pump function in exercised trained diabetic rats. J Appl Physiol 65: 187–193, 1988
10. Pieper GM, Gross GJ: Diabetes alters postischemic response to a prostacyclin mimetic. Am J Physiol 256: H1353–H1360, 1989
11. Feuvray D, Idell-Wenger JA, Neely JR: Effects of ischemia on rat myocardial function and metabolism in diabetes. Am J Physiol 44: 322–329, 1979

12. Khandoudi N, Bernard M, Cozzone P, Feuvray D: Intracellular pH and role of Na/H exchange during ischaemia and reperfusion on normal and diabetic rat hearts. Cardiovasc Res 24: 873–878, 1980
13. Tani M, Neely JR: Hearts from diabetic rats are more resistant to *in vitro* ischemia: possible role of altered calcium metabolism. Circ Res 62: 931–940, 1988
14. Paulson DJ, Traxler J, Schmidt M, Noonan J, Shug AL: Protection of the ischaemic myocardium by L-propionylcarnitine: effects on the recovery of cardiac output after ischaemia and reperfusion, carnitine transport and fatty acid oxidation. Cardiovasc Res 20: 536–541, 1986
15. Liedtke AJ, DeMaison L, Nellis SH: Effects of L-propionylcarnitine on mechanical recovery during reflow in intact hearts. Am J Physiol 255: H169–H176, 1988
16. Ferrari R, Ceconi C, Curello S, Pasini E, Visioli O: Protective effect of propionyl-L-carnitine against ischaemia and reperfusion-damage. Mol Cell Biochem 88: 161–168, 1989
17. Di Lisa F, Menabò R, Siliprandi N: L-Propionyl-carnitine protection of mitochondria in ischemic rat hearts. Mol Cell Biochem 88: 169–173, 1989
18. Leasure JE, Kordenat K: Effect of propionyl-L-carnitine on experimental myocardial infarction in dogs. Cardiovasc Drugs Ther 5 Suppl 1: 85–96, 1991
19. Aomine M, Nobe S, Arita M: Electrophysiologic effects of a short-chain acyl carnitine, L-propionylcarnitine, on isolated canine Purkinje fibers. J Cardiovasc Pharmacol 13: 494–501 1989
20. Aomine M, Arita M: Differential effects of L-propionylcarnitine on the electrical and mechanical properties of Guinea pig ventricular muscle in normal and acidic conditions. J Electrocardiol 20: 287–296, 1987
21. Aomine M, Arita M, Shimada T: Effects of L-propionylcarnitine on electrical and mechanical alterations induced by amphiphilic lipids in isolated Guinea pig ventricular muscle. Heart Vessel 4: 197–206, 1988
22. Di Biase M, Tritto M, Pitzalis MV, Favale S, Rizzon P: Electrophysiological evaluation of intravenous L-propionylcarnitine in man. Int J Cardiol 30: 329–333, 1991
23. Siliprandi N, Di Lisa F, Menabò R: Propionyl-L-carnitine: Biochemical significance and possible role in cardiac metabolism. Cardiovasc Drugs Ther 5 Suppl 1: 11–16, 1991
24. Hülsmann WC: Biochemical profile of propionyl-L-carnitine. Cardiovasc Drugs Ther 5 Suppl 1: 7–10, 1991
25. Wall SR, Lopaschuk GD: Glucose oxidation rates in fatty acid-perfused isolated working hearts from diabetic rats. Biochim Biophys Acta 1006: 97–103, 1989
26. Vary TC, Neely JC: A mechanism for reduced myocardial carnitine levels in diabetic animals. Am J Physiol 243: H154–H158, 1982
27. Shug A, Paulson D, Subramanian R, Regitz V: Protective effects of propionyl-L-carnitine during ischemia and reperfusion. Cardiovasc Drugs Ther 5 Suppl 1: 77–84, 1991
28. Packer L, Valenza M, Serbinova E, Starke-Reed P, Frost K, Kagan V: Free radical scavenging is involved in the protective effect of L-propionyl-carnitine against ischemia-reperfusion injury of the heart. Arch Biochem Biophys 288: 533–537, 1991
29. Randle PJ: Fuel selection in animals. Biochem Soc Trans 14: 799–806, 1986
30. Paulson DJ, Schmidt MJ, Traxler JS, Ramacci MT, Shug AL: Improvement of myocardial function in diabetic rats after treatment with L-carnitine. Metabolism 33: 358–363, 1984
31. Paulson DJ, Kopp SJ, Peace DG, Tow JP: Myocardial adaptation to endurance exercise training in diabetic rats. Am J Physiol 252: R1073–R1081, 1987
32. Pierce GN, Ramjiawan B, Dhalla NS, Ferrari R: Na^+-H^+ exchange in cardiac sarcolemmal vesicles isolated from diabetic rats. Am J Physiol 258: H255–H261, 1990

Molecular and Cellular Biochemistry **116**: 139–145, 1992.

Functional and metabolic effects of propionyl-L-carnitine in the isolated perfused hypertrophied rat heart

Roberto Motterlini, Michele Samaja, Massimo Tarantola, Rosella Micheletti and Giuseppe Bianchi
Istituto Scientifico San Raffaele, Dipartimento di Scienze e Tecnologie Biomediche dell'Università di Milano, and Prassis Sigma Tau Institute, Milano, Italy

Abstract

Aim of this study was to assess the effect of propionyl-L-carnitine (PLC), a naturally occurring derivative of L-carnitine, in cardiac hypertrophy induced by pressure overload in rats. The abdominal aorta was banded and the rats received one daily administration of PLC (50 mg/kg) or saline for four days. The hearts were excised 24 h after the last administration and were perfused retrogradely with oxygenated Krebs-Henseleit buffer containing 1.2 mM palmitate bound to 3% (w/v) albumin, 2.5 μM PLC and 25 μM L-carnitine. A saline-filled balloon was inserted into the left ventricle and the heart contractility was measured at three volumes of the balloon, corresponding to zero diastolic pressure and to increased volumes (110 and 220 μl) over the zero volume. At the end of the perfusion, the hearts were freeze-clamped, weighed and analyzed for adenine nucleotide and phosphocreatine (PCr) content by HPLC methods. No differences in the myocardial performance were found at zero diastolic pressure. In contrast, at high intraventricular volume, the maximal rate of ventricular relaxation was increased in PLC-treated with respect to saline-treated controls ($p < 0.05$). In addition, the increase of the end-diastolic pressure at increasing balloon volume was more marked in controls than in the PLC-treated hearts ($p < 0.02$). These data correlate well with the measured higher level of total adenine nucleotides ($p < 0.05$) and ATP ($p < 0.02$) in the PLC-treated hearts, while PCr was the same in both groups. Parallel experiments performed in the absence of palmitate in the perfusing media failed to show any effect of PLC. We conclude that PLC improves the diastolic function by increasing the fraction of energy available from fatty acid oxidation in the form of ATP. (Mol Cell Biochem **116:** 139–145, 1992)

Key words: isolated heart, propionyl-L-carnitine, palmitic acid, energetic metabolites

Introduction

L-Carnitine is essential in the transport of long-chain fatty acids from cytosol into the mitochondria [1, 2]. The mitochondrial inner membrane is not permeable to long-chain fatty acids preventing the access of these substrates to the fatty acid β-oxidation system, primary machinery for energy production in the cell. To overcome this barrier, L-carnitine combines with fatty acyl-CoA in the cytosol and shuttles the fatty acid into the mitochondrion by a reaction catalyzed by L-carnitine acyltransferase. Thus, reduction in cell carnitine levels results in depressed mitochondrial oxidation of long chain fatty acids. The consequent decrease in rate of

Address for offprints: M. Samaja, Dipartimento di Scienze e Tecnologie Biomediche, via Olgettina 60, 20132 Milano, Italy energy production is critical for those tissues whose function is heavily dependent on fatty acid oxidation, such as myocardium. Therefore, impaired fatty acid oxidation in the cardiomyopathies involving reduced myocardial level of L-carnitine [3–7], may further contribute to depress the cardiac function. Furthermore, the reduced utilization of long-chain fatty acids leads to the accumulation in the cytosol of potentially toxic substances that contribute to the pathogenesis of cardiac dysfunction [8, 9].

Propionyl-L-carnitine (PLC) is a naturally occurring derivative of L-carnitine to which the propionyl moiety seems to confer three potentially advantageous properties over L-carnitine, namely, a greater affinity for the cardiomyocyte sarcolemmal carrier [10], the ability to replenish mitochondria with intermediates of the citric acid cycle [11] and a greater positive inotropic activity demonstrated both in mechanical recovery after ischemia and in intact isolated hearts [10, 12–14].

Here, we tested the hypothesis that PLC improves the myocardial function secondary to specific metabolic adjustments. Aim of this study was: (1) to assess if free fatty acids are required for the cardiac action of PLC; (2) to study the relationship between changes in mechanical performance and high-energy phosphates content in the heart. We used the isolated perfused rat heart as experimental model to simulate the metabolic derangement due to carnitine deficiency. The hearts were made hypertrophic by pressure overload, a condition known to be associated with altered CoA and carnitine metabolism [15]. The suitability of this model to several biochemical and physiological measurements allows to relate metabolic changes to functional alterations. In addition, the response of the heart to PLC treatment is isolated from the response of the whole organism and can be studied in a greater detail.

Methods

Materials

PLC was synthesized at Sigma Tau Chemical Drug Dept. (batch 2649). Solutions for animal injection were prepared immediately before use dissolving PLC in saline and buffering pH to 7.4. The heart perfusion buffer contained 116 mM NaCl, 4.7 mM KCl, 1.2 mM KH_2PO_4, 0.5 mM EDTA, 1.2 mM Na_2SO_4, 28.5 mM $NaHCO_3$, 3 mM $CaCl_2$, 1.2 mM $MgCl_2$ and 16.6 mM glucose. Half of the hearts were perfused with 1.2 mM palmitate bound to 3% (w/v) bovine albumin added to the medium. All reagents were provided from Sigma Chemical (St. Louis, MO) and were analytical grade. In addition, 2.5 μM PLC and 25 μM L-carnitine were also present in the perfusion buffer.

Animals

Twenty male Wistar Kyoto rats (Charles River, Italy) weighing 200–250 g were used in this preliminary study. The animals were fed a standard diet (Altromin MT, Bolzano, Italy) containing approximately 55 nmol/g total L-carnitine. Cardiac hypertrophy was induced by banding the abdominal aorta above the renal arteries. The abdominal cavity was opened and a silver band (0.7 mm internal diameter) was fitted around the aorta. After 4 weeks recovery, the animals were randomly assigned to treatment consisting in one daily administration through an intraarterial catheter of either PLC (50 mg/kg) or saline (control group) for four days.

The studies were performed conforming to the guiding principles of the National Society for Medical Research.

Heart perfusion

Heart perfusion experiments were performed 24 h after the fourth administration of PLC or saline. Rats were anaesthetized with i.p. heparinized sodium thiopental (100 mg/kg). The chest was opened, the aorta was mounted onto a cannula and the perfusion was started retrogradely with the described buffer, pH 7.4, 37° C. The coronary flow rate was controlled by a peristaltic pump. The coronary sinus return was collected through an outflow cannula in the pulmonary artery. A latex balloon was introduced into the left ventricle through the left auricular appendage and connected to a pressure transducer to monitor end-diastolic pressure (EDP), left-ventricular developed pressure (LVDP), heart rate (HR), maximal rate of heart contraction ($+dP/dt_{max}$) and relaxation ($-dP/dt_{max}$).

The spontaneously contracting hearts were stabilized for 30 min at coronary flow rate = 15 ml/min, with the volume of the ventricular balloon adjusted to achieve EDP = 0.5 ± 0.5 Torr (V_0). At the end of the stabilization, the baseline values for the myocardial function were recorded and the intraventricular balloon was filled with accurately measured volumes of saline using a glass gas-tight 1 ml syringe (Hamilton Co, Reno, Nevada) and a micrometer. The selected volumes of the balloon were $V_0 + 110\,\mu L$ (V_{110}) or $V_0 + 220\,\mu l$ (V_{220}). The hearts were then allowed to stabilize and a complete series of measurements was taken again.

Metabolic measurements

The O_2 uptake (VO_2) was evaluated measuring the pO_2 in the arterial inflow and in the coronary sinus return (YSI mod. 5300 Oxygen Monitor, Yellow Springs Inc., OH). In the palmitate experiments, the myocardial content of high-energy phosphate compounds was also determined at the end of the perfusion. The hearts were freeze-clamped in liquid nitrogen, weighed and transferred to a tube containing 2 ml of 0.5 M cold perchloric acid. The content was homogenized using an OMNI 1000 (OMNI Int'l, Waterbury, Connecticut) operating at 20,000 RPM. After 15 min, the suspension was centrifuged, 0.5 ml of the supernatant was neutralized with 0.4 ml of 0.5 M KOH and 0.1 ml of 1 M KH_2PO_4, centrifuged again 30 min later, and filtered through $0.22\,\mu m$ pore size membrane (Nihon Millipore, Japan). ATP, ADP, AMP, creatine and creatine phosphate in myocardial extract were analyzed by HPLC. The equipment (Kontron Instruments, Milano, Italy) was composed of two mod. 420 pumps and a mod 432 UV/Vis detector set at 210 nm. The $3\,\mu m$ Supelcosil LC18 column (Supelco, Bellefonte, Pennsylvania) was equilibrated with 0.1 M KH_2PO_4 and 5 mM tetrabutylammonium sulphate, the sample ($20\,\mu l$) was injected and eluted using a composed gradient with a buffer containing 0.1 M KH_2PO_4, 4 mM tetrabutyl-ammonium sulphate and 90% (v/v) CH_3CN. The analysis was completed in 25 min and data were analyzed with the Kontron's dedicated software.

Statistics

All values are expressed as mean ± SEM. Between group comparison was by two-tailed unpaired Student's t-test. The significance was defined as a probability equal or less than 0.05.

Results

The described aortic constriction procedure induced a marked increase of the heart size with respect to normal (50%), indicating the presence of cardiac hypertrophy. As described earlier [16], this condition was associated with a reduction of left ventricle total carnitine content, that was decreased by about 40% with respect to control sham-operated animals.

Two series of heart perfusion experiments were performed. In the first series, the hearts were perfused with Krebs-Henseleit buffer with no palmitate added, and the two groups (PLC-treated and controls, $n = 5$ for each) were compared at V_0 and V_{110}. In the second series, ($n = 10$) buffer contained 1.2 mM palmitate bound to 3% (w/v) albumin, and measurements were made at three volumes of the balloon (0, 110, and $220\,\mu l$). Baseline data taken at V_0 for the two series are shown in Table 1. In both series, no differences were detected at V_0 between the controls and the PLC-treated hearts for any of the parameters evaluated including the double product LVDP × HR, index of the heart systolic function. Coronary pressure as well as contractility indexes were different in the two series of experiments probably because of the effects of fatty acid-bound albumin [17].

Figure 1 shows the diastolic function (EDP and $-dP/dt_{max}$) monitored at V_{110} in all the groups. No effect of PLC treatment was detected in the absence of palmitate. In contrast, in the presence of palmitate, EDP increased less and the maximal rate of relaxation ($-dP/dt_{max}$) was higher ($p < 0.05$ for both) in the PLC-treated hearts than in controls. The improvement of the diastolic function was not associated to higher systolic performance as shown by the values of LVDP × HR and $+dP/dt_{max}$.

In the second series of the experiments carried out in the presence of palmitate, more marked differences were detected between PLC-treated and control hearts when balloon volume was increased by $220\,\mu l$ over V_0 (V_{220}): EDP was 53.2 ± 5.0 vs 82.2 ± 8.4 Torr ($p < 0.02$) and $-dP/dt$ was 3060 ± 202 vs 2364 ± 184 Torr/s ($p < 0.05$) in PLC and saline groups, respectively. Again, the improvement of the diastolic function was not associated with higher mechanical work: LVDP × HR was 35.4 ± 3.0 vs 29 ± 3.6 Ktorr/min (NS) and $+dP/dt_{max}$ was 4510 ± 292 vs 4242 ± 194 Torr/s (NS) in PLC-treated and control hearts, respectively.

Figure 2 shows [ATP], [total adenine nucleotides] (TAN) and [PCr] in PLC-treated and control hearts perfused with palmitate containing buffer. All the phosphate high-energy compounds were significantly higher in the PLC-treated hearts than in controls. The higher [ATP] was accompanied by higher [TAN]. The energetic charge, i.e. the ratio ([ATP] + $^1/_2$[ADP])/([ATP] + [ADP] + [AMP]), was not significantly different in the two groups (0.78 ± 0.02 and 0.84 ± 0.02 in the controls and PLC-treated hearts, respectively) as well as the [PCr]/([PCr] + [Cr]) ratio (0.26 ± 0.05 and 0.36 ± 0.04).

Discussion

The present results indicate that PLC improves the diastolic performance of isolated hearts from rats with myocardial hypertrophy induced by pressure overload. The improvement was evident at high energy demand of the left ventricle, i.e, at high EDP, and was dependent on the presence of palmitate in the perfusion medium.

The hearts that were obtained from animals treated with PLC and were perfused with palmitate, displayed lower EDP for a given filling volume and faster relaxation rate compared to control hearts from saline-treated animals. PLC had no effect in hearts perfused at EDP = 0 Torr, i.e., at low energy demand. It can thus be inferred that PLC becomes effective when the energy demand of the left ventricle is increased and can be partly met through an increase in fatty acid oxidation. This is in essential agreement with the observed significant effect of EDP and VO_2 on the rate of palmitate oxidation [18] and with the reduced oxidation of fatty acids observed in cardiac hypertrophy caused by pressure overload [19].

In contrast to the marked effect on the diastolic performance, the systolic performance was not affected by PLC. However, as previously shown [16], in an *in vivo* situation the improved diastolic filling resulting from the lower EDP and the higher $-dP/dt_{max}$ might conceivably affect also the cardiac output.

PLC treated hearts display higher level of [TAN], [ATP] and [PCr] than controls. Associating metabolic and functional measurements helped to attribute the better diastolic distensibility of PLC-treated hearts to the higher level of high energy compounds. Irrespectively of the treatment, the sum [ATP] + [PCr] is correlated negatively with EDP ($r = -0.93$, $p < 0.001$) and positively with $-dP/dt_{max}$ ($r = 0.77$, $p < 0.05$) (Fig. 3), indicating that the mechanical effect of PLC is likely mediated by its metabolic effects. However, our data can not rule out a direct effect of PLC on the passive stiffness of the left ventricular wall. In fact, it was recently demonstrated that inhibition of fatty acid oxidation with 2-tedra-decylglycidic acid produced left ventricular hypertrophy and increased the left chamber stiffness constant [4, 20]. The stimulation of fatty acid oxidation by PLC, may have the opposite effect on diastolic function, improving it when depressed.

The higher [ATP] in the cell favors both the uptake of Ca^{++} into the sarcoplasmic reticulum at the end of the contraction, improving the active relaxation of the myocardium, and the total force output by interacting with the myosin-actin reaction. The finding that the [ATP] + [PCr] sum is related to diastolic function while apparently unrelated to systolic performance, is consistent with the greater sensitivity of diastolic function to myocardial levels of high-energy phosphate compounds.

Table 1. Baseline measurements taken at volume of the intraventricular balloon corresponding to end-diastolic pressure = 0 Torr (V_0) in the hearts (control and PLC-treated) perfused with Krebs-Henseleit buffer without and with 1.2 mM palmitate added

	No palmitate		1.2 mM palmitate	
	Controls	PLC-treated	Controls	PLC-treated
Heart weight, g	2.2 ± 0.2	2.2 ± 0.4	2.3 ± 0.4	2.3 ± 0.1
Heart rate, min^{-1}	212 ± 19	231 ± 19	217 ± 18	240 ± 10
Coronary pressure, Torr	65 ± 7	72 ± 9	90 ± 10	80 ± 9
End-diastolic pressure, Torr	0.7 ± 0.4	1.0 ± 0.2	0.1 ± 0.3	0.2 ± 0.2
Developed pressure, Torr	173 ± 14	168 ± 27	110 ± 10	111 ± 13
$+dP/dt_{max}$, Torr/s	5350 ± 550	5010 ± 580	3280 ± 270	3090 ± 330
$-dP/dt_{max}$, Torr/s	3360 ± 330	3410 ± 370	2220 ± 200	2200 ± 210
LVDP × HR, Torr/min/10^{-3}	36.6 ± 4.1	37.9 ± 5.3	24.1 ± 3.3	26.8 ± 3.9
O_2 uptake, μmoles/g/min	5.9 ± 0.6	5.7 ± 0.7	4.1 ± 0.4	4.0 ± 0.3

Data are mean ± standard error. PLC, propionyl-L-carnitine; $+dP/dt_{max}$, maximal rate of contraction; $-dP/dt_{max}$, maximal rate of relaxation; LVDP × HR, index of the heart systolic function.

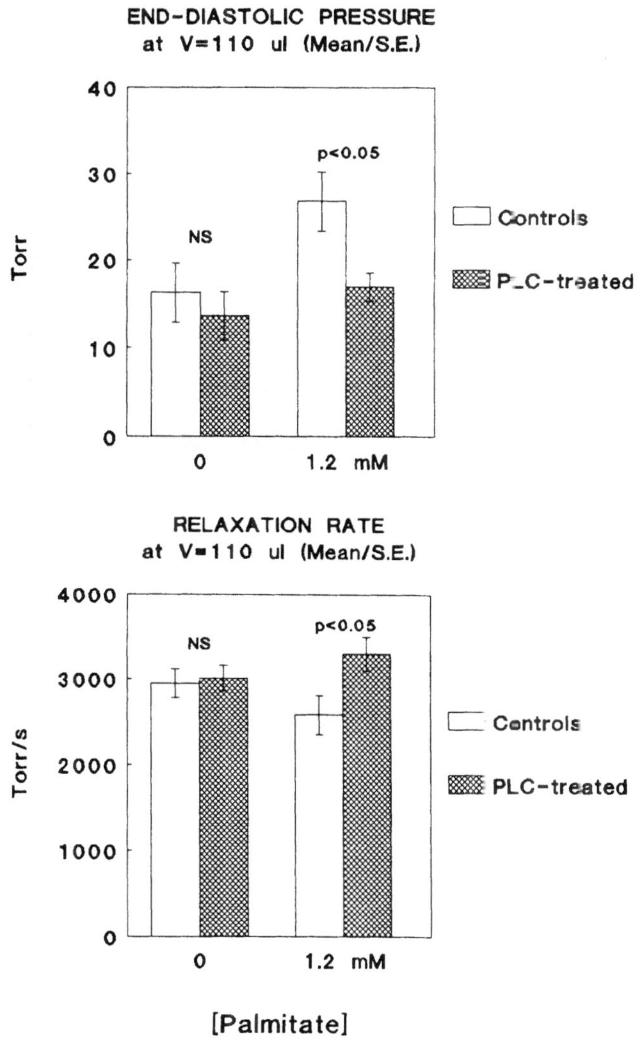

Fig. 1. Diastolic performance of the hearts perfused with Krebs-Henseleit buffer without and with 1.2 mM palmitate added, in controls and PLC-treated hearts (n = 5 for each group). End-diastolic pressure (upper panel) and relaxation rate (lower panel) were monitored at volume of the balloon increased by 110 μl (V_{110}) over the volume corresponding to end-diastolic pressure = 0 Torr. Data are mean ± standard error (S.E.). The significance tests (Student's unpaired t-test) refer to the comparison control vs PLC-treated hearts. PLC, propionyl-L-carnitine.

In fact, anomalies in diastolic relaxation rather than systolic performance are usually considered an earlier and more sensitive marker of a deficient energy state in heart failure [21].

The mechanism responsible for the PLC-driven increase of [ATP], and possibly [PCr], in the myocardial cell, appears mediated by increased [TAN] rather than increased energetic charge, which was virtually unchanged. It is probable that the larger availability of energy in the form of long-chain fatty acids increases oxidative phosphorylation and thus total production of

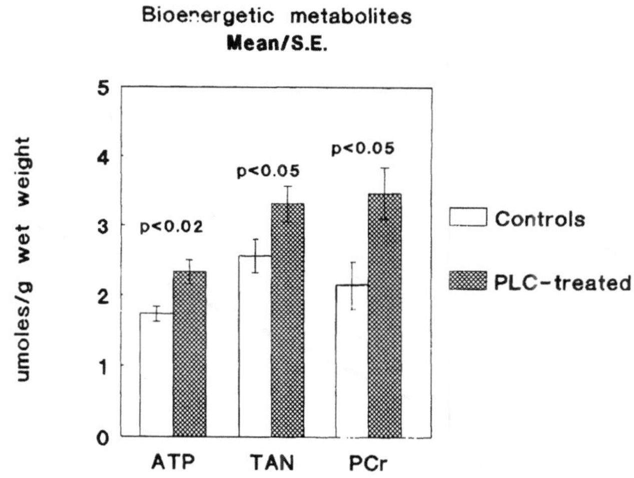

Fig. 2. Tissue level of ATP, total adenine nucleotides (TAN) and phosphocreatine (PCr) at the end of the perfusion with buffer containing 1.2 mM palmitate bound to 3% (w/v) albumin in control and PLC-treated hearts. Data are presented as mean ± standard error (S.E.), n = 5 for each group. PLC, propionyl-L-carnitine.

high energy compounds. Indeed, in the absence of palmitate no effect of PLC was observed, strongly suggesting a specific fatty acid-dependent effect of PLC.

It should be pointed out that the role of high energy phosphates in the pathogenesis of heart failure is controversial [22, 23]. The recent finding that significantly lower myocardial high energy phosphates (−20%) were detected at 2 and 4 weeks, but not at 6 weeks after aortic banding in the rat [24] adds uncertainty on the importance of ATP and PCr. Therefore, the tissue high energy phosphate level may not be the most appropriate parameter to study the relationship between the overall energy production by the heart and its contractile status. In fact during hypoxia, the myocardial high energy phosphate content decreases to much less extent than the high energy phosphate turnover and the changes of this last parameter paralleled much closer the contractile failure occurring under this condition [25].

In conclusion, the findings described suggest that the cardiac activity of PLC is linked to its metabolic effect, i.e., to the stimulation of fatty acid oxidation with consequent preservation of TAN and increase in ATP and PCr production. The described approach appears feasible to study the correlations existing between the myocardial function and the metabolic derangements in an induced cardiac hypertrophy model. Although the exact mechanism by which the high energy compounds are related to the myocardial function needs clarifying, it appears that PLC has an overall beneficial effect on both the function and metabolism of the isolated hyper-

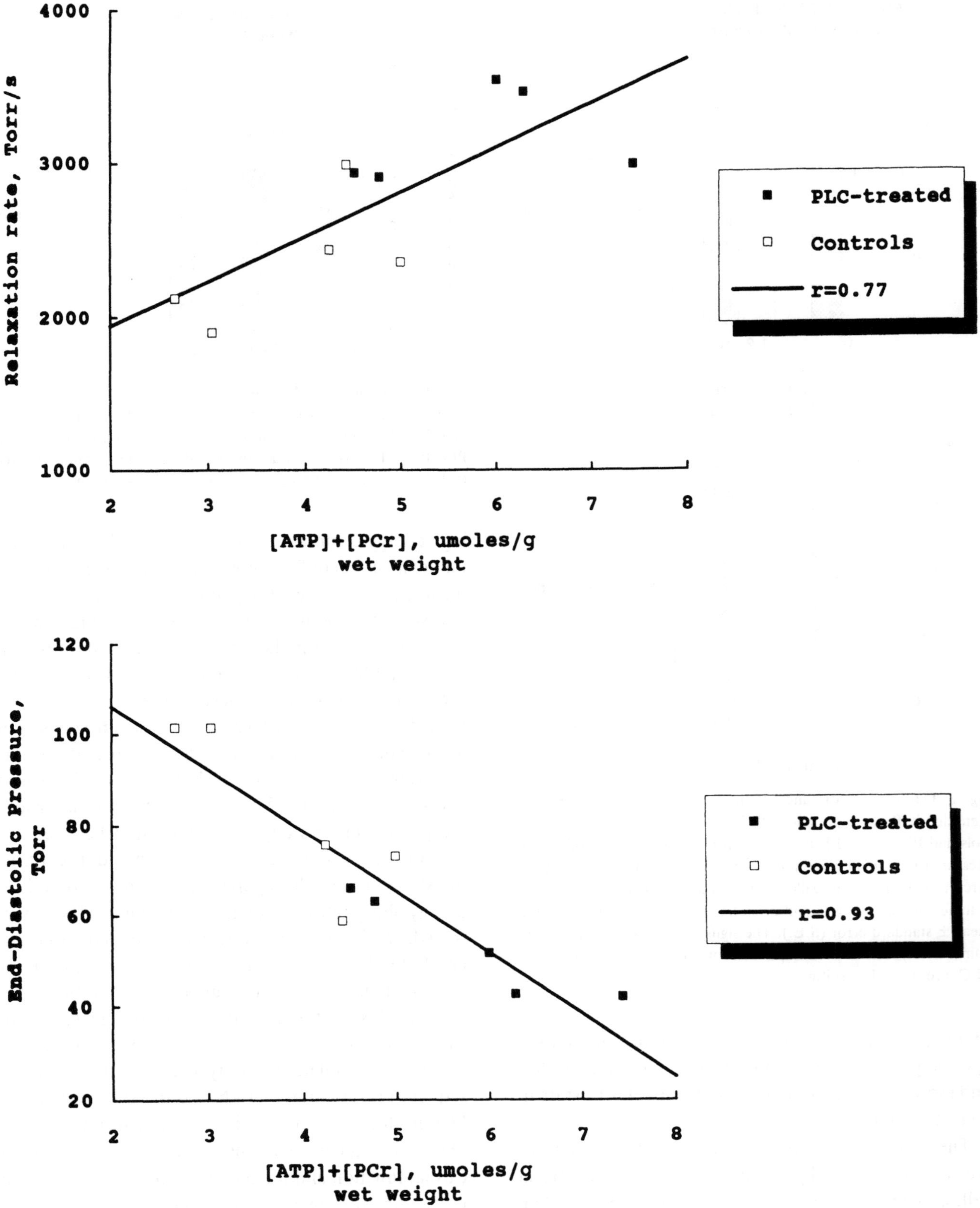

Fig. 3. Effect of the sum [ATP] + [PCr] on end-diastolic pressure (lower panel) and maximal relaxation rate ($-dP/dt_{max}$, upper panel) in PLC-treated (■) and control (□) hearts. The effect seems unrelated to the PLC treatment. The best fit line was evaluated by least square procedures. PLC, propionyl-L-carnitine.

trophied heart. As shown elsewhere [16], the findings described here on isolated heart are consistent with the cardiac stimulatory effect of PLC in conscious rats with a pressure overload heart hypertrophy. Further extension of these studies to other models is sought.

Acknowledgements

Roberto Matterlini is a recipient of a grant from Fondazione San Romanello del Monte Tabor, Istituto Scientifico San Raffaele. This work was supported by a grant from the Target Project Biotechnology and Bioinstrumentation, CNR, Italy.

References

1. Bremer J: Carnitine: metabolism and function. Physiol Rev 63: 1420–1480, 1983
2. Bieber LL: Carnitine. Ann Rev Biochem 57: 261–283, 1988
3. Regitz V, Shug AL, Fleck E: Defective myocardial carnitine metabolism in congestive heart failure secondary to dilated cardiomyopathy and to coronary, hypertensive and valvular heart disease. Am J Cardiol 65: 755–760, 1990
4. Bressler R, Gay R, Copeland JG, Bahl JJ, Bedotto J, Goldman S: Chronic inhibition of fatty acid oxidation: new model of diastolic dysfunction. Life Sci 44: 1897–1906, 1989
5. Spagnoli LG, Corsi M, Villaschi S, Palmieri G, Maccari F: Myocardial carnitine deficiency in acute myocardial infarction. Lancet i: 1419–1420, 1982
6. Suzuki Y, Masamura Y, Kobayashi Y, Yamazaki N, Harada Y, Osawa M: Myocardial carnitine deficiency in chronic heart failure. Lancet i: 116, 1982
7. Editorials: Carnitine deficiency. Lancet 335: 631–633, 1990
8. Corr PB, Saffitz JE, Sobel BE: Lysophospholipids, long-chain acylcarnitines and membrane dysfunction in the ischemic heart. Basic Res Cardiol 82, Suppl 1: 235–243, 1987
9. Corr PB, Creer MH, Yamada KA, Saffitz JE, Sobel BE: Prophylaxis of early ventricular fibrillation by inhibition of acylcarnitine accumulation. J Clin Invest 83: 927–936, 1989
10. Paulson DJ, Traxler J, Schmidt M, Noonan J, Shug AL: Protection of the ischaemic myocardium by L-propionylcarnitine: effects on the recovery of cardiac output after ischaemia and reperfusion, carnitine transport, and fatty acid oxidation. Cardiovasc Res 20: 536–541, 1986
11. Davis EJ, Spyderold O, Bremer J: Pyruvate carboxylase and propionyl-L-CoA carboxylase as anaplerotic enzymes in skeletal muscle mitochondria. Eur J Biochem 110: 255–262, 1980
12. Liedtke AJ, De Maison L, Nellis SH: Effects of propionyl-L-carnitine on mechanical recovery during reflow in intact hearts. Am J Physiol 255: H169–H176, 1988
13. Ferrari R, Ceconi C, Curello S, Pasini E: Protective effect of propionyl-L-carnitine against ischaemia and reperfusion damage. Mol Cell Biochem 88: 161–168, 1989
14. Pasini E, Ceconi C, Curello S, Cargnoni A, Ferrari R: Effect of acute and chronic treatment with propionyl-L-carnitine on mechanical performance of isolated heart. J Mol Cell Cardiol 22: S113, 1990
15. Reibel DK, Uboh CE, Kent RL: Altered coenzyme A and carnitine metabolism in pressure-overload hypertrophied hearts. Am J Physiol 244: H839–H843, 1983
16. Yang XP, English E, Micheletti R, Benatti P, Rossi R, Giacalone G, Bianchi G: Effect of propionyl-L-carnitine (PLC) on cardiac performance in awake WKY rats with cardiac hypertrophy (Abstr). FASEB J 5: A712, 1991
17. Van Beek JHGM, Bouma P, Westerhof N: Coronary resistance increase by nondefatted albumin in saline-perfused rabbit hearts. Am J Physiol 259: H1606–H1608, 1990
18. Ascuitto RJ, Ascuitto NT, Chen V, Downing SE: Ventricular function and fatty acid metabolism in neonatal piglet heart. Am J Physiol 256: H9–H15, 1989
19. Kagaya Y, Kanno Y, Takeyama D, Ishide N, Maruyama Y, Takahashi T, Ido T, Takishima T: Effects of long-term pressure overload on regional myocardial glucose and free fatty acid uptake in rats. A quantitative autoradiographic study. Circulation 81: 1353–1361, 1990
20. Litwin SE, Raya TE, Gay RE, Bedotto JB, Bahl JJ, Anderson PG, Goldman S, Bressler R: Chronic inhibition of fatty acid oxidation: new model of diastolic dysfunction. Am J Physiol 258: H51–H56, 1990
21. Grossman W: Diastolic dysfunction and congestive heart failure. Circulation 81 (suppl III): III-1, 1990
22. Katz AM: Cardiomyopathy of overload: a major determinant of prognosis in congestive heart failure. New Engl J Med 322: 100–110, 1990
23. Braunwald E: Pathophysiology of heart failure. In: E Braunwald (ed.) Heart Disease. Saunders, Philadelphia, 1988, pp 426–448
24. Tanonaka K, Sanbe A, Takeo S: Myocardial energy metabolism of pressure-overloaded rats *in vivo* state. J Molec Cell Cardiol 23: S31–S32, 1991
25. Bittl JA, Balshi A, Ingwall JS: Contractile failure and high-energy phosphate turnover during hypoxia: 31P-NMR surface coil studies in living rat. Circ Res 60: 871–878, 1987

Molecular and Cellular Biochemistry **116**: 147–153, 1992.

L-propionylcarnitine and myocardial performance in stunned porcine myocardium

L.M.A. Sassen, D.J. Duncker, A. Hogendoorn, E.O. McFalls, R. Krams, K. Bezstarosti,[1] J.M.J. Lamers[1] and P.D. Verdouw
Laboratory for Experimental Cardiology (Thoraxcenter) and [1] Department of Biochemistry, Erasmus University Rotterdam, Rotterdam, The Netherlands

Abstract

Recently, we showed that L-propionylcarnitine did not affect recovery of regional contractile function of porcine myocardium subjected to 1 h of low-flow ischemia followed by 2 hr of reperfusion. In that study, ischemia may have been too severe and/or the duration of reperfusion too short to detect a beneficial effect of the compound. Therefore, in the present study we investigated the effects of saline (control group; n = 14) or pretreatment with L-propionylcarnitine (3 days of 50 mg/kg p.o. b.i.d. + 50 mg/kg i.v. prior to the experiment; n = 13) on recovery of regional contractile function of the myocardium in open-chest anesthetized pigs, subjected to two cycles of 10 min of left anterior descending coronary artery (LADCA) occlusion, each followed by 30 min of reperfusion. In the control animals, at the end of the second reperfusion period, systemic vascular resistance had increased by 18%, which, however, was not observed in the L-propionylcarnitine-treated pigs. In the control group, during the first occlusion, systolic segment length shortening (SSLS) of the LADCA-perfused area decreased from 18.5 ± 5.5% to − 3.7 ± 3.2%. After 30 min of reperfusion, SSLS of the LADCA-perfused area had only partially recovered to 6.2 ± 5.9%. During the second occlusion-reperfusion cycle similar values for SSLS were observed. In the treated animals, SSLS of the LADCA-perfused area was slightly improved after the second occlusion-reperfusion cycle (p = 0.056). This effect did not result in an overall improvement in cardiac pump function. We conclude that in a model of myocardial stunning, L-propionylcarnitine prevents systemic vasoconstriction in response to ischemia and reperfusion and, possibly as a result of this effect, slightly ameliorates post-ischemic hypofunction. (Mol Cell Biochem **116:** 147–153, 1992)

Key words: stunned myocardium, systemic hemodynamics, regional contractile function, L-propionylcarnitine, pig

Introduction

L-propionylcarnitine, the naturally occurring derivate of L-carnitine, has been shown to enhance fatty acid utilization and to protect ischemic myocardium in the isolated working rat heart preparation [1, 2]. Additionally, a reduction in infarct size has been demonstrated in dogs by L-propionylcarnitine alone or in combination with the calcium-antagonist tiapamil [3]. Moreover, Liedtke *et al.* [4] demonstrated reversal of myocardial stunning by L-propionylcarnitine in a porcine model of mild myocardial ischemia. Similar improvement in regional contractility was also found with propionylcarnitine taurine [5].

Address for offprints: P.D. Verdouw, Laboratory for Experimental Cardiology, Thoraxcenter, Erasmus University Rotterdam, P.O. Box 1738, 3000 DR Rotterdam, The Netherlands

Recently, we investigated the effects of L-propionylcarnitine on changes in regional contractile function, high energy phosphates and calcium uptake by the sarcoplasmic reticulum after 60 min of low-flow myocardial ischemia in anesthetized swine but were unable to show a beneficial effect of the compound on these parameters, during the early phase (2 hr) of reperfusion [6]. Since in that study ischemia may have been too severe to detect a beneficial effect, we now wished to investigate the effects of L-propionylcarnitine on recovery of regional contractile function, in an open-chest porcine model of myocardial stunning. In this model, two cycles of 10 min of complete coronary artery occlusion each followed by 30 min of reperfusion result in approximately a 60% loss of myocardial contractile function, in the absence of myocardial necrosis [7].

One of the explanations for the beneficial effects of L-propionylcarnitine could be the repletion of intracellular carnitine. Therefore we also measured free carnitine and short chain acylcarnitine in transmural biopsies of the myocardium.

Materials and methods

General

Cross-bred Landrace × Yorkshire pigs (HVC, Hedel, The Netherlands) or either sex (25–45 kg, n = 27) were used to study the effects of no treatment (n = 14) or pretreatment with L-propionylcarnitine (n = 13) on changes in systemic hemodynamics and myocardial contractile function, induced by two sequences of myocardial ischemia (10 min) and reperfusion (30 min). L-propionylcarnitine pretreatment was started three days prior to the experiment and consisted of oral administration of 50 mg/kg L-propionylcarnitine twice a day (total dose: 100 mg/kg/day). On the day of the experiment the animals received 50 mg/kg/day orally in one dose two hours prior to the induction of anesthesia.

Preparation

Animals were sedated with intramuscular ketamine (20 mg/kg; AUV, Cuijk, The Netherlands), anesthetized with intravenous metomidate (5 mg/kg; Janssen Pharmaceutica, Beerse, Belgium), intubated and connected to a ventilator for intermittent positive pressure ventilation with a mixture of O_2 and N_2 (1:2). Respiratory rate and tidal volume were set to keep arterial blood gases, measured with an ABL3 (Radiometer, Copenhagen, Denmark) within the normal range: 7.35 $<$ pH $<$ 7.45; 35 mmHg $<$ pCO_2 $<$ 45 mmHg and 100 mmHg $<$ pO_2 $<$ 160 mmHg. Catheters (7F) were placed in the superior caval vein for an intravenous infusion of 10–15 mg/kg/h sodium pentobarbitone (Apharma, Arnhem, The Netherlands), haemaccel (Behringwerke A.G., Marburg, Germany) to replace blood withdrawn during sampling, and for administration of L-propionylcarnitine. The femoral arteries were cannulated with 7F catheters, which were positioned in the descending aorta for measurement of central aortic blood pressure and for withdrawal of blood samples for blood gas analysis. A 7F Sensodyn micromanometer-tipped catheter (B. Braun Medical B.V., Uden, The Netherlands) inserted via the left carotid artery, was used to measure left ventricular pressure and its first derivative (LV dP/dt). Rectal temperature was monitored throughout the experiment and maintained near 37° C with external heating pads.

In the animals that had been pretreated with L-propionylcarnitine, an intravenous bolus of 50 mg/kg L-propionylcarnitine was then administered over a 20 min period. Pancuronium bromide (4 mg; Organon Teknika B.V., Boxtel, The Netherlands) was given intravenously, and following a midline thoracotomy, the heart was suspended in a pericardial cradle. The left mammary vessels were ligated and the second left rib was removed for ease of further instrumentation. The adventitia surrounding the aorta was dissected free and an electromagnetic flow probe (15 mm; Skalar, Delft, The Netherlands) was placed for measurement of ascending aortic blood flow. On the proximal third of the left anterior descending coronary artery (LADCA), a small segment was dissected free of its adventitia for subsequent positioning of the arterial clamp.

Regional myocardial contractile function

To measure regional segment length changes in the area perfused by the LADCA, a pair of ultrasonic crystals (Triton, Technology Inc., San Diego, CA, USA) was placed into the subendocardium, approximately 10 mm apart. Another pair of crystals was placed in the subendocardium of the area perfused by the left circumflex coronary artery (non-LADCA), remote from the myocardium perfused by the LADCA. From the tracings the lengths at end-diastole (EDL; time of onset of positive LV dP/dt) and end-systole (ESL; time of peak negative LV dP/dt, reference [8]) were obtained and systolic segment length shortening (SSLS) calculated as:

$$SSLS\ (\%) = \frac{EDL - ESL}{EDL} \times 100$$

Experimental protocol

Following a 30–45 min stabilization period, baseline recordings were made of aortic and left ventricular pressures, LV dP/dt, ascending aortic blood flow and myocardial segment length in the two myocardial regions. All recordings were made at 50 mm/s. Furthermore, in 6 of the control animals and all the L-propionylcarnitine-treated pigs blood samples were drawn for the measurement of free and short chain fatty acyl carnitine. In the L-propionylcarnitine-treated animals another plasma sample was drawn after administration of the bolus of the compound. In 8 animals from the control group and 6 animals from the L-propionylcarnitine-treated group transmural myocardial biopsies were taken from the area perfused by the LADCA as well as from the area perfused by the left circumflex coronary artery, remote from the LADCA perfused area. Using a small clamp, the LADCA was then completely occluded for 10 min (occlusion-1) and reperfused for 30 min (reperfusion-1). This sequence was then repeated (occlusion-2 and reperfusion-2). At the end of reperfusion-2, recordings were repeated and biopsies taken from the LADCA- and the non-LADCA-perfused area. Furthermore, blood samples were drawn for determination of free and short chain acylcarnitine.

Carnitine and short-chain acylcarnitine levels

Left ventricular biopsies were homogenized in $HClO_4$, centrifuged, and the supernatant neutralized (as described above) for the determination of carnitine and short-chain acylcarnitine levels. The neutralized $HClO_4$ extracts and saline-treated plasma's were directly analyzed for free carnitine and short-chain acyl carnitine by the radioisotope procedure [9, 10]. Briefly, for determination of total carnitine, in one aliquot of the extract, short-chain acylcarnitine was first hydrolyzed to free carnitine by incubating in 0.1 M KOH at 37°C for 30 min and subsequently neutralized with 0.5 2-[4--(Hydroxyethyl)-1-piperazinyl ethansulfonic acid (HEPES). The other aliquot (not hydrolyzed) was used for determination of free carnitine. Carnitine was converted enzymatically with [^{14}C-]acetylCoA to [^{14}C-]acetylcarnitine. Unreacted [^{14}C-]acetylCoA was separated from the labeled carnitine derivative by passing through a Dowex AG1-X8 anion exchange resin (BioRad Laboratories, Richmond, VA). A sample of the [^{14}C]acetylcarnitine containing eluate was mixed with 8 ml Instagel and counted for radioactivity.

Protein

The protein content of SR was estimated according the method described by Lowry *et al.* [11]. In the needle biopsies, protein was estimated in the crude pellet obtained after centrifugation of the homogenate acidified with $HClO_4$. After dissolution of this pellet in 0.1 M KOH, protein was estimated with the Coomassie Briliant Blue assay [12] obtained from Bio-Rad (Bio-Rad Laboratories, Munich, Germany). For both methods bovine serum albumin was used as the standard.

Drugs

L-propionylcarnitine HCl (50 mg/kg) and $NaHCO_3$, in an equimolar ratio, were dissolved in 20 ml distilled water (pH of the final solution 6.8–7.3).

Data presentation and statistical analysis

Results have been expressed as arithmetic means ± standard errors of the mean. Changes were tested for significance at the $p < 0.05$ level (two-tailed) by analysis of variance (repeated measurements) followed by the Student-Newman-Keul test. Differences in the changes in myocardial blood flows of the LADCA- and non-LADCA-perfused myocardium were compared using the unpaired student's t-test.

Results

Systemic hemodynamics (Fig. 1)

During the first 10 min coronary artery occlusion mean arterial blood pressure in the control animals decreased

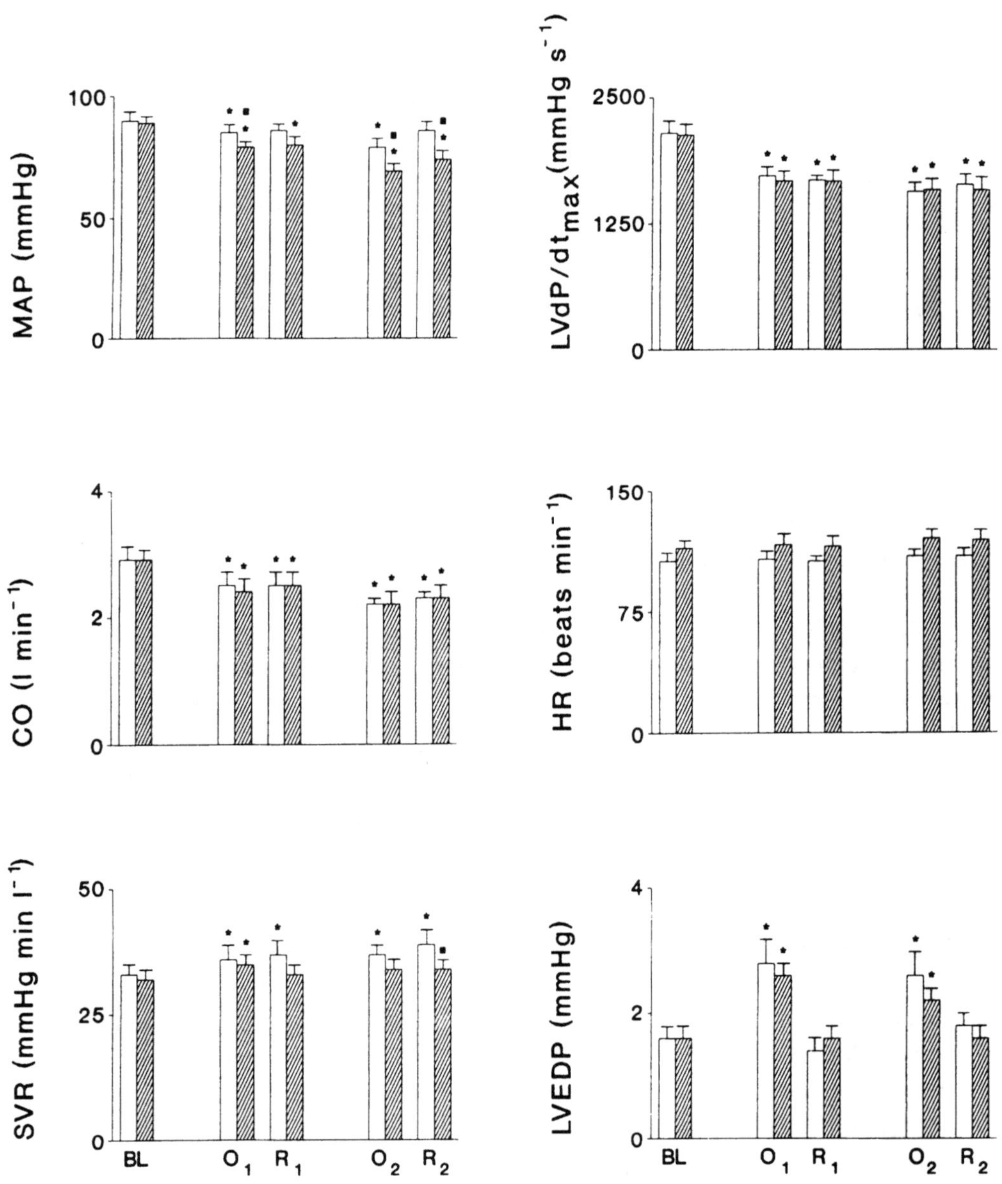

Fig. 1. Systemic hemodynamics in non-treated (open bars) and L-propionylcarnitine-treated (hatched bars) anesthetized pigs before (BL) and after two cycles of 10 min occlusion (O) of the left anterior descending coronary artery each followed by 30 min of reperfusion (R). MAP = mean arterial pressure, LV dP/dt_{max} = maximum rate of rise in left ventricular pressure, CO = cardiac output, HR = heart rate, SVR = systemic vascular resistance, LVEDP = left ventricular end diastolic pressure. ★ $p < 0.05$ versus baseline, ■ L-propionylcarnitine-induced changes versus baseline are significantly different from changes versus baseline in control swine.

from 90 ± 14 mmHg to 85 ± 13 mmHg (NS), while cardiac output fell from 2.9 ± 0.2 L/min to 2.4 ± 0.2 L/min ($p < 0.05$). Consequently, calculated systemic vascular resistance increased by approximately 10%. Heart rate did not change from its baseline value of 106 ± 15 beats/min, implying that the reduction in cardiac output was solely due to the attenuation of stroke volume. Finally, LV dP/dt_{max} decreased by 20%, while left ventricular end-diastolic pressure was elevated from 8 ± 1 mmHg to 14 ± 2 mmHg ($p < 0.05$). During the first reperfusion period, all ischemia-induced changes in hemodynamic parameters were maintained, except for left ventricular end-diastolic pressure which fully recovered. During the second cycle of coronary artery occlusion and reperfusion, changes in a similar direction as in the first cycle were observed although they were in general of a smaller dimension.

L-propionylcarnitine had no effect on systemic hemodynamics under baseline conditions (Fig. 1). Pretreatment with the drug, however, resulted in a drop in mean arterial blood pressure of 15% during the 2 cycles of occlusion and reperfusion, as an increase in systemic vascular resistance was prevented.

Regional myocardial segment length changes (Fig. 2)

The non-LADCA perfused segment showed a tendency towards enhanced shortening during occlusion from 14.9 ± 1.2% to 16.9 ± 1.5% ($p > 0.05$). During the first occlusion, the segment perfused by the LADCA became dyskinetic as SSLS decreased from 18.5 ± 1.5% to −3.7 ± 0.9% and recovered to 6.2 ± 1.6% after 30 min of reperfusion. During the second cycle of occlusion and reperfusion similar values for SSLS were observed.

Baseline value of SSLS of the non-LADCA area was significantly lower in the L-propionylcarnitine-treated animals (Fig. 2). L-propionylcarnitine induced a slight improvement in regional SSLS of the area perfused by the LADCA after 2 cycles of occlusion and reperfusion ($p = 0.056$).

Carnitine levels in plasma and myocardium (Tables 1 and 2)

In the control group plasma levels of carnitine and short chain acyl carnitine remained unchanged during the experimental period (Table 1). Two cycles of occlusion and reperfusion caused free carnitine of the stunned myocardium to decrease by 20%, while the level of short chain acyl carnitine was not affected (Table 2). Pretreatment with L-propionylcarnitine caused marked increases in plasmalevels of carnitine and short chain acylcarnitine (Table 1). After infusion of the bolus these

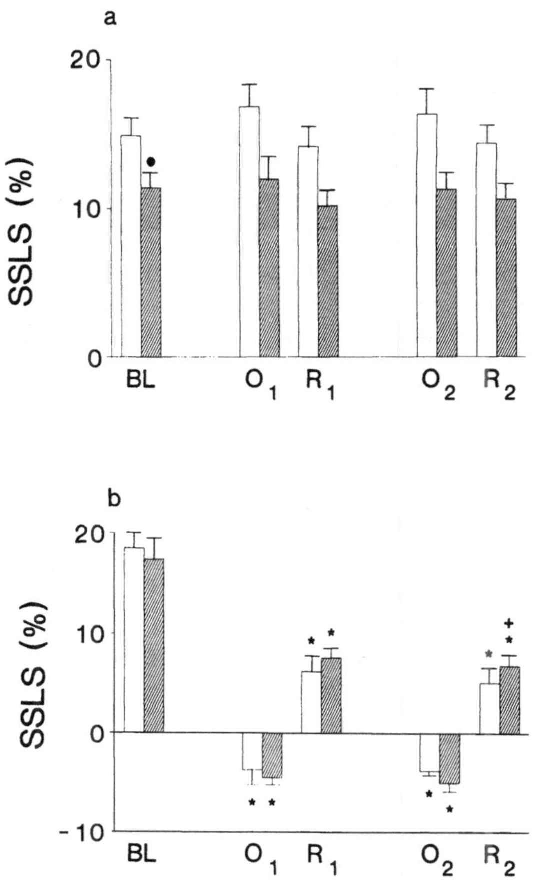

Fig. 2. Systolic segment length shortening (SSLS) of the area remote from that perfused by the left anterior descending coronary artery (non-LADCA, panel a) and of that perfused by the LADCA (panel b) in non-treated (open bars) and L-propionylcarnitine-treated (hatched bars) anesthetized pigs before (BL) and after two cycles of 10 min occlusion (O) of the left anterior descending coronary artery and 30 min of reperfusion (R). • $p < 0.05$ versus control animals, ★ $p < 0.05$ versus baseline, + $p = 0.056$ (L-propionylcarnitine-induced change from baseline versus change from baseline in control swine).

Table 1. Plasma levels of carnitine and short-chain acylcarnitine in non-treated and L-propionylcarnitine-pretreated anesthetized pigs before and at the end of two sequences of 10 min occlusion of the left anterior descending coronary artery and 30 min of reperfusion

	Group	n	Pre-baseline (before i.v. bolus)	Baseline (after i.v. bolus)	30 min reperfusion-2
Carnitine	C	6		8.8 ± 0.9	8.3 ± 0.9
	LPC	14	23.4 ± 1.7•	87.4 ± 4.0•+	75.8 ± 4.5*#
Short chain acyl carnitine	C	6		0.9 ± 0.2	0.6 ± 0.2
	LPC	14	5.4 ± 0.9•	163.3 ± 27.4•+	49.4 ± 7.5*#

C = control; LPC = L-propionylcarnitine. Date are in μM and have been expressed as mean ± SEM; • $p < 0.05$ versus Baseline in control swine; + $p < 0.05$ versus Pre-baseline in L-propionylcarnitine treated swine; * $p < 0.05$ versus Baseline; # L-propionylcarntine-induced changes versus Baseline after bolus are significantly different from those in the control animals.

levels increased even more noticeable. In the pretreated animals the myocardial level of carnitine at baseline was higher than in the control animals (Table 2). After two cycles of ischemia-reperfusion this difference was maintained in the control but not in the ischemic area. The myocardial level of short chain acyl carnitine was not affected by treatment. A decrease, as observed in the LADCA-perfused myocardium of the control animals after occlusion-reperfusion, was, however, prevented by L-propionylcarnitine treatment.

Discussion

The present study demonstrates that in a model of stunned myocardium L-propionylcarnitine is capable of modestly improving regional contractile function. Recently, we investigated the effects of L-propionylcarnitine in a model in which 60 min of low-flow ischemia was followed by 2 hr of reperfusion [6]. In that model no improvement of function by L-propionylcarnitine was observed, because the ischemic episode was too severe. It has been reported that in the model we used in the present study a loss of myocardial contractile function of approximately 60% occurs in the absence of myocardial necrosis [7]. We observed a 72% loss in the control group while in the treated animals the loss was 61%, which was borderline significantly different. A number of possible explanations can be forwarded.

Firstly, regional myocardial systolic segment shortening is a load-dependent index of myocardial contractile function. Therefore, the prevention by L-propionylcarnitine of peripheral systemic vasoconstriction, and the consequently lower arterial blood pressure (afterload) may have been responsible for the improvement in systolic segment shortening. Indeed, a vasodilatory action of L-propionylcarnitine has also been observed in anesthetized dogs, where L-propionylcarnitine (bolus: 20–200 mg/kg) dose-dependently decreased total peripheral resistance by up to 10% [13]. In parallel, we have found that in pigs after one hour of low-flow ischemia followed by 2 hr of reperfusion, L-propionylcarnitine caused vasodilation in the non-ischemic myocardium as well as in some peripheral organs, also revealing a vasodilatory effect [6]. However, we must keep in mind that in the latter study the higher myocardial work in the L-propionylcarnitine-treated animals could also have resulted in metabolic autoregulation secondary to the higher myocardial oxygen demand.

Secondly, L-propionylcarnitine may have caused partial reversal of myocardial stunning, similar to that found by Liedtke *et al.* [4] in a porcine model of mild myocardial ischemia. The latter group suggested that a positive inotropic action of L-propionylcarnitine might be responsible for this beneficial effect. A positive inotropic action of the compound has also been found in rabbits [14] as well as in man [15] although this was not confirmed by others in dogs [13]. In the present study we found in the pretreated animals a lower segment length shortening of the non-LADCA area at baseline (P = 0.034), which, however, was not noticed in the area perfused by the LADCA. We therefore believe that with this dose the positive inotropic effect on the stunned myocardium, if present, is only of minor importance.

The beneficial effect on regional contractile function of the stunned myocardium did not result in an overall improvement in cardiac pump function. This is in accordance with the results of the study by Liedtke *et al.* [4], where left ventricular developed pressure, heart rate and LV dP/dt_{max} were also not affected by L-propionylcarnitine.

Table 2. Myocardial tissue levels of carnitine and short-chain acylcarnitine in non-treated and L-propionylcarnitine-pretreated anesthetized pigs before and at the end of two sequences of 10 min occlusion of the left anterior descending coronary artery and 30 min of reperfusion

	Group	n	Baseline	30 min reperfusion-2	
				Control area	Ischemic area
Carnitine	C	8	4.0 ± 0.5	4.3 ± 0.4	3.2 ± 0.3^{+}
	LPC	6	5.7 ± 0.6$^{\bullet}$	5.8 ± 0.4$^{\bullet}$	4.4 ± 0.6
Short chain acyl carnitine	C	8	1.3 ± 0.1	1.2 ± 0.2	1.7 ± 0.2
	LPC	6	1.3 ± 0.2	1.2 ± 0.2	1.1 ± 0.2$^{\#}$

C = control; LPC = L-propionylcarnitine. Data are in μmol/g protein and have been expressed as mean ± SEM; $^{\bullet}p < 0.05$ versus control animals; $^{+}p < 0.05$ versus control area; $^{\#}p < 0.05$ L-propionylcarnitine-induced changes from Baseline versus changes from Baseline in the control pigs.

The present study was not designed to unravel the mechanism by which L-propionylcarnitine exerts its protective effect. However, one possibility deserved attention. L-propionylcarnitine could simply act by repleting carnitine, lost during ischemia. The decrease after ischemia-reperfusion was, however, not affected by L-propionylcarnitine. Similarly, in a former study on L-propionylcarnitine of our laboratory [6] we showed that ischemia followed by reperfusion indeed depleted the intracellular carnitine pool which, however, was not prevented by L-propionylcarnitine. On the other hand, baseline levels of myocardial free carnitine were higher in the L-propionylcarnitine-treated animals which might be beneficial by preventing a fall below critical levels. The increased plasma free carnitine after intravenous administration of L-propionylcarnitine has been reported before [6] and is most likely due to hydrolysation from L-propionylcarnitine to free carnitine in tissues or exchange for free carnitine.

In conclusion, in a model of myocardial stunning, L-propionylcarnitine prevents systemic vasoconstriction in response to ischemia and reperfusion and, possibly as a result of this effect, slightly ameliorates post-ischemic hypofunction. However, a protective effect in combination with a positive inotropic effect can not be excluded.

Acknowledgements

L-propionylcarnitine was generously supplied by Sigma Tau Pharmaceutical of Rome, Italy.

References

1. Paulson DJ, Traxler J, Schmidt M, Noonan J, Shug AL: Protection of the ischaemic myocarium by L-propionylcarnitine: effects on the recovery of cardiac output after ischaemia and reperfusion, carnitine transport, and fatty acid oxidation. Cardiovasc Res 20: 536–541, 1986
2. Shug A, Paulson D, Subraimanian R, Regitz V: Protective effects of propionyl-L-carnitine during ischemia and reperfusion. Cardiovasc Drugs Ther 5 (suppl 1): 77–83, 1991
3. Leasure JE, Kordenat K: Effect of propionyl-L-carnitine on experimental myocardial infarction in dogs. Cardiovasc Drugs Ther 5 (suppl 1): 85–95, 1991
4. Liedtke AJ, DeMaison A, Nellis DSH: Effect of L-propionylcarnitine on mechanical recovery during reflow in intact hearts. Am J Physiol 255 (Heart Circ Physiol 24): H169–H176, 1988
5. Molaparast-Saless F, Nellis SH, Liedtke AJ: The effects of propionylcarnitine taurine on cardiac performance in aerobic and ischemic myocardium. J Moll Cell Cardiol 20: 63–74, 1988
6. Sassen LMA, Bezstarosti K, van der Giessen WJ, Lamers JMJ, Verdouw PD: L-propionylcarnitine increases post-ischemic blood flow but does not affect recovery of the energy charge. Am J Physiol 261 (Heart Circ Physiol 30): H172–H180, 1991
7. Schott RJ, Rohmann S, Braun ER, Schaper W: Ischemic preconditioning slows energy metabolism and delays ultrastructural damage during a sustained ischemic episode. Circ Res 66: 913–931, 1990
8. Guth BD, Schultz R, Heusch G: Evaluation of parameters for the assessment of regional myocardial contractile function during asynchronous left ventricular contraction. Basic Res Cardiol 84: 227–246, 1989
9. Barth PG, Scholte HR, Berden JA, van der Klei-van Moorsel JM, Luyt-Houwen IEM, van het Veer-Korthof ETh, van der Harten JJ, Sobotka-Plojhar MA: An X-linked mitochondrial disease affecting cardiac muscle skeletal muscle and neutrophil leukocytes. J Neurol Sci 62: 327–355, 1983
10. Lamers JMJ, De Jonge-Stinis JT, Verdouw PD, Hülsmann WC: On the possible role of long chain fatty acycarnitine accumulation in producing functional and calcium permeability changes in membranes during myocardial ischaemia. Cardiovasc Res 21: 313–322, 1987
11. Lowry OH, Rosenbrough NJ, Fare AL, Randall RJ: Protein measurement with the folin phenol reagent. J Biol Chem 193: 265–275, 1951
12. Bradford MM: A rapid and sensitive method for the quantitation of microgram quantities of protein utilizing the principle of protein-dye binding. Anal Biochem 72: 248–254, 1976
13. Cevese A, Schena F, Cerutti G: Short-term hemodynamic effects of intravenous propionyl-L-carnitine in anesthetized dogs. Cardiovasc Drugs Ther 5 (suppl 1): 45–56, 1991
14. Ferrari R, Pasini E, Condorelli E, Boraso A, Lisciani R, Marzo A, Visioli O: Effect of propionyl-L-carnitine on mechanical function of isolated rabbit heart. Cardiovasc Drugs Ther 5 (suppl 1): 17–23, 1991
15. Chiddo A, Gaglione A, Musci S, Troito G, Grimaldi N, Locuratolo N, Rizzon P: Hemodynamic study of intravenous propionyl-L-carnitine in patients with ischaemic heart disease and normal left ventricular function. Cardiovasc Drugs Ther 5 (suppl 1): 107–111, 1991

Molecular and Cellular Biochemistry **116**: 155–162, 1992.

Release of heart fatty acid-binding protein into plasma after acute myocardial infarction in man

Appie H. Kleine, Jan F.C. Glatz, Frans A. Van Nieuwenhoven and Ger J. Van der Vusse
Department of Physiology, Cardiovascular Research Institute Maastricht, University of Limburg, Maastricht, The Netherlands

Abstract

The release of cytoplasmic heart fatty acid-binding protein (H-FABP) into the plasma of cardiac patients up to 38 hr after the onset of the first clinical symptoms of acute myocardial infarction (AMI) was studied, using a sensitive direct and noncompetitive Enzyme Linked Immunosorbent Assay of the antigen capture type (sandwich ELISA), newly developed for the measurement of small amounts of human H-FABP in plasma samples. Plasma levels of H-FABP were compared with plasma activity levels of the myocardial cytoplasmic enzymes creatine kinase MB (CK-MB) and alpha-hydroxybutyrate dehydrogenase (α-HBDH). Upper normal levels of H-FABP (19 μg/l), CK-MB (10 U/l) and α-HBDH (160 U/l) as determined in plasma from 72 blood donors served as threshold levels. H-FABP levels were significantly elevated above their threshold level within 3 hr after AMI. Peak levels of H-FABP, CK-MB and α-HBDH were reached 4.1 ± 0.9 hr, 8.4 ± 1.4 hr and 25.0 ± 9.5 hr (means ± S.D., n = 10) after acute myocardial infarction, respectively. Serial time curves of the plasma contents of H-FABP reveal that after myocardial infarction H-FABP is released in substantial amounts from human hearts. In 18 out of 22 patients with established AMI the plasma FABP level was at or above the threshold level in blood-samples taken within 3.5 hr after the first onset of symptoms of AMI, while for CK-MB this applied to 9 patients and for α-HBDH to 6 patients. These findings suggest that for an early indication of acute myocardial infarction in man cytoplasmic heart fatty acid-binding protein is more suitable than heart type creatine kinase MB and/or alpha-hydroxybutyrate dehydrogenase. (Mol Cell Biochem **116:** 155–162, 1992)

Key words: fatty acid-binding protein, ELISA, acute myocardial infarction

Abbreviations: H-FABP – (cytoplasmic) Heart Fatty Acid-Binding Protein; LDH – Lactate Dehydrogenase, α-HBDH – α-Hydroxybutyrate Dehydrogenase; CK–MB – Creatine Kinase-MB; AMI – Acute Myocardial Infarction; PBS – Phosphate Buffered Saline; BSA – Bovine Serum Albumin

Introduction

Cytoplasmic fatty acid-binding protein (FABP) is one of the most abundant cytoplasmic proteins in the heart, as in human and rat heart it comprises about 15 and 30%, respectively, of the number of protein molecules present in the aqueous cytoplasm [1]. In a human heart this amounts to 0.5 mg of FABP per gram wet weight. Cytoplasmic FABP is thought to be involved in the uptake, transport and metabolism of fatty acids, but its

Address for offprints: A.H. Kleine, Department of Physiology, P.O. Box 616, 6200 MD Maastricht, The Netherlands precise role remains to be determined [for reviews see 2 and 3]. Previously heart-type FABP (H-FABP) has been shown to be released from isolated rat hearts upon cellular injury at a rate and to an extent similar to that of lactate dehydrogenase (LDH). [4]. The *in vivo* release of H-FABP from myocardial tissue upon cellular damage has subsequently been demonstrated immunochemically in rats, using polyclonal as well as monoclonal mouse antibodies against rat H-FABP [5], and in cardiac patients using monoclonal mouse antibodies against rabbit H-FABP [6]. Recently Tanaka *et al.* [7], using anti-human H-FABP antibodies in a competitive enzyme immunoassay, also showed elevated serum levels of FABP after acute myocardial infarction in man. These findings suggest that in cardiac patients the plasma FABP level can be used as an indicator for damage to myocardial tissue. The above observations warrant further investigations on the use of plasma FABP level for the (early) assessment of an acute myocardial infarction (AMI) in man.

In the present study we explored the concept [4] that the plasma level of FABP is an indicator especially for the early diagnosis of AMI. For a proper study of this concept we had to develop a sensitive and relatively disturbance-insensitive analytical method for the precise quantitation of H-FABP in human plasma. For this, and Enzyme Linked Immuno-Sorbent Assay (ELISA) of the antigen capture type (sandwich ELISA), in which antigen is bound by immobilized IgG and quantified with a second antibody-enzyme conjugate, appears the appropriate choice. Measurements of H-FABP and of the activities of creatine kinase (CK-MB) and α-hydroxybutyrate dehydrogenase (α-HBDH) were made in serial plasma-samples obtained from 26 patients with initial diagnosis of AMI. Levels of the specific proteins, determined in non-haemolytic plasma from 72 healthy individuals, served as threshold levels. The data indicate that for the early indication of AMI the plasma H-FABP level may be more suitable than that of CK-MB or α-HBDH.

Materials and methods

Patients

Twenty-six patients with an initial diagnosis of AMI were included in this study. AMI was positively diagnosed when patients showed elevation of more than one millimetre of the ST-segment in the recorded cardiogram and typical presentation of chest pain (often combined with radiation of pain to the left arm), in combination with transpiration, nausea, and/or shortness of breath. Diagnosis of AMI was confirmed by data on enzyme release (i.e. significant rise in LDH, CK and α-HBDH activity within 8 hr after onset of first symptoms) in 22 patients, the remaining 4 patients were diagnosed having unstable angina pectoris.

Blood sampling

Blood-samples were collected in glass tubes coated with ethylenediaminetetraacetic acid (EDTA) and subsequently spun for 10 min at 1500 × g. The resulting plasma was divided into several aliquots in glass tubes. Samples were stored at − 80° C until further use. Starting from admission of the patients to the coronary care unit of the hospital (Academic Hospital, AZM, Maastricht) blood-samples were taken based on a time schedule of 0, 4, 10, 18, 28 and 38 hr from admission (hospital routine). A subgroup of ten patients was sampled hourly during the first ten hours of hospitalization. Thereafter blood-samples were again taken in accordance with hospital routine up to 38 hr after admission, together resulting in 11–14 blood-samples per patient in this subgroup.

Non-haemolytic blood-samples of healthy donors (n = 72), collected at the local bloodbank, were used for the exploration of the threshold (= upper normal) levels in plasma of FABP, CK-MB and α-HBDH. In this study the threshold level is defined as the mean value plus twice the standard deviation of the particular protein, as measured in normal human plasma [8]. In addition, equal aliquots of each of the 72 normal human plasma samples were pooled. This pool plasma served as inter-assay control in the assays.

Isolation and purification of H-FABP

Human H-FABP, needed for the immunization of rabbits and for calibration of the immunoassay, was purified by gel permeation and anion-exchange chromatography as previously described [9]. Briefly, homogenates of human heart tissue (25%, w/v) were prepared in buffer, consisting of 10 mM Tris-HCl (pH 8.0), 150 mM KCl and 1 mM dithiothreitol. After centrifugation at 2,600 × g for 10 min and the supernatant at 105,000 × g for 90 min, the final supernatant (containing cytosolic proteins) was concentrated by ultra-filtration using a Diaflo YM5 membrane (Amicon, Danvers, MA, USA), and applied to a Sephacryl S200-SF column (Parmacia/LKB, Uppsala, Sweden) equilibrated with homogenization buffer. Elution of proteins was monitored spectrophotometrically at 280 nm. FABP containing fractions, identified by Lipidex 1000 assay [10], were pooled, concentrated, dialyzed overnight at 4° C against 5 mM Tris-HCl (pH 8.0), and then applied to a Sepharose-Q fast flow column (Pharmacia/LKB, Uppsala, Sweden) equilibrated with dialysis-buffer, H-FABP was collected after stepwise gradient-elution (0–30 mM NaCl in 5 mM Tris-HCl, pH 8.0). Purity of H-FABP was assessed by SDS-PAGE and isoelectric focusing. Finally, the protein was dialysed against 10 mM K-phosphate (pH 7.4), containing 150 mM NaCl and stored in aliquots at − 20° C.

Production of antibodies

Rabbits (Flemish giant) were immunized with 200 μg of pure human H-FABP each in a mixture of 2 ml PBS and Freund's Complete Adjuvant (FCA) (1:1, v/v), and were boosted in weeks 4 and 8 with antigen in a mixture of 2 ml PBS and Freund's Incomplete Adjuvant (FIA) (1:1, v/v). Two weeks after the last boosting blood was collected in glass tubes and spun for 10 min at 1500 × g. From the resulting polyclonal antiserum of IgG fraction was isolated by means of protein-A chromatography (Pharmacia/LKB, Uppsala, Sweden). Subsequently monospecific rabbit IgG against human H-FABP (αP-IgG) was isolated from this fraction by affinity chromatography, using a Sepharose column containing covalently bound human H-FABP. Monospecific rabbit IgG from one rabbit was conjugated with Horse Radish Peroxidase (αP-IgG-HRP) [11] to serve as the detector antibody in the sandwich ELISA.

Sandwich ELISA

In order to estimate H-FABP in plasma accurately we developed a sensitive direct noncompetitive Enzyme Linked Immunosorbent Assay of the antigen capture type (sandwich ELISA) essentially similar to that described by Börchers *et al.* [12] for the assay of H-FABP in subcellular fractions of bovine cardiac tissue. Our assay had to be more sensitive because, unlike subcellular tissue fractions, human plasma contains a relatively large amount of proteins (cf. albumin) in comparison to the amount of H-FABP. Thus, we developed an assay that would meet the requirements for measurement of minute amounts of human H-FABP in plasma samples. The assay was carried out in Falcon 3912 Micro Test III polystyrene microtiter plates (Becton Dickinson Labware, Oxnard, CA, USA). Each well was coated with 200 ng of αP-IgG (see above) in 50 μl 0.1 M ammonium carbonate (pH 8.8). Subsequently the wells were washed with PBS containing 0.5% BSA and 0.05% Tween-20 (PBT) to remove all unbound IgG and at the same time saturate the wells, thereby preventing unspecific binding during the following steps of the assay. Next the standards and serially diluted (PBT) plasma samples were incubated in the wells for 1.5 hr at room temperature. After washing with PBT to remove unbound FABP the wells were incubated (1.5 hr at room temperature) with 70 ng of αP-IgG-HRP in 50 μl PBT for specific detection of bound FABP. Plates were developed with 100 μl substrate solution containing 10 mM ortho-phenylenediamine (OPD) and 0.008% H_2O_2, dissolved in 0.1 M sodium citrate (pH 5.0). Development was stopped with 50 μl 4N H_2SO_4 after 3–5 min and extinction was read at 492 nm using a MkII Plus microplate reader (Flow Lab, Bioggio, Switzerland).

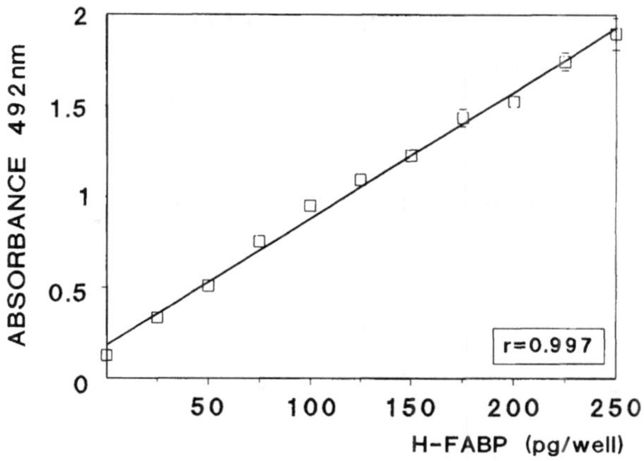

Fig. 1. Standard calibration curve for the sandwich enzyme linked immunosorbent assay (sandwich ELISA) for H-FABP. See text for details.

Measurement of CK-MB and α-HBDH activity

Activity of CK-MB was measured by an enzyme immunoinhibition assay kit (# 418234, Boehringer, Mannheim, Germany). Activity of α-HBDH, which is predominantly LDH-isoenzyme-1, was measured spectrophotometrically with α-ketobutyrate as the substrate [13] using a commercially available test kit (# 161080, Boehringer, Mannheim, Germany).

Results

Sandwich ELISA

The calibration curve for the assay leveled beyond 500 pg H-FABP, but was found to be strictly linear

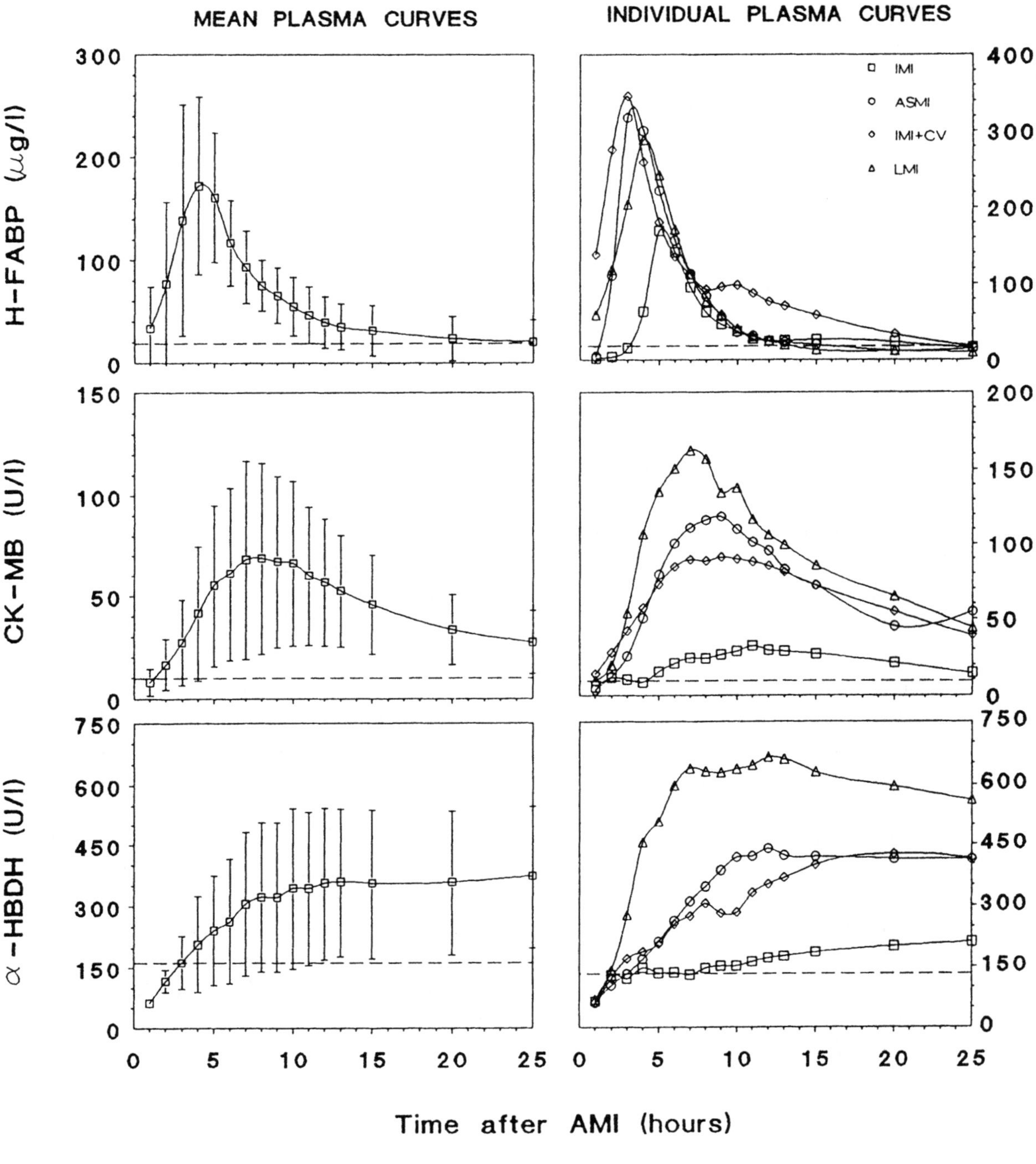

Fig. 2. Left-hand side panels: Mean plasma time-content curves of H-FABP and plasma time-activity curves of CK-MB and α-HBDH after acute myocardial infarction (n = 10). Right-hand side panels: Four typical examples of plasma curves. Each separate curve represents data from one patient. Curve IMI: Inferior myocardial infarction; Curve ASMI: Anteroseptal myocardial infarction; Curve IMI + CV: Inferior myocardial infarction in combination with Cardioversion; Curve LMI: Lateral myocardial infarction. Dotted lines in all panels indicate threshold levels of H-FABP, CK-MB and α-HBDH in normal human plasma (see text for details).

(R = 0.9970) within the range of 0 pg to 250 pg H-FABP (Fig. 1). All plasma samples studied were serially diluted with PBT, so as to assure that their H-FABP content would fall into the linear portion of the calibration curve. Recovery experiments (n = 11) using normal human plasma spiked with purified H-FABP yielded an average recovery of 93.5%. The interassay coefficient of variation was on the order of 6.5%.

Threshold levels

The plasma threshold levels for the three proteins studied, measured in plasma from 72 healthy volunteers,

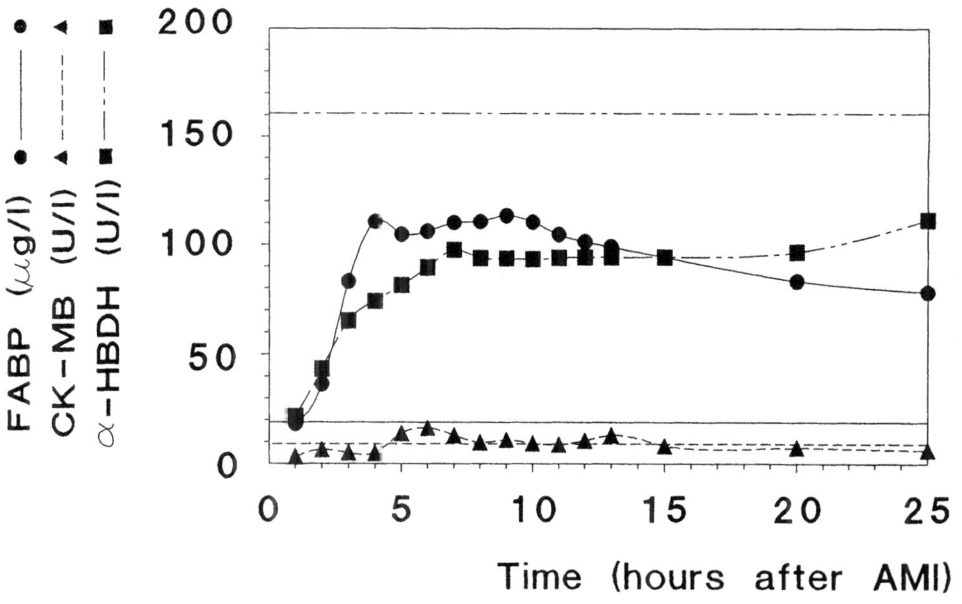

Fig. 3. Plasma levels of H-FABP, CK-MB and α-HBDH, and their respective thresholds (same line pattern) in a patient with inferior myocardial infarction in combination with instable angina pectoris and severe renal insufficiency, during 25 hr after the first onset of symptoms.

were calculated to amount to 19 μg/l for H-FABP, 10 U/l for CK-MB, and 160 U/l for α-HBDH.

Plasma curves

The time curves of the plasma content of FABP and the plasma activities of CK-MB and α-HBDH (Fig. 2) are the resultants of both the release into and the clearance from the circulation of these cytoplasmic proteins. In the left-hand side panels of Fig. 2 the mean release curves for FABP, CK-MB and α-HBDH of the subgroup of frequently sampled cardiac patients are presented. The plasma levels of FABP, CK-MB, and α-HBDH peaked at 4.1 ± 0.9 hr, 8.4 ± 1.4 hr and 25.0 ± 9.5 hr (means ± S.D., n = 10), respectively, after the onset of the first symptoms. In general relatively high plasma H-FABP levels in blood samples taken early are followed by relatively high levels of CK-MB and α-HBDH in the blood samples taken later, as is illustrated by typical release curves of individual patients (Fig. 2, right-hand side panels, which shows four typical release curves). The release curves for all 26 patients studied revealed that within 20 hr after the onset of the first symptoms, plasma H-FABP levels had returned to the threshold level (cf. Fig. 2). The disappearance rate constant of FABP from plasma (Kd) was calculated to range between 0.1 and 0.4 h^{-1}. In contrast, plasma activities of CK-MB and α-HBDH remained well above the threshold level until at least 25 hr after the onset of the first symptoms. Interestingly, one patient, who suffered from AMI in combination with unstable angina pectoris and severe renal insufficiency, showed a heightened (4 to 8 times the threshold level) FABP plasma level during the whole course of the sampling time (Fig. 3). The CK-MB plasma level in this case was only elevated 2 times the threshold level during a short period of the sampling time, while the α-HBDH plasma level stayed normal during sampling time.

Early diagnosis of AMI

In Fig. 4 the FABP plasma content and the CK-MB and α-HBDH plasma activities of the first blood-samples obtained (between 0.5 and 3.5 hr after AMI) from the 26 patients monitored in this study are plotted on a relative scale against their respective threshold levels. In 16 out of 22 patients with established AMI (for establishment of AMI see Material and methods), i.e. 73%, the plasma FABP level was already at or above the threshold level in these early blood-samples taken, while for CK-MB this was the case in only 9 (= 41%) and for α-HBDH in 6 (= 27%) patients (Fig. 4, closed symbols). Furthermore, of the six patients who showed no significant rise in plasma FABP level in the first blood-sample taken, two showed a significantly elevated level of FABP (not of CK-MB or α-HBDH) in a second blood-sample taken within 3.5 hr after AMI. The remaining four patients with established AMI showed significant rises in the plasmalevels of FABP, CK-MB, and α-HBDH in a next blood-sample taken

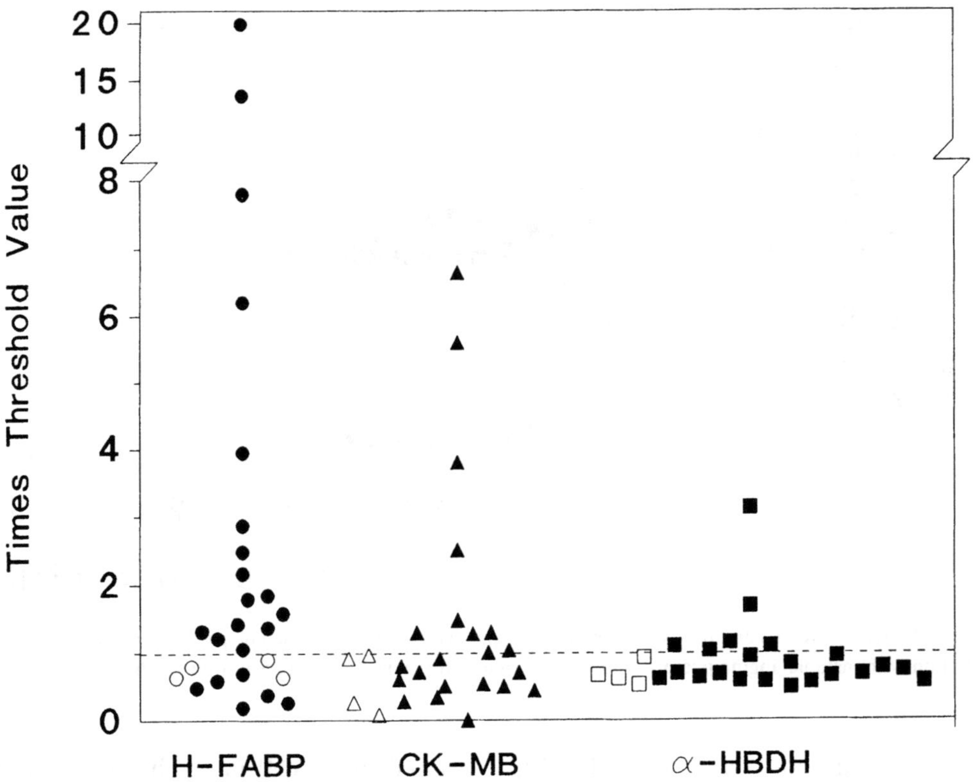

Fig. 4. Plasma levels of H-FABP, CK-MB and α-HBDH in *first* blood samples obtained (within 3.5 hr after onset of symptoms) from in total 26 cardiac patients in relative relation to their threshold level in plasma. Dotted line indicates threshold levels of H-FABP, CK-MB and α-HBDH in plasma. Solid symbols represent patients with established acute myocardial infarction (AMI, for establishment see Material and methods), open symbols refer to patients eventually diagnosed having instable angina pectoris.

between 3.75 and 7.75 hr after onset of the first symptoms. The four patients, from the total of 26, diagnosed as unstable Angina pectoris (Fig. 4, open symbols) showed no plasma levels of FABP, CK-MB or α-HBDH above their respective threshold level during sampling time. Thus, for the early (within 3.5 hr) diagnosis of AMI the plasma level of FABP appears to have been a useful extra tool in 18 out of the 22 established AMI patients (82%), that of CK-MB in 9 patients (41%) and that of α-HBDH in 6 patients (27%).

Discussion

In this study we demonstrate the accurate and reproducible measurement of H-FABP in serial plasma samples of cardiac patients with initial diagnosis of acute myocardial infarction (AMI), using a highly sensitive enzyme linked immunosorbent assay (ELISA) of the sandwich type, newly developed for the measurement of human H-FABP in plasma samples. Sensitivity and accuracy of the assay were evident from a linear calibration curve (between 0 and 250 pg H-FABP) and an almost complete recovery of purified H-FABP added to plasma samples. Comparison of the release of H-FABP into the circulation with that of CK-MB and α-HBDH revealed that the plasma level of the cytoplasmic heart type FABP can be used especially for the early diagnosis of AMI in man.

Plasma threshold levels

Measurement of H-FABP content in human plasma of 72 healthy individuals revealed a normal plasma level of $9 \pm 5\,\mu g/l$ (mean ± S.D.), leading to a threshold level of $19\,\mu g/l$ (mean + 2 S.D.). Assuming a total plasma volume of 3 litres, this means there is a constant pool of some $60\,\mu g$ of FABP, at the most, present in the bloodstream. Because FABPs from human heart and skeletal muscle are identical [14], and FABP from heart is not likely to be found in the bloodstream under normal conditions, the normal plasma level of FABP most probably stems for the greatest part from the continuous release of FABP from damaged skeletal myocytes. Although the mean FABP content of skeletal muscle is relatively low, i.e. 0.2 mg per gram wet weight [14], the large total mass of striated muscle (about 40% of the body weight) makes that a human body contains about 6 grams of skeletal muscle FABP. Thus, the normal plasma level of FABP would represent less than 0.001% of this total amount.

The normal plasma FABP level found is of similar magnitude as the value (0–2.8 μg/l) reported by Tanaka *et al.* [7], but 3 to 4 orders of magnitude lower than the plasma levels of 15–60 mg/l given by Knowlton *et al.* [6]. However, the latter value seems very unlikely since it would imply that under normal circumstances the plasma pool of FABP equals that of 225–900 grams of skeletal muscle or 90–370 grams of cardiac muscle.

FABP as indicator for AMI

Comparison of the time curves of the plasma content of FABP and of the plasma activities of CK-MB and α-HBDH reveals that FABP is released into and cleared from the circulation more rapidly than is CK-MB and α-HBDH. The rapid clearing will relate to the fact that small molecules like H-FABP (15 kDa) are easily cleared from the circulation by urinary excretion [15–17] unlike larger molecules such as CK-MB (80 kDa) or α-HBDH (138 kDa) which stay in the circulation for a longer period of time. With our immunoassay we were indeed able to qualitatively establish the presence of FABP in urine collected from some patients during their hospitalization, thus confirming urinary excretion of FABP, similar to findings of Tanaka *et al.* [7]. In addition, the patient with renal insufficiency showed elevated plasma FABP levels for a prolonged period of time. Nevertheless, our preliminary and qualitative data on renal excretion of FABP and also bearing the probable inter-patient variation in mind, do not justify any conclusions about clearance rates of FABP from the bloodstream as yet. The relatively short half-life time of FABP in plasma is helpful in maintaining a low normal c.q. low threshold plasma level. In turn, as demonstrated in this study, the relatively low threshold plasma level will permit the discrimination of elevated plasma levels at an earlier point in time after AMI than the discrimination of elevated CK-MB or α-HBDH levels from their respective threshold values. The rapid release of FABP into the circulation seems to be independent of the site of myocardial damage in the heart (see Fig. 2, right-hand side panels).

Furthermore, its low threshold level and rapid clearing from plasma makes the plasma level of FABP not only useful for the early assessment of AMI, but presumably also for the demonstration of the occurrence of a possible recurrent infarction.

Myoglobin, the oxygen-carrying protein inside muscle cells, has also been introduced as a plasma marker for early indexation and quantification of acute myocardial infarction [18–25]. However, the discrimination between skeletal muscle and cardiac tissue damage will be more difficult with myoglobin than with H-FABP, since the skeletal muscle myoglobin content is approximately twice that of the heart [26, 27] while the FABP concentration is striated muscle is only 10–50% of that in human cardiac muscle [28]. Hence, interference on an assay from skeletal muscle damage will be 4-20 times less severe in case of an FABP-measuring assay, compared to a myoglobin-measuring assay. On the other hand, for both indicators significant interference will take place only if AMI and extensive skeletal muscle injury occur simultaneously within a few hours.

Concluding remarks

The plasma level of H-FABP was shown to be a useful parameter for the assessment or exclusion of acute myocardial infarction in man, in addition to the commonly used parameters. Furthermore, as a plasma marker for damage to myocardial tissue H-FABP was found to have advantages over those of other protein markers (i.e. CK-MB and α-HBDH) as it appears to be suitable especially for the early (within a few hours) detection of AMI and presumably also for the monitoring of a possible second infarction. The reported sandwich ELISA, when adapted for more rapid clinical application, could become a valuable tool for the early assessment of AMI. In addition to its use for clinical diagnostic purposes, the use of the plasma FABP level as early marker of myocardial infarction will also be of interest for various types of experimental studies in which cardiac tissue damage (may) occur(s).

In view of the existence of other tissue specific types of cytoplasmic FABP, i.e. in kidney, liver, intestine, brain, and adipose tissue [1, 3], cellular injury in these tissues may be specifically and accurately assessed by a similar approach.

Acknowledgements

We would like to thank the Coronary Care Unit of the Academic Hospital, AZM, Maastricht (Head: Dr. A.P.M. Gorgels), The Netherlands for their help in collecting the blood samples for this investigation, and Dr. Torsten Börchers (University of Münster, Münster, Germany) for his help with the development of the immunochemical assay, and Dr. George M. Willems for stimulating discussions. This study was supported by StiPT, Executive Agency for Technology Policy, grant MTR88002. Jan F.C. Glatz is an Established Investigator of the Netherlands Heart Foundation.

References

1. Glatz JFC, Van der Vusse GJ: Cellular fatty acid-binding proteins: Current concepts and future directions. Mol Cell Biochem 98: 237–251, 1990
2. Bass NM: The cellular fatty acid binding proteins: Aspects of structure, regulation, and function. Int Rev Cytol 111: 143–184, 1988
3. Veerkamp JH, Peeters RA, Maatman RGHJ: Structural and functional features of different types of cytoplasmic fatty acid-binding proteins. Biochim Biophys Acta 1081: 1–24, 1991
4. Glatz JFC, van Bilsen M, Paulussen RJA, Veerkamp JH, van der Vusse GJ, Reneman RS: Release of fatty acid-binding protein from isolated rat heart subjected to ischemia and reperfusion or to the calcium paradox. Biochim Biophys Acta 961: 148–152, 1988
5. Knowlton AA, Apstein CS, Saouf R, Brecher P: Leakage of heart fatty acid binding protein with ischemia and reperfusion in the rat. J Mol Cell Cardiol 21: 577–583, 1989
6. Knowlton AA, Burrier RE, Brecher P: Rabbit heart fatty acid-binding protein: Isolation, characterization, and application of a monoclonal antibody. Circ Res 65: 981–998, 1989
7. Tanaka T, Hirota Y, Sohmiya K-I, Nishimura S, Kawamura K: Serum and urinary human heart fatty acid-binding protein in acute myocardial infarction. Clin Biochem 24: 195–201, 1991
8. Ingelfinger JA, Mosteller F, Thibodeau LA, Ware JH. In: Biostatistics in clinical medicine. MacMillan Publishing Co., Inc., New York, 1983, pp 85–113
9. Kleine AH, Glatz JFC, van Nieuwenhoven FA, Vallinga MIJ, Salden MHL, Bosman FT, Boersma WJA, Zegers ND, Van der Vusse GJ: Type-specific immunodetection of human heart fatty acid-binding protein with polyclonal anti-peptide antibodies. Mol Cell Biochem 98: 41–48, 1990
10. Glatz JFC, Veerkamp JH: A radiochemical procedure for the assay of fatty acid binding by proteins. Anal Biochem 132: 89–95, 1983
11. Wilson MB, Nakane PK. In: W Knapp, K. Holubar, G Wick (eds) Immunofluorescence and Related Techniques. Elsevier/North-Holland, Amsterdam, 1978, pp 215–224
12. Börchers T, Unterberg C, Rüdel H, Robenek H, Spener F: Subcellular distribution of cardiac fatty acid-binding protein in bovine heart muscle and quantitation with an enzyme-linked immunosorbent assay. Biochim Biophys Acta 1002: 54–61, 1989
13. Rosalki SB, Wilkinson JH: Reduction of alpha-ketobutyrate by human serum. Nature 188: 1110–1111, 1960
14. Peeters RA, Veerkamp JH, Kanda T, Ono T, Van Kessel AD: Cloning of the cDNA encoding human skeletal muscle fatty acid-binding protein, its peptide sequence and chromosomal localization. Biochem J 276: 203–207, 1991
15. Waldman TA, Strauber W, Mogielnicki RP: The renal handling of low molecular weight proteins. II. Disorders of serum protein catabolism in patients with proteinuria, the nephrotic syndrome, or uremia. J Clin Invest 51: 2162–2174, 1972
16. Hall CL, Hardwicke J: Low molecular weight proteinuria. Ann Rev Med 30: 199–211, 1979
17. Klocke FJ, Copley DP, Krawczyk JA, Reichlin M: Rapid renal clearance of immunoreactive canine plasma myoglobin. Circulation 65: 1522–1528, 1982
18. Stone MJ, Waterman MR, Harimoto D, Murray G, Wilson N, Platt MR, Blomqvist G, Willerson JT: Serum myoglobin level as diagnostic test in patients with acute myocardial infarction. Br Heart J 39: 375–380, 1977
19. Varki AP, Roby DS, Watts H, Zatuchni J: Serum myoglobin in acute myocardial infarction: A clinical study and review of the literature. Am Heart J 96: 680–688, 1978
20. Drexel H, Dworzak E, Kirchmair W, Milz MM, Puschendorf B, Dienstl F: Myoglobinemia in the early phase of acute myocardial infarction. Am Heart J 105: 642–651, 1983
21. McComb JM, McMaster EA, MacKenzie G, Adgey AAJ: Myoglobin and creatine kinase in acute myocardial infarction. Br Heart J 51: 189–194, 1984
22. Isakov A, Shapira I, Burke M, Almog C: Serum myoglobin levels in patients with ischemic myocardial insult. Arch Intern Med 148: 1762–1765, 1988
23. Ellis AK, Saran BR: Kinetics of myoglobin release and prediction of myocardial depletion after coronary artery reperfusion. Circulation 80: 676–683, 1989
24. Ohman EM, Casey C, Bengtson JR, Pryor D, Tormey W, Horgan JH: Early detection of acute myocardial infarction: additional diagnostic information from serum concentrations of myoglobin in patients without ST elevation. Br Heart J 63: 335–338, 1990
25. Driessen-Kletter MF, Amelink GJ, Bär PR, van Gijn J: Myoglobin is a sensitive marker of increased muscle membrane vulnerability. J Neurol 237: 234–238, 1990
26. Sylvén C, Jansson E, Böör K: Myoglobin content in human skeletal muscle and myocardium: relation to fibre size and oxidative capacity. Cardiovasc Res 18: 443–446, 1984
27. Lin L, Sylvén C, Sotonyi P, Somogyi E, Kaijser L, Jansson E: Myoglobin content and citrate synthase activity in different parts of the normal human heart. J Appl Physiol 69: 899–901, 1990
28. Paulussen RJA, Van Moerkerk HTB, Veerkamp JH: Immunochemical quantitation of fatty acid-binding proteins. Tissue distribution of liver and heart FABP types in human and porcine tissues. Int J Biochem 22: 393–398, 1990

Molecular and Cellular Biochemistry **116**: 163–169, 1992.

Regulatory functions of the coronary endothelium

Victor W.M. van Hinsbergh
Gaubius Laboratory IVVO-TNO, P.O. Box 430, 2300 AK Leiden, The Netherlands

Abstract

In this brief review three functions of the coronary endothelium are surveyed: (a) its barrier and exchange function, (b) the prevention of coagulation and platelet aggregation, and (c) its role in vasoregulation. Impairment of these functions can occur in ischemia, hypertension, arteriosclerosis and inflammation. (Mol Cell Biochem **116:** 163–169, 1992)

Key words: endothelium, permeability, vasoregulation, antithrombotic, arteriosclerosis

Introduction: the endothelium and its functions

The coronary circulation provides oxygen, nutrients and hormones to the myocardial tissue. The exchange between blood and the tissues occurs in the capillaries and postcapillary venules of the microvasculature. The endothelium is the main component of the wall of capillaries and postcapillary venules and represents an active and selective barrier for fluid, solutes, macromolecules and white blood cells between the blood and the tissue. In the past fifteen years the insight has grown that the endothelium represents an important and multifaceted organ with many different functions and a maximal exposition to the blood. The functions of the endothelium include: (1) the regulation of vascular permeability, (2) the prevention of coagulation and platelet deposition, (3) a contribution to the regulation of the vascular tonus in muscular blood vessels by the production of factors and the conversion of vasoactive substances, (4) the expression of many, often specifically localized, receptors and binding sites, e.g. albumin and insulin receptors and binding sites for lipoprotein lipase, (5) the secretion of many specific proteins, including basal membrane proteins, (6) the margination and extravasation of white blood cells, (7) a contribution to local and systemic response to inflammatory stimuli, and (8) angiogenesis. In this brief review three main functions of the coronary endothelium are surveyed: its barrier and exchange function, the prevention of thrombosis and its role in vasoregulation. Subsequently, several alterations in endothelial functioning in pathophysiological conditions are mentioned.

The barrier and exchange function of the endothelium

The major function of the coronary microvessels is the exchange of nutrients, hormones and waste products between the myocardium and the blood. This exchange is actively regulated by the endothelial cells of capillaries and postcapillary venules. Nutrients, macromolecules and hormones can pass across the endothelium via several routes, which are schematically depicted in Fig.

Address for offprints: V.W.M. van Hinsbergh, Gaubius Laboratory IVVO-TNO, P.O. Box 430, 2300 AK Leiden, The Netherlands

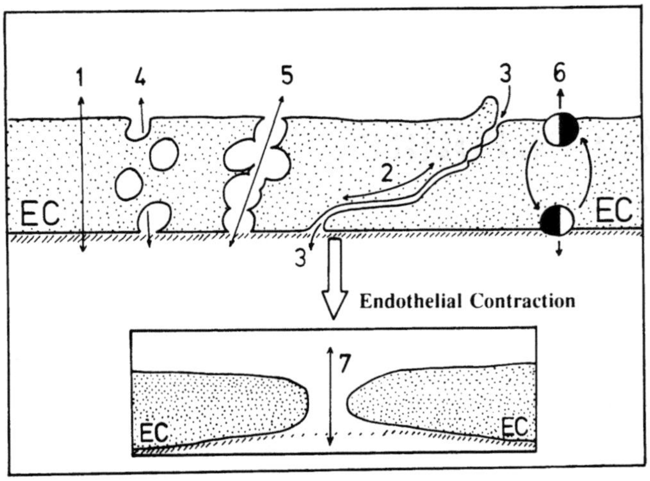

Fig. 1. Schematic representation of various exchange pathways which may be involved in the permeation of various nutrients, fluid and macromolecules across the endothelium. 1. diffusion through the cell membranes and the cytoplasm; 2. lateral diffusion of small lipophilic molecules through the lipid bilayers of the cellular junctions; 3. extracellular passage through the narrow slits of endothelial junctions; 4. vesicular exchange; 5. transendothelial channels which consist of fused vesicles; 6. carrier-mediated exchange and receptor mediated transport. The figure depicts the junctional area between two endothelial cells (EC). Inset: gaps are observed between endothelial cells in activated postcapillary venules [7].

1. The very large lipoproteins VLDL and chylomicrons cannot pass; their triglycerides are hydrolyzed by lipoprotein lipase (LPL) which is bound to glycosaminoglycans on the surface of the endothelium. The pathway(s) by which nutrients cross the endothelium depend on the physicochemical properties and size of the molecules [1, 2]. Oxygen diffuses through the plasma membranes and the cytoplasm, whereas it is generally believed that small lipophilic molecules pass, by means of lateral movement through the lipid bilayers of the cell membranes in junctional areas and, if present, transendothelial channels [3]. The exchange of solutes predominantly proceeds through the narrow slits of intercellular junctions. In postcapillary venules and relatively big capillaries the tightness of the junctional complex can be regulated. Several lines of evidence have indicated that elevation of the cytoplasmic calcium concentration in endothelial cells results in cellular contraction around the junctional areas [4–6]. This contraction is caused by the interaction of actin and non-muscle myosin [5]. Calcium and magnesium ions, ATP, calmodulin and myosin light chain kinase are needed to induce this contraction [4]. In postcapillary venules the contraction can cause a separation of the cells, which results in temporary gaps between the cells (Fig. 1, inset). It is of interest to note that norepinephrine and various agents that increase the cellular cAMP concentration can counteract this process and hence can improve the barrier function of the endothelium]7–9]. Although it is generally accepted that pericellular passage of macromolecules occurs in inflammation [7, 10], the significance of pericellular extravasation of macromolecules in healthy capillaries has been debated. In a series of experiments with electron dense markers, evidence was obtained that vesicles are the predominant site of transcytosis of macromolecules in capillaries [11, 12]. Another proposition, which is based on carefully conducted serial ultrathin sectioning of mesenteric and cardiac muscle capillaries, suggests that nearly all vesicles are connected to the upper or lower leaflet of the plasma membrane and hence provide extreme extrusions of the plasma membrane [13, 14]. At most 2% of the structures that are observed in cross sections as vesicles appear to be true vesicles. However, these observations need not to be as conflicting as they seem. First, the number of vesicles is two orders of magnitude larger than is necessary to explain the exchange of albumin by vesicles. A small subpopulation of specific vesicle structures may explain the apparent difference. Secondly, receptor-mediated transcytosis, as is known for insulin and transferrin in blood brain barrier capillaries [15], may be a more general phenomenon in other capillaries. Evidence has been provided for the high affinity binding and transcytosis of (glyco)albumin to/via vesicular structures for various types of capillary endothelia [12, 16]. On the other hand, whereas vesicular passage of LDL has been suggested by morphological data [17], no evidence for the involvement of a specific carrier in the passage of LDL across the endothelium could be demonstrated [18–20]. One might speculate that in the microvascular bed a gradient exists from the small capillaries, where carrier- or receptor-mediated transport plays a role in the exchange of glucose, insulin, albumin and other molecules, to the postcapillary venules, where the exchange of nutrients and macromolecules is predominantly governed at the cell junctions.

In addition to actively regulating the access of nutrients to the tissue, the endothelium prevents the influx of noxious particles into the heart tissue and the arterial wall. In normal coronary arteries, cholesterol-containing lipoproteins are prevented from crossing the endothelium in significant amounts. In arteriosclerotic regions, however, the rate of influx of macromolecules is increased [21, 22], which may contribute to the accumulation of lipids into the vessel wall.

Prevention of blood coagulation and platelet deposition

Another important function of the endothelium in the coronary circulation is the prevention of thrombus formation. The body has been equipped with a complex defence mechanism to prevent blood loss by wounding or vascular leakage. This is achieved by the interplay between platelet aggregation, fibrin deposition and vasoconstriction [23–25]. If the endothelium is damaged, coagulation and platelet deposition are triggered by the exposed extracellular matrix. However, (spontaneous) coagulation and platelet aggregation within the blood vessels can result in the formation of a mural or occluding thrombus with severe obstruction of the blood flow and ischemia of the distal tissue. The intact endothelium prevents thrombus formation by three mechanisms: (1) the production of proteins that retard or inhibit the coagulation pathway, (2) the production of factors that interfere with platelet aggregation and vasoconstriction, and (3) the release of tissue-type plasminogen activator, which triggers fibrinolysis. Each of these processes is regulated by the endothelium at several levels.

Endothelial cells expose two major anticoagulant proteins: antithrombin III and thrombomodulin [26]. Antithrombin III bound to glycosaminoglycans rapidly inactivates thrombin and factor Xa. The endothelial surface protein thrombomodulin immediately catalyses the activation of protein C upon thrombin generation. This results in a shutdown of thrombin generation, because activated protein C inactivates the cofactors factor Va and factor VIIIa, which are necessary for the activation of factor X and prothrombin. Endothelial cells *in vitro* also produce an inhibitor of extrinsic pathway (EPI), identical to lipoprotein-associated coagulation inhibitor (LACI) [28]. This inhibitor slows down the extrinsic coagulation pathway by the formation of an inhibitory complex with tissue factor and factor Xa. Recent studies in rabbits, in which plasma EPI levels were depleted by injections of anti-EPI IgG, point to a physiological role of this inhibitor [29]. The *in vivo* significance of a lipocortin-related vascular anticoagulant protein, which is produced by cultured endothelial cells and which inhibits thrombin generation is not yet known.

With respect to the inhibition of platelet deposition, the endothelium produces prostacyclin, which suppresses the activation of blood platelets [30]. This suppression is enhanced by another endothelial product: nitric oxide [31]. The surface of the endothelium contains ectonucleotidases including ADPase activity, which readily convert ADP released from blood platelets into AMP and adenosine [32]. This limits the extent of platelet aggregation, because ADP activates other adjacent platelets.

Fibrinolysis is the third mechanism by which the endothelium ensures the patency of blood vessels. Fibrinolysis, the dissolution of fibrin, is achieved by the serine protease plasmin, which is formed from its zymogen plasminogen by the action of plasminogen activators [33]. In the blood compartment this action is performed by tissue-type plasminogen activator (t-PA) [34], which is synthesized in the endothelium [35]. Various vasoactive substances, including PAF, bradykinin, endothelin and thrombin can induce a rapid release of t-PA from the endothelium [36]. This results in an increase of a hundred-fold or more in the local t-PA concentration. This mechanism permits the cells to generate a high t-PA concentration at a site where coagulation is initiated, so that the generating fibrin threads can be dissolved immediately. The synthesis rate of t-PA itself and the constitutive secretion of t-PA by the endothelium also contribute to the plasma concentration of t-PA [37]. Under normal resting conditions, almost equal amounts of t-PA and an efficient t-PA inhibitor, named PAI-1, are produced. The endothelial production of PAI-1 increases dramatically upon exposure of the endothelium to inflammatory mediators.

Role of the endothelium in the regulation of the vascular tone

As mentioned above, the endothelium is involved in the conversion and catabolism of vasoactive substances. ATP and ADP are converted to AMP and adenosine, which is a potent vasodilator in coronary arteries. Bradykinin is converted by angiotensin converting enzyme [38] into the potent constrictor angiotensin II, which can be degraded by angiotensinase to angiotensin III. Histamine is methylated to form its competitive inhibitor methyl-histamine or inactivated by decarboxylation [39]. The endothelium further modulates the vascular response to vasoactive mediators by the uptake and degradation of these mediators.

Since the late seventies it has been recognized that the endothelium plays a crucial role in the regulation of the vascular tonus [40]. Today it is appreciated that the

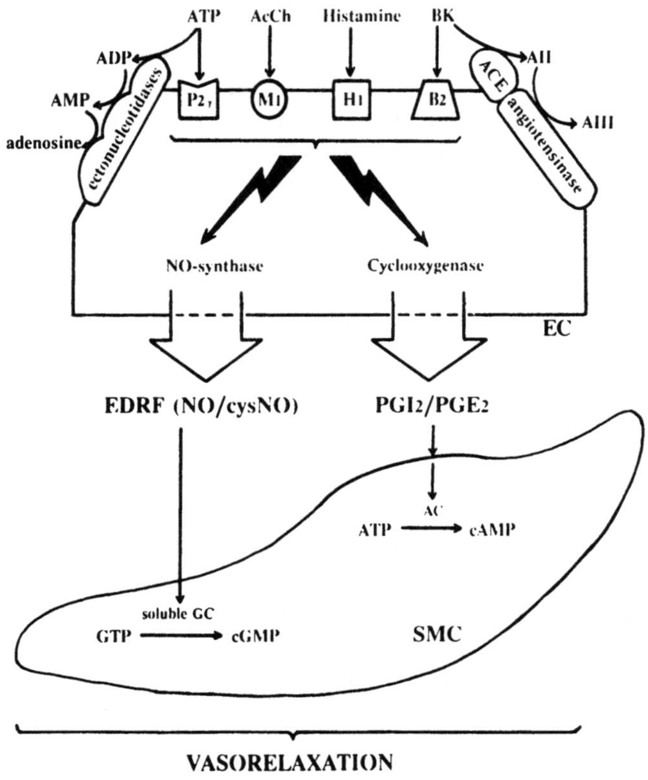

Fig. 2. Schematic representation of the stimulation of the release of vasorelaxing agents by endothelial cells. Various endothelial receptors are indicated by M_1, H_1, P_{2Y} and B_2. Abbreviations: EC: endothelial cell; SMC: smooth muscle cell; AcCh: acetylcholine; BK: bradykinin; ACE: angiotensin converting enzyme; AII: angiotensin II; AIII: angiotensin III; EDRF: endothelium-derived relaxing factor; NO: nitric oxide; cysNO: cysteinylated NO; PGI_2: prostacyclin; PGE_2: prostaglandin E_2; GC: guanylate cyclase.

production of endothelium-derived relaxing factor (EDRF) plays a regulatory role in coronary and other arteries as well in resistance vessels [41–43]. EDRF production can be induced by many vasoactive substances including acetylcholine, histamine, bradykinin, substance P and ATP (Fig. 2). These agents activate receptors, which results in the activation of phospholipase C and the generation of inositoltriphosphate and diacylglycerol. These mediators respectively increase the cytoplasmatic calcium concentration and stimulate protein kinase C. As the calcium ionophore A23187 also activates EDRF production, a general role of calcium ions in the release of EDRF has been accepted. EDRF has recently been identified as nitric oxide (NO) [31]. Although some debate still exists about the actual nature of EDRF(s) – whether it is secreted as NO or as cysteinyl NO [44] – it is generally agreed that NO is the active component that activates soluble guanylate cyclase in smooth muscle cells and thus causes vascular relaxation. Nitric oxide is synthesized from L-arginine via the enzyme NO-synthase. Two NO-synthases have been recognized thus far: a NO-synthase that is constitutively present in endothelial cells, and a second NO synthase that can be induced by inflammatory mediators in macrophages and endothelial cells [31, 45].

Prostaglandins, prostacyclin, and PGE_2 also have a relaxing effect on muscular vessels, which is caused via an increase in cAMP in the smooth muscle cells. Local variation in the production of prostaglandins exists. Data on rabbit coronary blood vessels have indicated that prostacyclin plays a dominant role in the larger coronary arteries, whereas in the smaller vessels the production of PGE_2 predominates over prostacyclin [46]. The production of prostaglandins is regulated by the activity of the enzyme cyclo-oxygenase. Cyclo-oxygenase activity is constitutively present in endothelial cells [46] and can be additionally induced by activation of the cells [47, 48]. It would be interesting to know whether – similar to NO synthase – two cyclo-oxygenases can be produced in endothelial cells: one that is constitutively synthesized and another that can be induced.

Additional endothelial factors that influence vasoregulation have been described [see refs. 42 and 43 for review]. They include several vasoconstrictor factors which may be involved in the autoregulation of blood flow after rapid elevation of the intraluminal blood pressure in arteries and resistance vessels. The very potent vasoconstricting protein endothelin was discovered in the supernatant of cultured endothelial cells a few years ago [49], and has drawn much attention during the last few years. Endothelin appears to play an important role in vasoregulation and in the regulation of water and salt balance. It has been demonstrated in a precursor form ('big endothelin'), which has relatively little activity and in three processed forms (endothelin 1, 2 and 3). It is likely that endothelin plays a role in the heart, as endothelin-1 mRNA has been demonstrated in heart ventricle [50].

Changes in endothelial functions

Regional differences in the various functions of the endothelium along the vascular tree exist: e.g. exchange of nutrients occurs predominantly in the thin flattened continuous endothelium of the microvasculature. In addition during development and upon stimulation or activation, the endothelial cell acquires additional properties which enables the cell to cope with local or temporary demands. Cell proliferation and reorganisation require an altered responsiveness during heart development and collateral formation. Furthermore, endothelial functioning is influenced by physicochemical forces and in various diseases.

Haemodynamic forces

Alteration in blood flow accompanied by a change in shear forces on the surface of the endothelium can induce vasorelaxation by the release of EDRF and prostacyclin [51, 52]. It is still uncertain whether the response of the endothelium to changes in shear forces is mediated by specific mechano-receptors [53], or is the result of an altered access of intact vasoactive mediators to endothelial receptors.

Ischemia

In addition to a direct relaxation of smooth muscle cells, endothelium-dependent vasorelaxation has been observed during hypoxia in isolated coronary arteries [54]. Release of prostacyclin plays an important role in this relaxation. Whether endothelium-dependent vasoconstriction, which has been observed in anoxic conditions, plays a role *in vivo* is less certain. *In vitro* studies have suggested that at low oxygen pressure the permeability and anticoagulant properties of endothelial cells may decrease [55]. Further studies have to establish whether these properties of the endothelium are also markedly affected by hypoxia *in vivo,* or whether the endothelial-metabolism is predominantly impaired by the reoxygenation that occurs after an ischemic period.

Arteriosclerosis

In early arteriosclerotic lesions the endothelium is intact, but has altered characteristics. An obvious one is the binding of monocytes to the endothelium. Studies in rabbits suggest that a specific monocyte-binding protein is expressed on the endothelium [56]. The monocytes extravasate into the arterial wall and form lipid-loaded foam cells together with smooth muscle cells which have migrated from the media into the intima. The response of the endothelium to several vasoactive mediators to release EDRF is impaired in the endothelium overlying arteriosclerotic human coronary arteries [57, 58]. This impairment probably plays a role in angina pectoris. It has been suggested that the cause of the defective EDRF induction is in the transmembrane signal transduction pathway [59] and that addition of cod liver oil or 3-ω-polyunsaturated fats can partly reverse the defect [60]. The impaired signal transduction probably also affects the release of other components, like prostacyclin and t-PA. Other altered properties of the endothelium in arteriosclerotic areas include cell enlargement [61, 62], the production of PDGF-A mRNA [63] and a decreased barrier function [21].

Inflammation

Activation of the endothelium by inflammatory mediators results in a marked alteration of endothelial functioning. This activation is induced by inflammatory mediators, in particular interleukin-1, tumour necrosis factor and interferon-gamma. Interleukin-1 and TNF have similar effects on the endothelium, which usually involve the synthesis of new proteins [see ref. 64 for review]. Among these induced proteins are the leukocyte adhesion molecules ELAM-1, VCAM-1 and ICAM-1, interleukin-6, plasminogen activator inhibitor type-1, tissue factor, MHC class 1 antigens and inducible forms of NO-synthase and cyclooxygenase. Local inflammatory reactions may affect endothelial functions in arteriosclerotic coronary arteries, as suggested by the presence of the inflammatory mediator TNFα in arteriosclerotic plaques. Furthermore, inflammation and endothelial activation appear to contribute to the accelerated arteriosclerosis that occurs in coronary bypass vessels and in blood vessels of transplanted hearts [65].

References

1. Curry FE: Determinants of capillary permeability: a review of mechanisms based on single capillary studies in the frog. Circ Res 59: 367–380, 1986
2. Renkin EM: Capillary transport of macromolecules: pores and other endothelial pathways. J Appl Physiol 58: 315–325, 1985
3. Blanchette-Mackie EJ, Scow RO: Lipolysis and lamellar structures in white adipose tissue of young rats: lipid movement in membranes. J Ultrastruc Res 77: 295–318, 1981
4. Wysolmerski RB, Lagunoff D: Involvement of myosin light-chain kinase in endothelial cell retraction. Proc Natl Acad Sci USA 87: 16–20, 1990
5. Schnittler H-J, Wilke A, Gress T, Suttorp N, Drenckhahn D: Role of actin and myosin in the control of paracellular permeability in pig, rat and human vascular endothelium. J Physiol 431: 379–401, 1990
6. He P, Pagakis SN, Curry FE: Measurement of cytoplasmic calcium in single microvessels with increased permeability. Am J Physiol 258: H1366–H1374, 1990
7. Grega GJ, Persson CGA, Svensjö E: Endothelial cell reactions to inflammatory mediators assessed *in vivo* by fluid and solute

flux analysis. In: US Ryan (ed.) Endothelial Cells Vol III. CRC Press Incorporated, Boca Raton, 1988, pp 103–119
8. Stelzner TJ, Weil JV, O'Brien RF: Role of cyclic adenosine monophosphate in the induction of endothelial barrier properties. J Cell Physiol 139: 157–166, 1989
9. Langeler EG, Van Hinsbergh VWM: Norepinephrine and iloprost improve barrier function of human endothelial cell monolayers: role of cAMP. Am J Physiol 260: C1052–C1059, 1991
10. Majno G, Shea SM, Leventhal M: Endothelial contraction induced by histamine-type mediators. An electron microscopic study. J Cell Biol 42: 647–672, 1969
11. Palade GE: The microvascular endothelium revisited. In: N Simionescu and M Simionescu (eds) Endothelial Cell Biology in Health and Disease. Plenum Publishing Corporation, New York, 1988, pp 3–21
12. Predescu D, Simionescu M, Simionescu N, Palade GE: Binding and transcytosis of glycoalbumin by the microvascular endothelium of the murine myocardium: evidence that glycoalbumin behaves as a bifunctional ligand. J Cell Biol 107: 1729–1738, 1988
13. Frøkjær-Jensen J: Three-dimensional organization of plasmalemmal vesicles in endothelial cells. An analysis by serial sectioning of frog mesenteric capillaries. J Ultrastruct Res 73: 9–20, 1980
14. Bundgaard M, Hageman P, Crone C: The three-dimensional organization of plasmalemmal vesicular profiles in the endothelium of rat heart capillaries. Microvasc Res 25: 358–368, 1983
15. Cornford EM: The blood-brain barrier, a dynamic regulatory interface. Mol Physiol 7: 219–260, 1985
16. Milici AJ, Watrous NE, Stukenbrok H, Palade GE: Transcytosis of albumin in capillary endothelium. J Cell Biol 105: 2603–2612, 1987
17. Vasile E, Simionescu M, Simionescu N: Visualization of the binding, endocytosis, and transcytosis of low-density lipoprotein in the arterial endothelium *in situ*. J Cell Biol 96: 1677–1689, 1983
18. Wiklund O, Carew TE, Steinberg D: Role of the low density lipoprotein receptor in penetration of low density lipoprotein into rabbit aortic wall. Arteriosclerosis 5: 135–141, 1985
19. Langeler EG, Snelting-Havinga I, Van Hinsbergh VWM: Passage of low density lipoproteins through monolayers of human arterial endothelial cells. Effects of vasoactive substances in an *in vitro* model. Arteriosclerosis 9: 550–559, 1989
20. Rutledge JC, Curry FE, Lenz JF, Davis PA: Low density lipoprotein transport across a microvascular endothelial barrier after permeability is increased. Circ Res 66: 486–495, 1990
21. Bremmelgaard A, Stender S, Lorentzen J, Kjeldsen K: *In vivo* flux of plasma cholesterol into human abdominal aorta with advanced atherosclerosis. Arteriosclerosis 6: 442–452, 1986
22. Stemerman MB: Effects of moderate hypercholesterolemia on rabbit endothelium. Arteriosclerosis 1: 25–32, 1981
23. Colman RW, Hirsch J, Marder VV, *et al.*: Hemostasis and Thrombosis: Basic Principles and Clinical Practice. Lippincott, Philadelphia, 1987
24. Harker LA: Endothelium and hemostasis. In: US Ryan (ed.) Endothelial Cells, Vol 1. CRC Press, Boca Raton, 1988, pp 167–177
25. Mann KG, Jenny RJ, Krishnaswamy S: Cofactor proteins in the assembly and expression of blood clotting enzyme complexes. Ann Rev Biochem 57: 915–956, 1988
26. Preissner KT: Anticoagulant potential of endothelial cell membrane components. Haemostasis 18: 271–306, 1988
27. Esmon CT: The role of protein C and thrombomodulin in the regulation of blood coagulation. J Biol Chem 264: 4743–4746, 1989
28. Rapaport SI: The extrinsic pathway inhibitor: a regulator of tissue factor-dependent blood coagulation. Thromb Haemostas 66: 6–15, 1991
29. Sandset PM, Warn-Cramer BJ, Rao LVM, Maki SL, Rapaport DI: Depletion of extrinsic pathway inhibitor (EPI) sensitizes rabbits to disseminated intravascular coagulation induced with tissue factor: evidence supporting a physiologic role for EPI as a natural anticoagulant. Proc Natl Acad Sci USA 88: 708–712, 1991
30. Sixma JJ: Platelet adhesion in health and disease. In: M Verstraete, J Vermylen, R Lijnen and J Arnout (eds) Thrombosis and Haemostasis 1987. Leuven University Press, Leuven, 1987, pp 127–146
31. Moncada S, Palmer RMJ, Higgs EA: Nitric oxide: physiology, pathophysiology, and pharmacology. Pharmacol Rev 43: 109–142, 1991
32. Gordon JL: Extracellular ATP: effects, sources and fate. Biochem J 233: 309–319, 1986
33. Bachmann F: Fibrinolysis. In: M Verstraete, J Vermylen, R Lijnen and J Arnout (eds) Thrombosis and Haemostasis 1987. Leuven University Press, Leuven, 1987, pp 227–265
34. Wun T-C, Capuano A: Spontaneous fibrinolysis in whole human plasma: identification of tissue activator-related protein as the major plasminogen activator causing spontaneous activity *in vitro*. J Biol Chem 260: 5061–5066, 1985
35. Kristensen P, Larsson L-I, Nielsen LS, Grøndahl-Hansen J, Andreasen PA, Danø K: Human endothelial cells contain one type of plasminogen activator. FEBS Lett 168: 33–37, 1984
36. Emeis JJ: Mechanisms involved in short-term changes in blood levels of t-PA. In: C Kluft (ed.) Tissue-Type Plasminogen Activator (t-PA): Physiological and Clinical Aspects, Vol II. CRC Press, Boca Raton, 1988, pp 21–35
37. Van Hinsbergh VWM, Kooistra T, Emeis JJ, Koolwijk P: Regulation of plasminogen activator production by endothelial cells: role in fibrinolysis and local proteolysis. Int J Radiat Biol 60: 261–272, 1991
38. Erdos EG: Angiotensin-I converting enzyme. Circ Res 36: 247–255, 1975
39. Haddock RC, Mack P, Fogerty FJ, Lewis Baenziger N: Role of receptors in metabolic interaction of histamine with human vascular endothelial cells and skin fibroblasts. J Biol Chem 262: 10220–10228, 1987
40. Furchgott RF, Zawadzki JV: The obligatory role of endothelial cells in the relaxation of arterial smooth muscle by acetylcholine. Nature 288: 373–376, 1980
41. Bassenge E, Busse R: Endothelial modulation of coronary tone. Progr Cardiovasc Disease 30: 349–380, 1988
42. Furchgott RF, Vanhoutte PM: Endothelium-derived relaxing and contracting factors. FASEB J 3: 2007–2018, 1989
43. Rubanyi GM: Endothelium-derived relaxing and contracting factors. J Cell Biochem 46: 27–36, 1991
44. Meyers PR, Minor RL Jr, Guerra R Jr, Bates JN, Harrison DG: Vasorelaxant properties of the endothelium-derived relaxing factor more closely resemble S-nitrosocysteine than nitric oxide. Nature 345: 161–163, 1990
45. Radomski MW, Palmer RMJ, Moncada S: Glucocorticoids inhibit the expression of an inducible, but not the constitutive, nitric oxide synthase in vascular endothelial cells. Proc Natl Acad Sci USA 87: 10043–10047, 1990

46. Gerritsen ME: Functional heterogeneity of vascular endothelial cells. Biochem Pharmacol 36: 2701–2711, 1987
47. Dejana E, Breviario F, Balconi G, Rossi V, Remuzzi G, de Gaetano G, Mantovani A: Stimulation of prostacyclin synthesis in vascular cells by mononuclear cell products. Blood 64: 1280–1283, 1984
48. Maier JA, Hla T, Maciag T: Cyclooxygenase is an immediate-early gene induced by interleukin-1 in human endothelial cells. J Biol Chem 265: 10805–10808, 1990
49. Yanagisawa M, Kurihara H, Kimura S, Tomobe Y, Kobayashi M, Mitsui Y, Yazaki Y, Goto K, Masaki T: A novel potent vasoconstrictor peptide produced by vascular endothelial cells. Nature 332: 411–415, 1988
50. Nunez DJR, Brown MJ, Davenport AP, Neylon CB, Schofield JP, Wyse RK: Endothelin-1 mRNA is widely expressed in porcine and human tissues. J Clin Invest 85: 1537–1541, 1990
51. Holtz J, Giesler M, Bassenge E: Two dilatory mechanisms of anti-anginal drugs on epicardial coronary arteries *in vivo:* indirect, flow-dependent, endothelium-mediated dilation and direct smooth muscle relaxation. Z Kardiol 72: 98–106, 1983
52. Pohl U, Holtz J, Busse R, *et al.:* Crucial role of endothelium in the vasodilator reponse to increased flow *in vivo*. Hypertension 7: 37–44, 1986
53. Lansman JB, Hallam TJ, Rink TJ: Single stretch-activated ion channels in vascular endothelial cells as mechanotransducers? Nature 325: 811–813, 1987
54. Busse R, Förstermann U, Matsuda H, Pohl U: The role of prostaglandins in the endothelium-mediated vasodilatory response to hypoxia. Pflügers Arch 401: 77–83, 1984
55. Ogawa S, Gerlach H, Esposito C, Pasagian-Macaulay A, Brett J, Stern D: Hypoxia modulates the barrier and coagulant function of cultured bovine endothelium. Increased monolayer permeability and induction of procoagulant properties. J Clin Invest 85: 1090–1098, 1990
56. Cybulsky MI, Gimbrone Jr MA: Endothelial expression of a mononuclear leukocyte adhesion molecule during atherogenesis. Science 251: 788–791, 1991
57. Ludmer PL, Selwyn AP, Shook TL, Wayne RR, Mudge GH, Alexander RW, Ganz P: Paradoxical vasoconstriction induced by acetylcholine in atherosclerotic coronary arteries. New Engl J Med 315: 1046–1051, 1986
58. Bossaller C, Habib GB, Yamamoto H, Williams C, Wells S, Henry PD: Impaired muscarinic endothelium-dependent relaxation and cyclic guanosine 5′-monophosphate formation in atherosclerotic human coronary artery and rabbit aorta. J Clin Invest 79: 170–174, 1987
59. Shimokawa H, Flavahan NA, Vanhoutte PM: Natural course of the impairment of endothelium-dependent relaxations after balloon endothelium removal in porcine coronary arteries. Possible dysfunction of a pertussis toxin-sensitive G-protein. Circ Res 65: 740–753, 1989
60. Shimokawa H, Vanhoutte PM: Dietary cod-liver oil improves endothelium-dependent responses in hypercholesterolemic and atherosclerotic porcine coronary arteries. Circulation 78: 1421–1430, 1988
61. Repin VS, Dolgov VV, Zaikina OE, Novikov ID, Antonov AS, Nikolaeva MA, Smirnov VN: Heterogeneity of endothelium in human aorta. A quantitative analysis by scanning electron microscopy. Atherosclerosis 50: 35–52, 1984
62. Tokunaga O, Fan J, Watanabe T: Atherosclerosis- and age-related multinucleated variant endothelial cells in primary culture from human aorta. Am J Pathol 135: 967–976, 1989
63. Wilcox JN, Smith KM, Williams LT, Schwartz SM, Gordon D: Platelet-derived growth factor mRNA detection in human atherosclerotic plaques by *in situ* hybridization. J Clin Invest 82: 1134–1143, 1988
64. Pober JS, Cotran RS: Cytokines and endothelial cell biology. Physiol Rev 70: 427–451, 1990
65. Salomon RN, Hughes CCW, Schoen FJ, Payne DD, Pober JS, Libby P: Human coronary transplantation-associated arteriosclerosis. Evidence for a chronic immune reaction to activated graft endothelial cells. Am J Pathol 138: 791–798, 1991

Molecular and Cellular Biochemistry **116**: 171–179, 1992.

Lipid metabolism of myocardial endothelial cells

Kees Schoonderwoerd and Hans Stam
Department of Biochemistry I, Faculty of Medicine and Health Sciences, Erasmus University Rotterdam, The Netherlands

Abstract

The vascular endothelium can be regarded as a widely distributed organ, interposed between the intravascular and extravascular spaces, with a pluripotent function in the regulation of capillary diameter, vascular homeostasis, lipoprotein metabolism and the vascular response to injury. In the basal physiological state these processes provide a non-thrombotic, non-inflammatory vascular lining preventing uncontrolled inflammation and coagulation. Endothelial cells respond to potential harmful conditions (mechanical stress, anoxia, ischemia and oxidative stress) and a variety of hormones and vasoactive mediators by inducing coagulation and production of inflammatory mediators through the production of 'bioactive' lipids. Although the number of studies in isolated myocardial endothelial cells is limited, from the presumed metabolic analogy with endothelial cells isolated (and cultured) from other organs, one may conclude that the bioactive lipids include oxygenated arachidonate metabolites (eicosanoids) and the platelet activating factor (1--O-alkyl-2-acetyl-sn-glycerol-3-phosphocholine; PAF). All aspects of lipid metabolism, related to the production of eicosanoids and PAF, are present within myocardial endothelial cells. There is uptake and incorporation of fatty acids by endothelial cells and liberation from endogenous triacylglycerol and (membrane) phospholipid stores by (phospho)lipases. Endothelial cells oxidize fatty acids in a carnitine-dependent, mitochondrial, pathway. Endothelial cells actively interact with high density lipoprotein (HDL) and low density lipoprotein (LDL) leading to uptake of cholesterol(esters) that undergo intracellular hydrolysis, and re-esterification to phospho- and neutral lipids, and leaving the LDL-particle modified in a way that makes them bind to the scavenger receptor on macrophages. Extravascular triacylglycerols in lipoproteins (very low density lipoprotein (VLDL), chylomicrons) are handled by endothelial cell lipoprotein lipase, providing substrate fatty acids for the underlying muscle tissue. Eicosanoid production from (membrane)phospholipids and PAF synthesis from alkylphospholipids are tightly coupled and interrelated to the flow of arachidonic acid between cellular lipid pools. (Mol Cell Biochem **116:** 171–179, 1992)

Key words: endothelium, myocardium, lipid metabolism, eicosanoids, platelet activating factor (PAF)

Address for offprints: K. Schoonderwoerd, Department of Biochemistry I, Faculty of Medicine and Health Sciences, Erasmus University Rotterdam, P.O. Box 1738, 3000 DR Rotterdam, The Netherlands

Abbreviations: AA – Arachidonic Acid; ACAT – Acyl-Cholesterolacyltransferase; Alkyl-PL – Alkyl(ether)-Phospholipids; C – Cholesterol; CE – Cholesterolester; DG – Diacylglycerol; E – Eicosanoids; EDRF – Endothelium Derived Relaxing Factor; G – Glycerol; G-3-P – Glycerol-3-Phosphate; G-3-PC – Glycerol-3-Phosphocholine; HDL – High Density Lipoprotein; 15-HETE – 15-Hydroxy-5,8,11,13-Eicosatetraenoic Acid; 13-HODE – 13-Hydroxy-Octadecadienoic Acid; LDL – Low Density Lipoprotein; LPL – Lipoprotein Lipase; Lyso PAF – 1-Alkyl-2-0-3-Phosphocholine; Lyso PL – Lysophospholipids; MG – Monoacylglycerol; PA – Phosphatidic Acid; PAF – Platelet Activating Factor; PGE_2 – Prostaglandin E_2; PGI_2 – Prostacyclin; PL – Phospholipids; PMN – Polymorphonuclear Leukocytes; TG – Triacylglycerol; VLDL – Very Low Density Lipoprotein

Introduction

Over the last twenty years it has become apparent that the endothelial layer covering the vascular and lymphatic system not only functions as a selective barrier for plasma components to reach the underlying tissue but also plays an active role in the regulation of vascular homeostasis, inflammation, lipoprotein metabolism and vascular diameter [1, 2]. The active processes of the endothelium therefore include specific interactions with hormones, lipoproteins, neutrophils, platelets and other inflammatory cells, modulation of coagulation components and synthesis and/or secretion of a number of bioactive compounds such as the endothelium derived relaxing factor(s) (EDRF's), platelet activating factor (PAF), eicosanoids, interleukines, tissue plasminogen activator, endotheline(s), various growth factors and of the heparan sulphate proteoglycans, involved in the binding of lipoprotein lipase to the luminal surface of the endothelial cell.

There is increasing evidence that the so-called bioactive lipids produced by the endothelium play a role in the inflammatory response to injury and that endothelial damage is involved in atherogenesis [2–6]. Therefore knowledge of endothelial lipid metabolism is of great importance. In this overview we will focuss on endothelial lipid transport, the metabolism of triacylglycerol, phospholipids and cholesterol(esters) in general and the specific synthesis and role of bioactive lipids such as eicosanoids and PAF in particular.

Studies concerning lipid metabolism in endothelial cells isolated from the myocardial vasculature are very limited. From this limited information, however, it can be concluded that with respect to metabolic properties in general there exists clear similarity between cultured endothelial cells from different origin and species. Moreover it is conceivable that lipid metabolic pathways do not differ fundamentally from those in other mammalian cell types in culture. A differential function in this respect is the capacity of endothelial cells to synthesize (and release) eicosanoids and other 'bioactive' lipids, albeit that this property is not unique.

Transport of lipids through the endothelium

Fatty acids

An important function of the endothelium is the transport of fatty acids from the plasma compartment to the underlying tissue. Albumin-bound fatty acids are thought to be transported across the endothelial cell lining by a number of mechanisms i.e. penetration via interendothelial clefts, by lateral diffusion in the plasmalemma, by simple diffusion as fatty acid across the endothelial cell or by a carrier mediated process. Bassingthwaighte *et al.* [7] have proposed and underpinned the following transport mechanism: 1. Binding of the albumin-fatty acid complex to a specific receptor on the luminal membrane of endothelial cells, 2. Rapid release of albumin and translocation of the fatty acid across the plasmalemma, cytosol and abluminal cell membrane by passive diffusion, and 3. Delivery of the fatty acid to albumin present in the interstitial space. Scow and coworkers [8] on the other hand have presented experimental evidence for the mass transport of fatty acids through the outer layer of endothelial membrane systems (see also this volume). In addition to the uptake of fatty acids derived from the albumin-fatty acid complex, fatty acids can be released from triacylglycerol-rich lipoproteins (chylomicrons and very low density lipoproteins (VLDL) after hydrolysis by lipoprotein lipase (LPL) present on the luminal site of the endothelial cells [9]. Local high concentrations of fatty acids, generated by LPL activity, have been shown to increase the transfer of the atherogenic low density lipoprotein (LDL) across the endothelium [10]. Stress hormones stimulate the functional LPL activity bound at the vascular endothelium [11]. Therefore during stress the myocardial endothelium is exposed to higher concentrations of fatty acids which may lead to an increased transfer of LDL.

Cholesterol

In addition to the receptor-mediated interaction of the endothelium with chylomicrons and VLDL, the endothelium contains specific receptors for LDL and high density lipoproteins (HDL) [9]. In addition the endothelium also contains β-VLDL- and scavenger-receptors [12, 13]. The number of receptors for chylomicrons, VLDL and the scavenger-receptors are not down regulable in contrast to the number of LDL- and β-VLDL-receptors [14]. Contact-inhibited endothelial cells have a low LDL-receptor activity [15], but they can still take up cholesterol from VLDL, chylomicrons and by means of the scavenger-receptor, which, however, may lead to intracellular cholesterol accumulation. The cholesterol taken up by the endothelial cell can be removed by HDL, after its binding to the HDL-receptor. A so called up regulation of the number of HDL-receptors has been described with an increased number of HDL-receptors in response of an increase in the cellular cholesterol content [16]. The number of HDL-receptors on the endothelium is much higher than the number of HDL-receptors on smooth muscle cells [17]. Inside the cell cholesterolesters are hydrolyzed by cholesterolesterase and free cholesterol can be esterified to cholesterolester, dependent on the amount of free cholesterol, by acyl-CoA-cholesterol-acyltransferase (ACAT).

Lipoprotein particles

Transport of lipoproteins across the endothelium may take place by so-called transendothelial channels, originating from the fusion of a series of plasmalemmal vesicles [18, 19]. Myocardial endothelium contains a large number of these plasmalemmal vesicles, indicating a high activity of this potential transport mechanism. In addition, transport of lipoproteins can occur through the endothelial gap-junctions. Under physiological conditions the gaps are not large enough for transport of the atherogenic lipoproteins. However, under conditions of hypertension, injury or a reduced energy state during myocardial ischemia, the size of the endothelial gaps may increase and allow penetration of atherogenic lipoproteins [20]. On the other hand endothelial cells may be involved in the process of atherosclerosis by modifying LDL. Endothelium and/or endothelium-attached monocytes or neutrophils readily oxidize LDL [21–23]. A role of endothelial cell lipoxygenase in this process has been presented [24]. Oxidized LDL enhances the binding of monocytes to the endothelium and inhibits the synthesis of platelet derived growth factor [25, 26]. Modified LDL also binds to the scavenger-receptor present on macrophages [21, 22]. The scavenger-receptor is not down regulated and therefore this uptake of oxidized LDL leads to cholesterol and lipid accumulation in the macrophages. Oxidized cholesterol is suggested to be an initiator of atherosclerotic plaque-formation. In addition to LDL also β-VLDL is modified after interaction with the endothelium, as follows from the increased binding of β-VLDL to smooth muscle cells after incubation with endothelial cells [26].

Triacylglycerol and phospholipid metabolism

(Phospho)lipolysis and the salvage of arachidonic acid

Circulating fatty acids, fatty acids derived from triacylglycerol-rich lipoproteins and fatty acids that originate from hydrolysis of smooth muscle or myocyte (phospho)lipids are reesterified in the endothelium to triacylglycerol and phospholipids. At low fatty acid concentrations, incorporation in phospholipids takes place and at higher concentrations the fatty acids are reesterified and stored as triacylglycerol [27–29]. In addition to reesterification of exogenous fatty acids, the endothelium contains the enzymes for *de novo* fatty acid, cholesterol, triacylglycerol and phospholipid synthesis [15, 29–31]. Contact inhibited endothelial cells, however, have a very low *de novo* rate of cholesterol and phospholipid synthesis [30, 32]. The route of fatty acid reesterification does not differ from the destination of fatty acids in other mammalian cells. In endothelial cells reacylation and transacylation (Land's pathway) of phospholipids is more important than *de novo* synthesis [34, 36]. Reacylation of phospholipids starts with its degradation by phospholipase A_2 and subsequent reacylation of the lysophospholipid catalyzed by specific

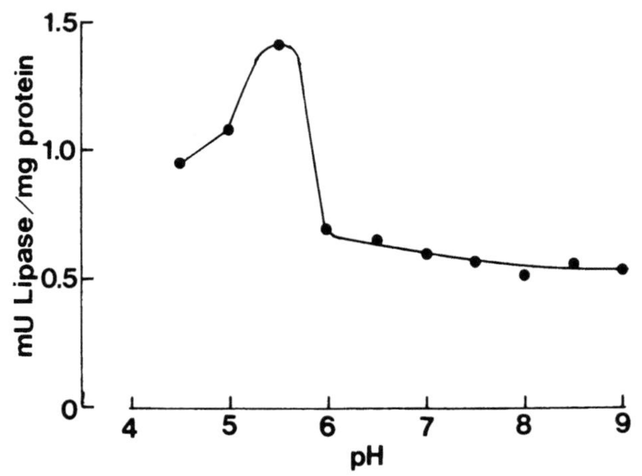

Fig. 1. pH dependency of triacylglycerolhydrolase activity in homogenates of vascular endothelial cells freshly prepared from rat heart according to Linssen *et al.* [48]. Triacylglycerolhydrolase activity was measured at various pH by the release of ^{3}H-oleic acid from a uniformly labeled trioleate suspension at 37° C as described in [50]. Results are the mean of two separate experiments in duplicate.

acyltransferases or transacylases [36, 37]. A consequence of the low *de novo* synthesis of phospholipids in endothelial cells is the relatively high incorporation of arachidonic acid and other polyunsaturated fatty acids in endogenous triacylglycerol when compared to other mammalian cells [27, 33]. In contrast, in adipose tissue, for instance, polyunsaturated fatty acids are mainly incorporated in phospholipids, probably because adipose tissue lacks a specific and high arachidonyl-CoA synthetase activity [34, 35]. The presence of arachidonyl-CoA synthetase in endothelial cells may specifically enhance the incorporation of arachidonic acid in triacylglycerol above the incorporation of other fatty acids in combination with the observed stimulation of diacylglycerol-acyltransferase by polyunsaturated fatty acids [34]. Moreover, at high fatty acid levels product inhibition of choline-P-cytidyl-transferase, the rate limiting step in phospholipid synthesis, by phosphatidylcholine may occur [37]. The high incorporation of arachidonic acid in endogenous triacylglycerol may be very useful because the endothelium lacks the Δ 6 desaturase activity and is not able to elongate and desaturate linolenic acid (18:3) to arachidonic acid (20:4) [27]. Therefore endogenous triacylglycerols may function as a pool of arachidonic acid for the reacylation of lysophospholipids. In addition to phospholipase A_2, also phospholipase D activity has been suggested to play an important role in the interconversion of different phospholipid-classes in the endothelium, especially the flow of fatty acids from phosphatidylcholine to phosphatidylethanolamine and triacylglycerol [38–41]. Protein kinase C is probably involved in the stimulation of the reacylation since phorbolesters stimulate both choline-P-cytidyl--transferase, the rate limiting enzyme in the synthesis of phosphatidylcholine, and phospholipase D involved in the degradation of phospholipids [38, 39, 42–44]. The alteration in fatty acid composition of endothelial cell triacylglycerols and phospholipids is very fast and its rate decreases in the order of triacylglycerol > phosphatidylcholine > phosphatidylethanolamine > phosphatidylinositol > phosphatidylserine [34]. In cultured endothelial cells oleic acid and linolenic acid are likely to stimulate flow of fatty acids from phosphatidylcholine to triacylglycerol [45–47]. As presented in Fig. 1, triacylglycerol in the endothelium is hydrolyzed by an acid triacylglycerol-hydrolase since no neutral triacylglycerol hydrolase activity was found in homogenates from rat myocardial endothelial cells, freshly isolated according to Linssen *et al.* [48]. It has been described for other cells that corticosteroids stimulate the turnover of fatty acids and subsequently the flow of fatty acids through the different lipid classes by stimulation of an acid triacylglycerol-hydrolase [34, 49]. Triacylglycerol is hydrolyzed by the acid lipase to diacylglycerol [50] and diacylglycerol can subsequently be converted to phospholipids or reacylated to triacylglycerol.

Endothelial fatty acid oxydation

A reduced energy state in endothelial cells, as may occur during myocardial ischemia, leads to endothelial injury [51]. Knowledge about energy metabolism in myocardial endothelial cells, however, is very limited. Using cultured rat heart endothelial cells, Spahr *et al.* [52] only found a low β-oxydation rate of palmitate which, moreover, was inhibited by more than 80% in the presence of 5 mM glucose. It was concluded that endothelial cells mainly use glucose as substrate. However, carnitine, necessary for the transport of fatty acids to its site of oxidation, the mitochondria, was not added in the experimental medium that contained fetal calf serum as only source of this cofactor, while the amount of carnitine endogenously present in the (cultured) endothelial cells may not be sufficient, as followed from the observations of Hülsmann *et al.* [53], who demonstrated an increased rate of fatty acid oxidation in aortic endothelial cells after addition of carnitine, and from the experiments of Van Hinsbergh *et al.* [54], who described a disappearance of accumulated triacylglycerols

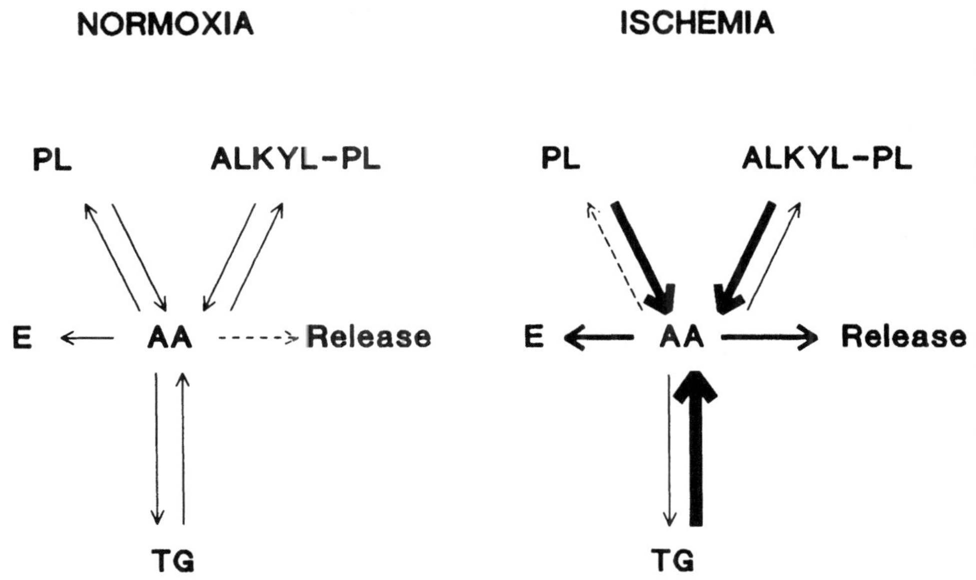

Fig. 2. Proposed pathways of arachidonic acid (AA) production during normoxia and ischemia/reperfusion. During ischemia (phospho)lipolytic activities are enhanced and (re)esterification activities depressed, leading to a rise in AA level in endothelial cells followed by AA release, especially during reperfusion. Abbreviations: PL = phospholipids; TG = triacylglycerol; Alkyl-PL = alkyl(ether)-phospholipids; E = eicosanoids.

in endothelial cells after incubation with carnitine. Moreover, endothelial cells possess a high carnitine acyl-CoA transferase activity, indicative for a high capacity for β-oxydation [55]. Mertens *et al.* [58], concluded that in the absence of exogenous substrates the endothelium mainly oxidizes endogenous fatty acids for their energy supply. Experiments with oleate instead of palmitate indeed showed a high endothelial rate of β-oxydation [53, 56, 57]. The difference between palmitate and oleate may be due to either a low transport of palmitate or a low activation rate of palmitate to palmitoyl-CoA in the endothelium since the incorporation of palmitate in various endothelial lipid classes is also very low when compared to oleate [29, 34].

Lipid metabolism of endothelial cells during ischemia/reperfusion

In myocardial endothelial cells phosphorylase activity is very low in comparison with its activity in cardiac myocytes [59]. As a consequence, the ischemic supply of glycerol-3-phosphate (from endogenous glycogen), the intracellular acceptor of fatty acids for reesterification, is hampered. Intracellular lipidosis, occurring as a consequence of fatty acid reesterification with glycerol-3-phosphate during the drop in pH from 7.5 to 7.0 is present in a number of tissues [60]. It can be regarded as a protecting mechanism, preventing fatty acid accumulation in, for instance, cardiac myocytes. The limited capacity of endothelial cells with respect to (re)esterification makes these cells more prone for fatty acid-induced deleterious processes. The release of arachidonate during ischemia [62] and reperfusion [61] may therefore originate from vascular endothelial cells (Fig. 2). Indeed, Figard *et al.* [63] have demonstrated the ability of isolated endothelial cells to release arachidonate from their triacylglycerol pool when external fatty acid concentration is low. The ischemia-induced rise in intracellular Ca^{2+}-level will promote phospholipase A_2 activity as well as phospholipase C and D activity [38, 64–66]. Since lyso-acyltransferase is hampered during ischemia, the released arachidonate can no longer be reesterified with lysophospholipids [62]. Instead arachidonate may be reesterified with diacylglycerol and stored in the endogenous triacylglycerol pool, when energy supply is still sufficient. Acidosis during ischemia may activate endothelial lysosomal acid lipase and hereby induce the degradation of endogenous triacylglycerols [67]. At a low energy status, fatty acid-activation and (re)esterification are depressed and release of fatty acids, in particular arachidonate, will ensue.

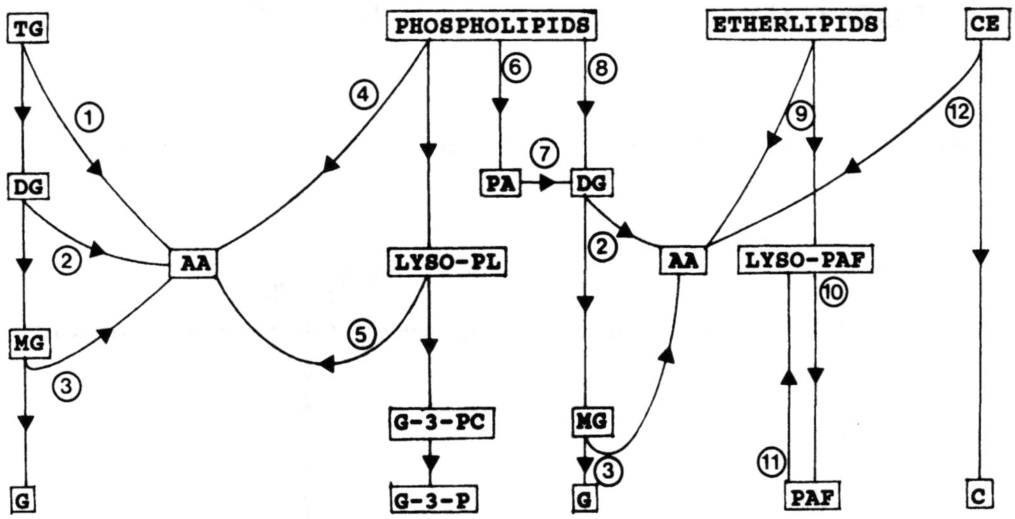

Fig. 3. Sources of arachidonic acid (AA) in endothelial cells. AA can be derived from triacylglycerol (TG), phospholipids, etherlipids and cholesterolesters. The enzymes involved are: 1. Triacylglycerolhydrolase; 2. Diacylglycerolhydrolase; 3. Monoacylglycerolhydrolase; 4. Phospholipase A_2; 5. Lyso-phospholipase; 6. Phospholipase D; 7. Phosphatidic acid phosphatase; 8. Phospholipase C; 9. Phospholipase A_2; 10. Acetyltransferase; 11. PAF acetylhydrolase; 12. Cholesterolesterhydrolase. Abbreviations: TG = Triacylglycerol; DG = Diacylglycerol; MG = Monoacylglycerol; G = Glycerol; AA = Arachidonic acid; LYSO-PL = Lysophospholipid; G-3-PC = Glycerophosphocholine; G-3-P = Glycerol-3-phosphate; PA = phosphatidic acid; LYSOPAF = 1-O-alkyl-2-hydroxyl-sn-glycero-3-phosphocholine; PAF = Platelet activating factor; CE = (Lipoprotein)- cholesterolester; C = Cholesterol.

Synthesis of bioactive lipids in the endothelium

Eicosanoids

The most important bioactive lipids synthesized by the endothelium are eicosanoids like prostacyclin (PGI_2), prostaglandin E_2 (PGE_2), thromboxane A_2, 15--hydroxy-5,8,11,13-eicosatetraenoic acid (15-HETE), 13-hydroxy-ocadecadienoic acid (13-HODE) and the so-called platelet activating factor (PAF) [2, 3]. The production of these substances is activated by numerous factors such as thrombin, histamine, bradykinin, shear stress and the interaction of endothelial cells with platelets and neutrophils [1–6, 65, 66, 68]. Whereas eicosanoids have a definite physiological importance in a variety of processes (local regulation of vessel tone, intravascular platelet activation, leukocyte traffic, endothelial cell mobility, the release of growth factors, cellular proliferation during angiogenesis and mediation of inflammation), PAF is involved in the aggregation and degranulation of platelets, polymorphonuclear leukocytes (PMN's) and monocytes in areas of inflammation [1, 3, 4].

The synthesis of the bioactive lipids starts with the release of arachidonic acid from its intracellular lipid sources like triacylglycerol, phospholipids, etherlipids and (lipoprotein)-cholesterolesters (see Fig. 3). Most of the released arachidonic acid ($\approx 80\%$) is liberated from phosphatidylcholine by a specific phospholipase A_2. A small part of arachidonic acid is released by the concerted action of phospholipase C and diacylglycerol-hydrolase [64, 69]. PGI_2 production from arachidonic acid released during shear-stress is mainly derived from the combined action of phospholipase C and diacylglycerol-hydrolase [66]. After stimulation of endothelial cells with thrombin, release of arachidonic acid from endogenous triacylglycerols was found [64], probably to replenish the arachidonic acid fraction in the phospholipid pool. Thrombin also activates phospholipase C leading to the production of diacylglycerol and to the activation of protein kinase C which subsequently stimulates phospholipase D [38, 41, 44]. As stated before, activation of phospholipase D stimulates the flow of fatty acids from triacylglycerol to phospholipids, leading to a release of arachidonic acid from triacylglycerol and a replenishment of arachidonic acid in the phospholipids. It has been proposed that the regulation of free arachidonic acid levels involves changes in the rate of reesterification [70].

The production of PGI_2 is also stimulated by HDL and LDL. Arachidonic acid esterified to cholesterol is an important component of HDL and LDL. Upon internalization of the lipoprotein and after release by the lysosomal cholesterolesterase, arachidonic acid is used for the biosynthesis of PGI_2 and/or incorporated in phospholipids and endogenous triacylglycerols [71, 72]. Arachidonic acid may also be hydrolysed from phosphatidylethanolamine or phosphatidylinositol. When endothelial cells are stimulated by the Ca^{2+}-ionophore A23187, release and subsequent reesterification of arachidonic acid in phosphatidylethanolamine and phosphatidylinositol have been found [64]. Especially the decline in arachidonic acid in the phosphatidylinositol fraction was very rapid. The Ca^{2+} that stimulates the various phospholipase activities is mainly derived from the extracellular fluid [3, 73]. The combined action of phospholipase C, producing diacylglycerol, and phospholipase D, that is activated by protein kinase C, may be necessary for reacylation of lysophospholipids, preventing the accumulation of this amphiphilic membrane perturbating agent in the endothelial cell. As stated before during myocardial ischemia, reacylation of lysophospholipids may be hampered, leading to a release of arachidonic acid from the endothelium. The main eicosanoid derivative of arachidonic acid in myocardial endothelium is PGI_2 [1–5]. Since myocardial endothelial cells lack the Δ 6 desaturase enzyme linoleate is not a precursor of arachidonic acid.

Platelet activating factor

Another important bioactive lipid produced by endothelial cells is PAF (1-O-alkyl-2-acetyl-sn-glycerol-3--phosphocholine). (For a review see [3].) Upon appropriate stimulation PAF is synthesized from alkylphospholipids (etherlipids) in the plasmamembrane by a combined action of a specific phospholipase A_2 and acetyltransferase. PAF is therefore locally formed, has only a surface expression and is short-lived. PAF production by endothelial cells is stimulated by the same compounds that cause PGI_2 synthesis and release. It is likely that the mechanism leading to the production of arachidonic acid and lyso-PAF are the same. This mechanism includes increased entry of extracellular Ca^{2+} after hormone-receptor binding by the opening of specific Ca^{2+}-channels, mediated by G-proteins [44, 74, 75]. The G-proteins may also couple the receptor binding to phosphatidylinositol turnover leading to Ca^{2+} influx and stimulation of the protein kinase C coupled phospholipase D. These pathways have been described in endothelial cells, the relative activity of each pathway, however, is still unknown. The elevation of intracellular Ca^{2+} leads to an activation of phospholipase A_2, liberating arachidonic acid from the membrane phospholipid pool as well as formation of lyso-PAF from etherlipids. The PAF formation from alkylphospholipid breakdown by phospholipase A_2 and subsequent acetyltransferase action on lyso-PAF is called the 'remodelling pathway' and is the main route of PAF synthesis in the endothelium. A small part of PAF is synthesized *de novo* from alkyldihydroxyacetonphosphate. The biological actions of PAF are ended after degradation of PAF by PAF-acetylhydrolase and resynthesis of alkyl phosphatidylcholine. The regulation of PAF-acetylhydrolase activity, largely unknown yet, may be of great importance for the physiological duration of PAF-mediated processes.

Acknowledgements

We thank dr. M.C.J.G. Linssen, Department of Physiology, University of Limburg, The Netherlands, for the preparation of myocardial vascular endothelial cells.

References

1. Jaffe EA: Cell biology of endothelial cells. Hum Pathol 18: 234–239, 1987
2. Vane JR, Gryglewski RJ, Botting RM: The endothelial cell as a metabolic and endocrine organ. TIPS 8: 491–496, 1987
3. Whatley RE, Zimmerman GA, McIntyre TM, Prescot SM: Lipid metabolism and signal transduction in endothelial cells. Prog Lipid Res 29: 45–63, 1990
4. Munro JM, Cotran RS: The pathogenesis of atherosclerosis; atherogenesis and inflammation. Lab Invest 58: 249–261, 1988
5. Gimbrone MA: Endothelial dysfunction and the pathogenesis of atherosclerosis. In: AM Gotto, LC Smith and B Allen (eds) Atherosclerosis V. Proceedings of the Fifth International Symposium. Springer Verlag, New York, Heidelberg, Berlin, 1979, pp 415–425
6. Reidy MA: Biology of disease. A reassesment of endothelial injury and arterial lesion formation. Lab Invest 53: 513–520, 1985
7. Bassinthwaighte JB, Noodleman L, Van der Vusse G, Glatz JFC: Modeling of palmitate transport in the heart. Mol Cell Biochem 88: 51–59, 1989
8. Scow RO, Blanchette-Mackie EJ, Smith LC: Transport of lipids across capillary endothelium. Fed Proc 39: 2610–2617, 1980
9. Cryer A: The role of the endothelium in myocardial lipoprotein dynamics. Mol Cell Biochem 88: 7–16, 1988
10. Hennig RB, Shasby DM, Spector AA: Exposure to fatty acids increases human low density lipoprotein transfer across cultured endothelial monolayers. Circ Res 57: 761–786, 1985
11. Hülsmann WC, Dubelaar ML: Lipoprotein lipases and stress

hormones: Studies with glucocorticoids and choleratoxin. Biochim Biophys Acta 875: 69–75, 1986
12. Stein O, Stein Y: Bovine aortic endothelial cells display macrophage-like properties towards ^{125}I-labelled low density lipoprotein. Biochim Biophys Acta 620: 631–635, 1980
13. Tanimura N, Asada Y, Hayashi T, Kisanuki A, Sumiyoshi A: Aortic endothelial cell damage induced by β-VLDL and macrophages *in vitro*. Atherosclerosis 85: 161–167, 1990
14. Baker DP, Van Lenten BJ, Fogelman AM, Edwards PA, Kean C, Berliner JA: LDL, scavenger, and β-VLDL receptors on aortic endothelial cells. Arteriosclerosis 4: 248–255, 1984
15. Fielding CJ: The endothelium, triglyceride-rich lipoproteins and atherosclerosis. Insights from cell biology and lipid metabolism. Diabetes 30 (suppl 2): 19–23, 1981
16. Tauber JP, Goldminz D, Gospodarowicz D: Up-regulation in vascular endothelial cells of binding sites of high density lipoprotein induced by 25-hydroxycholesterol. Eur J Biochem 119: 327–339, 1981
17. Savion N, Kotev-Emeth S: Cholesterol efflux from and high density lipoprotein binding to cultured bovine vascular endothelial cells are higher than with vascular smooth muscle cells. Eur J Biochem 183: 363–370, 1989
18. Palade GE: Transport in quanta across endothelium of blood capillaries. Anat Rev 136: 254–262, 1960
19. Williams SK: Vesicular transport of proteins by capillary endothelium. In: AL Copley, GV Seaman (eds) Surface phenomena in hemorheology. Their theoretical, experimental and clinical aspects. Vol 416 NY Acad Sci, New York, 1985, pp 457–465
20. Constantidines P: The arterial endothelium. Present state of knowledge. Wien Klin Wochenschr 94: 627–630, 1982
21. Hendriksen T, Mahoney EM, Steinberg D: Enhanced macrophage degradation of low density lipoproteins previously incubated with cultured endothelial cells: recognition by receptors for acetylated low density lipoprotein. Proc Natl Acad Sci USA 78: 6499–6503, 1981
22. Parthasarathy S, Printz DJ, Boyd D, Joy L, Steinberg D: Macrophage oxidation of low density lipoprotein generates a modified from recognized by the scavenger receptor. Arteriosclerosis 6: 505–510, 1986
23. Cathcart MK, Morel D, Chisolm GM III: Monocytes and neutrophils oxidize low density lipoprotein making it cytotoxic. J Leukocyte Biol 38: 341–350, 1985
24. Parthasarathy S, Wieland E, Steinberg D: A role for endothelial cell lipoxygenase in the oxidative modification of low density lipoprotein. Proc Natl Acad Sci USA 86: 1046–1050, 1989
25. Quinn MT, Parthasarathy S, Fong LG, Steinberg D: Oxidatively modified low density lipoproteins: A potential role in recruitment and retention of monocyte/macrophages during atherogenesis. Proc Natl Acad Sci USA 84: 2995–2998, 1987
26. Horrigan S, Campbell JH, Campbell GR: Effect of endothelium on β-VLDL metabolism by cultured smooth muscle cells of differing phenotype. Atherosclerosis 71: 57–69, 1988
27. Spector AA, Kaduce TL, Hoak JC, Fry GL: Utilization of arachidonic and linoleic acids by cultured human endothelial cells. J Clin Invest 68: 1003–1011, 1981
28. Spector AA, Mathur SN, Kaduce TL, Hyman BT: Lipid nutrition and metabolism of cultured mammalian cells. Prog Lipid Res 19: 155–186, 1981
29. Hoak JC, Czervionke RL, Lewis LJ: Uptake and utilization of free fatty acids (FFA) by human endothelial cells. Thromb Res 4: 879–883, 1974
30. Spector AA, Hoak JC, Fry GL, Stoll LL, Tanke CT, Kaduce TL: Essential fatty acid availability and prostacyclin production by cultured human endothelial cells. Prog Lip Res 20: 471–477, 1981
31. Roux FS, Mokni R, Hughes CC, Clouet PM, Lefauconnier JM, Bourre JM: Lipid synthesis by rat brain microvessel endothelial cells in tissue culture. J Neuropathol Exp Neurol 48: 437–447, 1989
32. Fielding PE, Vlodavsky I, Gospodarowicz D, Fielding CJ: Effect of contact inhibition on the regulation of cholesterol metabolism in cultured vascular endothelial cells. J Biol Chem 244: 749–755, 1979
33. Denning GM, Figard PH, Kaduce TL, Spector AA: Role of triglycerides in endothelial cell arachidonic acid metabolism. J Lipid Res 24: 993–1001, 1983
34. Rosenthal MD: Fatty acid metabolism of isolated mammalian cells. Prog Lipid Res 26: 87–124, 1987
35. Neufeld EJ, Wilson DB, Sprecher H, Majerus PW: High affinity esterification of eicosanoid precursor fatty acids by platelets. J Clin Invest 72: 214–220, 1983
36. Lands WEM, Crawford CG: Biosynthesis of cell components. In: A Martonosi (ed.) The enzymes of Biological Membranes Vol 2. Plenum Press, New York, 1976, pp 3–85
37. Bell RM, Coleman RA: Enzymes of triacylglycerol formation in mammals. In: PD Boyer (ed.) The enzymes. Acad Press Inc Vol XVI, 1983, pp 87–111
38. Huang CF, Chabot MC: Phorbol diesters stimulate the accumulation of phosphatidate, phosphatidylethanol and diacylglycerol in three cell types. Evidence for the indirect formation of phosphatidylcholine-derived diacylglycerol by a phospholipase D pathway and direct formation of diacylglycerol by a phospholipase C pathway. J Biol Chem 265: 14858–14863, 1990
39. Martin TW: Formation of diacylglycerol by a phospholipase-D-phosphatidate phosphatase pathway specific for phosphatidylcholine in endothelial cells. Biochim Biophys Acta 962: 282–296, 1988
40. Exton JH: Signaling through phosphatidylcholine breakdown. J Biol Chem 265: 1–4, 1990
41. Wey HE, Jakubowski JA, Deykin D: Incorporation and redistribution of arachidonic acid in diacyl and ether phospholipids of bovine aortic endothelial cells. Biochim Biophys Acta 878: 380–386, 1986
42. Kolesnick RN, Paley AE: 1,2-Diacylglycerols and phorbolesters stimulate phosphatidylcholine metabolism in GH3 pituitary cells. Evidence for separate mechanisms of action. J Biol Chem 262: 9204–9210, 1987
43. Daniel LW, Waite M, Wykle RL: A novel mechanism of diglyceride formation. 12-O-tetradecanoylphorbol-13-acetate stimulates the cyclic breakdown and resynthesis of phosphatidylcholine. J Biol Chem 261: 9128–9132, 1986
44. Fuse I, Tai HH: Regulation of arachidonate release by G-proteins and protein kinase C in human platelets. Adv Prostaglandin Thromboxane Leukotriene Res 19: 574–579, 1989
45. Tsai PY, Geyer RP: Fatty acid synthesis and metabolism of phospholipid acyl groups in strain L mouse fibroblasts. Biochim Biophys Acta 489: 381–389, 1977
46. Tsai PY, Geyer RP: Effect of exogenous fatty acids on the retention of phospholipid acyl groups by mouse L fibroblasts. Biochim Biophys Acta 528: 344–354, 1978

47. Hennig B, Shasby DM, Fulton AB, Spector AA: Exposure to free fatty acid increases the transfer of albumin across cultured endothelial monolayers. Arteriosclerosis 4: 489–497, 1984
48. Linssen MCJG, Vork MM, De Jong YF, Glatz JFC, Van der Vusse GJ: Fatty acid oxydation capacity and fatty acid binding protein content of different cell types isolated from rat heart. Mol Cell Biochem 98: 19–25, 1990
49. Tsai PY, Geyer RP: Stimulating effect of hydrocortisone on the catabolism of endogenous fatty acyl groups by fatty acid – supplemented mouse L fibroblasts. J Biol Chem 253: 5087–5089, 1978
50. Stam H, Hülsmann WC: Regulation of lipases involved in the supply of substrate fatty acids for the heart. Eur Heart J 6: 158–167, 1985
51. Stam H, Hülsmann WC, Jongkind JF, Van der Kraay AMM, Koster JF: Endothelial lesions, dietary composition and lipid peroxidation. Eicosanoids 2: 1–14, 1989
52. Spahr R, Krützfeldt A, Mertens S, Siegmund B, Piper HM: Fatty acids are not an important fuel for coronary microvascular endothelial cells. Mol Cell Biochem 88: 59–64, 1989
53. Hülsmann WC, Dubelaar ML: Aspects of fatty acid metabolism in vascular endothelial cells. Biochimie 70: 681–686, 1988
54. Van Hinsberg VWM, Emeis JJ, Havekes L: Interaction of lipoproteins with cultured endothelial cells. In: The Endothelial Cells – a pluripotent control of the vessel wall. 1st Int Endothelial Cell Symp of the ETCS, Paris, Karger, Basel, 1982, pp 99–112
55. Leighton B, Curi R, Hussein A, Newsholme EA: Maximum activities of some key enzymes of glycolysis, glutaminolysis, Krebs cycle and fatty acid utilization in bovine pulmonary endothelial cells. FEBS Lett 225: 93–96, 1987
56. Chace KV, Odessey R: The utilization by rabbit aorta of carbodydrates, fatty acids, ketone bodies and amino acids as substrates for energy production. Circ Res 48: 850–858, 1981
57. Kohane D, Cunningham L, Cohen R, Brecher P: Fatty acid utilization by vascular endothelium (Abstract). J Cell Biol 107: 863a, 1988
58. Mertens S, Noll T, Spahr R, Krützfeldt A, Piper HM: Energetic response of coronary endothelial cells to hypoxia. Am J Physiol 258: H689–H694, 1990
59. Krützfeldt A, Spahr R, Mertens S, Siegmund B, Piper HM: Metabolism of exogenous substrates by coronary endothelial cells in culture. J Mol Cell Cardiol 22: 1393–1404, 1990
60. McKenzie CG, McKenzie JB, Reiss OK: Increase in cell lipid and cytoplasmatic particles in mammalian cells cultured at reduced pH. J Lipid Res 8: 642–645, 1967
61. Van der Vusse GJ, Roemen THM, Prinzen FW, Coumans WA, Reneman RS: Uptake and tissue content of fatty acids in dog myocardium under normoxic and ischaemic conditions. Circ Res 50: 538–546, 1982
62. Chien KR, Han A, Sen A, Buja LM, Willerson JT: Accumulation of unesterified arachidonic acid in ischemic canine myocardium. Circ Res 54: 313–322, 1984
63. Figard PH, Hejlik DP, Kaduce TL, Stoll LL, Spector AA: Free fatty acid release from endothelial cells. J Lipid Res 27: 771–780, 1986
64. Ragab-Thomas JM, Hullin F, Chap H, Douste-Blazy L: Pathways of arachidonic acid liberation in thrombin and calcium-ionophore A23187 stimulated human endothelial cells: Respective roles of phospholipids and triacylglycerol and evidence for diacylglycerol formation from phosphatidylcholine. Biochim Biophys Acta 917: 388–397, 1987
65. Hong SL, McLaughin NJ, Tzeng CY, Patton G: Prostacyclin synthesis and deacylation of phospholipids in human endothelial cells: Comparison of thrombin, histamine and ionophore A23187. Thromb Res 38: 1–10, 1985
66. Bhagyalakshmi A, Frangos JA: Mechanism of shear-induced prostacyclin production in endothelial cells. Biochem Biophys Res Commun 158: 31–37, 1989
67. Schoonderwoerd K, Broekhoven-Schokker S, Hülsmann WC, Stam H: Involvement of lysosome-like particles in the metabolism of endogenous myocardial triglycerides during ischemia/reperfusion. Uptake and degradation of triglycerides by lysosomes isolated from rat heart. Basic Res Cardiol 85: 153–163, 1990
68. Hong SL: Effect of bradykinin and thrombin on prostacyclin synthesis in endothelial cells from calf and pig aorta and human umbilical cord vein. Thromb Res 18: 787–795, 1980
69. Martin TW, Wysolmerski RB, Lagunoff D: Phosphatidylcholine metabolism in endothelial cells: Evidence for phospholipase A and a novel calcium-independent phospholipase C. Biochim Biophys Acta 917: 296–307, 1987
70. Irvine RF: How is the level of free arachidonic acid controlled in mammalian cells? Biochem J 204: 3–16, 1982
71. Pomerantz KB, Fleisher LN, Tall AR, Cannon PJ: Enrichment of endothelial cell arachidonate by lipid transfer from high density lipoproteins: Relationship to prostaglandin I_2 synthesis. J Lipid Res 26: 1269–1276, 1985
72. Spector AA, Scanu HM, Kaduce TC: Effect of human plasma lipoproteins on prostacyclin production by cultured endothelial cells. J Lipid Res 26: 288–297, 1985
73. Hallam TJ, Jacob R, Merritt JE: Evidence that agonists stimulate bivalent – cation influx into human endothelial cells. Biochem J 255: 179–184, 1988
74. Burch RM: G protein regulation of phospholipase A2. Mol Neurobiol 3: 155–171, 1989
75. Irving HR, Exton JH: Phosphatidylcholine breakdown in rat liver plasma membranes. Roles of guanine nucleotides and P_2-purinergic agonists. J Biol Chem 262: 3440–3443, 1987

Molecular and Cellular Biochemistry **116**: 181–191, 1992.

Endothelium, the dynamic interface in cardiac lipid transport

Robert O. Scow and E. Joan Blanchette-Mackie
Endocrinology Section, Laboratory of Cellular and Developmental Biology,
National Institute of Diabetes and Digestive and Kidney Diseases, National Institutes of Health, Bethesda, MD 20892, USA

Abstract

Vascular endothelium is the dynamic interface in transport of lipid from blood to myocytes in heart and arteries. The luminal surface of endothelium is the site of action of lipoprotein lipase on chylomicrons and VLDL and the site of uptake of fatty acids from albumin. Fatty acids and monoacylglycerols are transported from the lumen in an interfacial continuum of endothelial and myocyte membranes. Lipoprotein lipase is transferred from myocytes to the vascular lumen, and is anchored there, by proteoheparan sulfate in cell membranes. Insulin, needed for synthesis of lipoprotein lipase and esterfication of fatty acids, is captured from the blood stream and delivered to myocytes by endothelial insulin receptors. Fatty acids, monoacylglycerols, lipoprotein lipase and insulin are transported along the same route, but by different mechanisms. The route involves the plasma membrane of endothelium and myocytes, the membrane lining transendothelial channels, and intercellular contacts. (Mol Cell Biochem **116:** 181–191, 1992)

Key words: fatty acid transport, monoacylglycerol transport, lipoprotein lipase, insulin, lipolysis, cell membranes, plasma triacylglycerol

Introduction

Vascular endothelium has several functions in the regulation of lipid transport from blood to myocytes in heart and arteries [1–6]. The principal form of lipid taken up is fatty acids for oxidation [7–9]. Fatty acids are delivered to these tissues as acyl components of triacylglycerol in chylomicrons and very low density lipoproteins (VLDL) [3, 4, 10–12], or as fatty acids bound to albumin molecules in the blood stream [12–14]. Uptake of fatty acids from albumin probably involves transient binding of albumin to the endothelium and direct transfer of fatty acids from albumin to endothelial cells [15, 16]. The interaction of low density and high density lipoproteins with endothelium of heart has been discussed elsewhere [6].

Uptake of acyl components from chylomicrons and VLDL requires the action of lipoprotein lipase at the luminal surface of vascular endothelium [1, 2, 10, 17]. Lipoprotein lipase is synthesized by myocytes and transferred to the luminal surface of vascular endothelium where it acts [18]. The lipase hydrolyzes triacylglycerol to fatty acids and 2-monoacylglycerols which are transported to myocytes [2, 5, 11]. Phosphatidylcholine and cholesterol are also transferred from chylomicrons to tissues during uptake of lipolytic products [19]. Fatty acids and 2-monoacylglycerols transported to myocytes are either esterified and stored temporarily as triacylglycerol, or oxidized immediately. Fatty acids are released from triacylglycerol stores in

Address for offprints: R.O. Scow, National Institutes of Health, Building 6, Room 137, Bethesda, MD 20892, USA

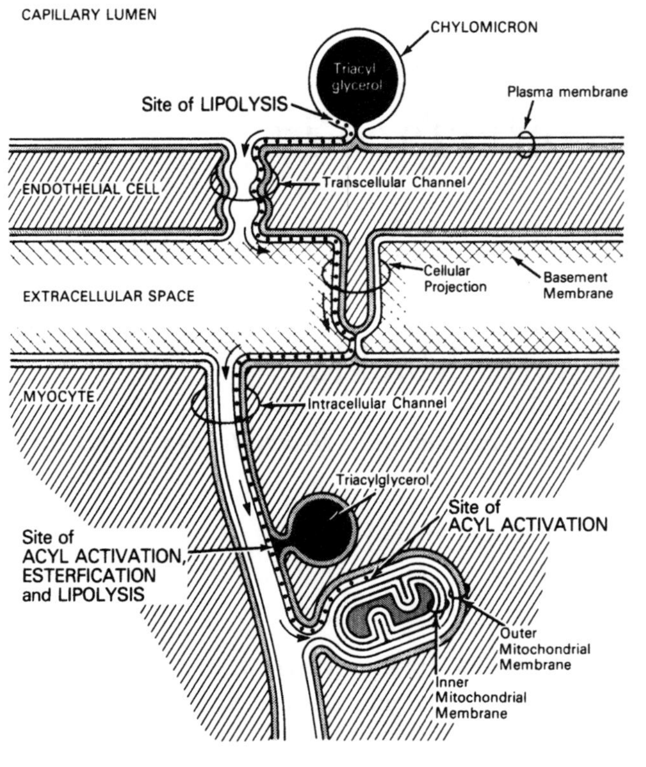

Fig. 1. A drawing showing the interfacial continuum of cell membranes (indicated by black dots) for transport of lipolytic products from triacylglycerol droplets in the capillary lumen and myocytes of heart to sites of triacylglycerol synthesis and fatty acid oxidation in myocytes. The interfacial continuum is composed of the outer (luminal) leaflets of plasma and intracellular membranes of endothelial cells and myocytes. Lipolytic products (fatty acids and monoacylglycerols) enter the continuum at sites of activity of lipoprotein lipase in capillaries and hormone-sensitive lipase in myocytes, and leave the continuum when they are activated and reesterified in sarcoplasmic reticulum for storage, or activated in mitochondria for transfer to the inner matrix for oxidation. Cellular projections of endothelium (as shown) and myocytes extend through the basement membrane to make contact with other cells. (From Scow and Blanchette-Mackie [5].)

myocytes by action of hormone-sensitive lipase when needed for oxidation [20].

Synthesis of lipoprotein lipase, as well as triacylglycerol, requires insulin [21, 22]. Insulin is delivered to the tissue by the blood stream, taken up from blood by insulin receptors in endothelial cells, and transported to myocytes [23, 24].

Transport of fatty acids, monoacylglycerol, lipoprotein lipase and insulin between the vascular lumen and myocytes requires crossing a continuous endothelium, the extracellular space between endothelial cells and myocytes, and the basement membrane of both kinds of cells [3, 5, 18, 24]. Of these substances, only insulin is soluble in aqueous media at physiological pH. The mechanisms by which these substances are transported between the blood stream and myocytes, and the role of endothelial cells in these processes are the subject of this review.

Transport of fatty acids and monoacylglycerols

Physical state of fatty acids and monoacylglycerol in aqueous media

Long chain fatty acids are insoluble in aqueous media at pH less than 9.6 [5, 25]. At pH below 7.0, fatty acids are mostly protonated (un-ionized) and usually locate in lipid droplets, and sometimes in lipid-water interfaces [5, 25]. Fatty acids form at pH 7.4–9.0 acid-soap complexes which locate solely in lipid-water interfaces [5, 25]. When fatty acid-soaps overcrowd a lipid-water interface, such as a monolayer or a leaflet of a bilayer, they expand the interface by making bilayered extensions of the interface, in the form of multilamellar structures, into the aqueous media [3, 5, 26–29]. Lowering pH of the aqueous media from 7.4–9.0 to below 7.0 rapidly changes the lamellar structures to oil droplets. Long chain fatty acids form micelles and disperse in the aqueous media only at or above pH 9.6, when $>90\%$ of the fatty acids are ionized [25].

Long chain monoacylglycerols are insoluble in aqueous media and locate only in lipid-water interfaces [5]. Monoacylglycerols and fatty acid-soaps have the same high affinity as phosphatidylcholines for lipid-water interfaces (i.e. lipid bilayers) [5].

Role of carrier proteins

Long chain fatty acids become 'solubilized' in blood when they bind to plasma albumin [5, 13, 14]. Albumin molecules in blood usually bind less than 2 fatty acid molecules, but under certain conditions (e.g. diabetes) they can bind up to 5 [13, 14]. Monoacylglycerols bind very poorly to albumin, with a molar ratio less than 0.33 [30], and thus are not transported in blood by albumin.

It is widely proposed that fatty acids may be transported in cells by fatty acid-binding proteins [16, 31–33]. These proteins belong to a family of low-molecular weight cytosolic proteins (14–16 kDa) which bind hydrophobic ligands and are found in a wide variety of tissues in many mammalian species [31]. While fatty

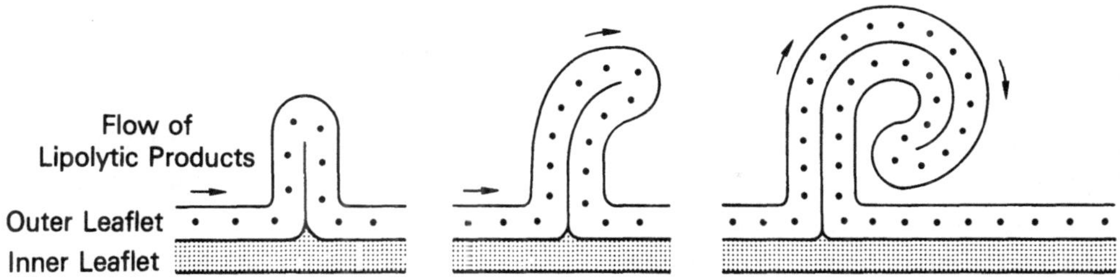

Fig. 2. Formation of lamellar structures associated with cell membranes. Overcrowding by fatty acids in the outer leaflet of a membrane causes lamellar extensions of the leaflet into the adjoining aqueous compartment. (From Blanchette-Mackie and Scow [26].)

acid-binding proteins appear to have some role in modulation of long chain fatty acid metabolism in tissues, their ability to enhance directly fatty acid uptake by cells is still debated [31] and their involvement in fatty acid transport is still hypothetical [33]. Furthermore, there is no direct evidence demonstrating specific fatty acid binding to fatty acid-binding protein *in situ* or *in vivo* [32].

The possible role of a specific binding protein in the metabolism of monoacylglycerols in cells and tissues has not been addressed.

Transport of fatty acids and monoacylglycerols in a continuum of cell membranes

We have studied movement of long chain fatty acids and monoacylglycerols in tissues and model membranes [2, 5, 34]. Our findings indicate that fatty acids and monoacylglycerols can be transported in cells and tissues by lateral movement in an interfacial continuum of cell membranes (Fig. 1) [1–5, 26–29].

We postulate that fatty acids and monoacylglycerols enter the continuum at sites of hydrolysis of triacylglycerol in capillaries or parenchymal cells, and flow in the continuum to sites in parenchymal cells where fatty acids and monoacylglycerols are removed for esterification and storage in sarcoplasmic (endoplasmic) reticulum, or fatty acids are removed for oxidation in mitochondria (Fig. 1) [1, 2, 5]. The interfacial continuum consists of the outer (luminal) leaflets of the plasma membrane and membranes of intracellular channels of endothelial cells, and the corresponding leaflets of plasma membrane, sarcoplasmic reticulum and outer mitochondrial membrane of parenchymal cells. The route of transport between capillaries and the interior of parenchymal cells has been marked by lipolytic products which overcrowded the continuum in fixed tissues and formed lamellar structures in aqueous compartments adjoining the continuum (Fig. 2).

Interfacial continuum of cell membranes

Evidence for an interfacial continuum of cell membranes between the luminal surface of capillaries and the outer leaflet of plasma membrane of parenchymal cells was obtained primarily in electron microscopic studies of adipose tissue [1, 2, 5, 29, 35, 36]. The vascular endothelium in heart and adipose tissue is continuous, and the tight junctions between endothelial cells preclude passage of substances between or around endothelial cells. Capillaries in these tissues, however, have transcellular channels which connect the capillary lumen with extracellular space (Fig. 1)]1, 26, 35]. The membranes lining these channels provide an interfacial continuum across endothelial cells. Cell processes of endothelial and parenchymal cells are often found extending through the basement membrane to make contact with other cells (Fig. 1) [5, 26, 29, 35]. Evidence for fusion of external leaflets of plasma membranes of adjoining cells was observed between capillary endothelium and a parenchymal cell in white adipose tissue [35].

Studies made with extracellular markers (tannic acid and lanthanum) showed that intracellular channels in adipocytes and liver cells extend from the extracellular space to the surface of intracellular lipid droplets [35, 36]. Thus, the membranes lining these intracellular channels are continuous with plasma membrane. Tannic acid and lanthanum surrounded lipid droplets which appeared to project into the lumen of intracellular channels [35, 36]. Because the primary site of triacylglycerol synthesis is endoplasmic reticulum, the intracellular channels associated with lipid droplets were identified as endoplasmic reticulum. The very thin electron opaque line seen between the lumen and the droplet suggests that a monolayered structure, probably a leaflet of the reticulum, separates the triacylglycerol droplet from the lumen of endoplasmic reticulum [36]. Long chain triacylglycerol molecules would be expected to accumulate between leaflets of endoplasmic reticulum because their affinity for lipid-water interfaces is much less than that of fatty acids, monoacylglycerols and phospholipids [1, 2, 5, 36]. Thus, the luminal leaflet of endoplasmic reticulum provides an interfacial continuum in adipocytes between the outer leaflet of plasma membrane and the surface of intracellular lipid droplets (Fig. 1).

Membrane continuity has been observed between endoplasmic reticulum and outer mitochondrial membrane in several tissues [5]. Fatty acids are activated in the luminal leaflet of outer mitochondrial membrane before being transferred to the mitochondrial matrix for oxidation (see [3, 37] for references). Long chain fatty acids lose their strong affinity for membrane leaflets and become hydrophilic when bound to CoA [5]. We postulate that these changes result in immediate displacement of activated fatty acids from the luminal leaflet of outer mitochondrial membrane to the aqueous phase of the outer mitochondrial chamber where the activated fatty acids react with carnitine for transfer across inner mitochondrial membrane to the matrix for oxidation [5].

These observations constitute the morphological evidence for an interfacial continuum of membrane leaflets between the lumen of capillaries, intracellular lipid droplets, and mitochondria in parenchymal cells (Fig. 1).

Fatty acid flow in membranes

Evidence for fatty acid flow in cell membranes is derived from experimental observations showing the formation of lamellar structures continuous with membrane leaflets in tissues in which fatty acids were produced by lipolysis. Studies showed that lamellar structures developed in glutaraldehyde-fixed adipose tissue incubated at 38° C in medium at pH 8.3 [37]. Lamellar structures were found in vesicles and vacuoles of capillary endothelium and in extracellular space of adipose tissue which had been perfused prior to fixation with chylomicrons. Biochemical studies showed that lipoprotein lipase was active and could hydrolyze chylomicron-triacylglycerol to fatty acids in glutaraldehyde-fixed specimens.

Electron microscopic study of latacting mammary gland, which has a very high level of lipoprotein lipase activity, showed the presence of lamellar structures in capillary endothelium and extracellular space in tissue taken from rats injected with chylomicrons [2]. Continuity between cell membranes and leaflets of lamellar structures was found. This suggested that the lamellar structures were lipolytic products which formed bilayered extensions of cell membranes (Fig. 2) [2].

We then studied several tissues to determine conditions which might affect formation and distribution of lipolytic products in fixed tissues [3, 26–29]. Lipolytic products formed from plasma triacylglycerol were studied in heart, muscular artery, white adipose tissue and brown adipose tissue from rats injected with chylomicrons or emulsified triacylglycerol. Lipolytic products formed from intracellular triacylglycerol were studied in heart from fed rats, white adipose tissue from fasted rats, brown adipose tissue from rats exposed to cold, and mouse adipose tissue treated with lipolytic hormone. Lipoprotein lipase and hormone-sensitive lipase hydrolyze triacylglycerols to fatty acids and 2-monoacylglycerols [30, 38]. Whether monoacylglycerol lipase was active in fixed tissue was not determined. Nonetheless, 2-monoacylglycerols could be isomerized in fixed tissues to 1(3)-isomers and hydrolyzed by lipoprotein lipase or hormone-sensitive lipase to fatty acids and glycerols [1, 30]. In most experiments, the tissues were fixed at 4° C with glutaraldehyde at pH 7.4, and incubated at 25° C in medium at pH 7.4. Glutaraldehyde fixation stopped reesterification and fatty oxidation but not lipolysis, and thereby caused accumulation of lipolytic products in the tissues. Visualization of lamellar structures in tissues was enhanced by adding tannic acid to the incubation medium.

Findings in heart and muscular artery were similar to those in adipose tissue. Lamellar structures were found in capillaries, extracellular space and myocytes of the left ventricle of rat heart incubated for 30 min in tannic acid solution, pH 7.4, at 25° C after being fixed with glutaraldehyde at 4° C [3]. About 7% of the lamellar structures found in heart from young rats injected with chylomicrons were in capillaries, 10% were in extracellular space, and 83% were in myocytes. The lamellar structures in capillaries were located in the lumen near chylomicrons and in intracellular channels of endothelial cells (Fig. 3-A). The lamellar structures in extracellular space were sometimes continuous with lamellar structures located in intracellular channels of endothelial cells (Fig. 3-A).

Some of the lamellar structures in myocytes were located in intracellular channels (sarcoplasmic reticulum) near cell surfaces (Fig. 3-B), in channels near intracellular lipid droplets, at the periphery of lipid

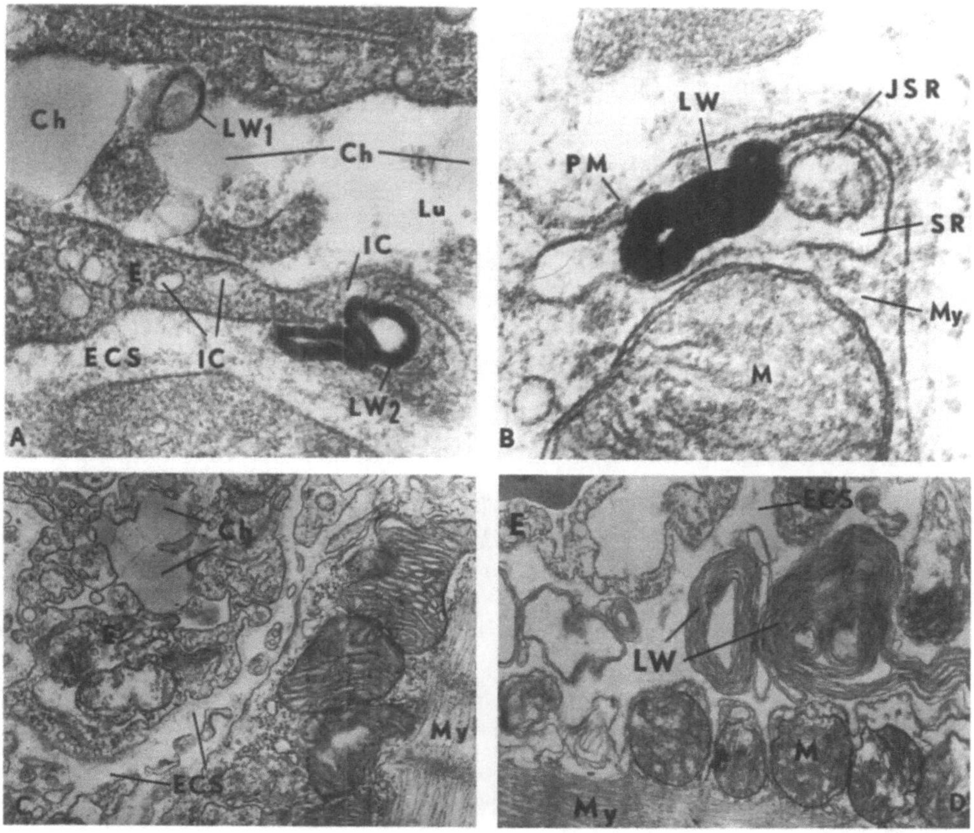

Fig. 3. Sections of heart from young rats injected with chylomicrons. The rats were injected 1.5 min before the tissues were taken for study. A and B. These tissues were fixed with glutaraldehyde for 30 min at 4° C, incubated in sodium cacodylate solution containing tannic acid at pH 7.4 for 30 min at 25° C, and then fixed in osmium tetroxide at 4° C. Note that the lamellar structures (LW) in these tissues (Figs. 3-A and 3-B), which were incubated with tannic acid, are tightly packed in contrast to the widely spaced lamellar structures in tissues incubated without tannic acid (Fig. 3-D). In panel A, a lamellar structure (LW_1) is associated with a chylomicron (Ch) in the capillary lumen (Lu), and another lamellar structure (LW_2) extends from the lumen of an intracellular channel (IC) of the endothelium (E) into the extracellular space (ECS). In panel B, a lamellar structure is present in the lumen of sarcoplasmic reticulum (SR) in a myocyte. C. This tissue was taken 1.5 min after injection of chylomicrons and immediately fixed with osmium tetroxide at 4° C. Note the presence of chylomicrons (Ch) in the capillary lumen and the absence of lamellar structure in the extracellular space (ECS). D. This tissue was incubated in sodium cacodylate solution (without tannic acid) at pH 7.4 for 30 min at 25° C before being fixed in osmium tetroxide at 4° C. Note the large number of widely spaced, concentrically arranged lamellar structures (LW) in the extracellular space (ECS) between the capillary endothelium and myocyte, and the absence of chylomicrons in the capillary lumen. JSR, junctional sarcoplasmic reticulum; M, mitochondrion; My, myocyte; PM, plasma membrane. A, × 50,000; B, × 102,000; C, × 21,000; D, × 10,000. (From Wetzel and Scow [3].)

droplets, and in the nuclear envelope [3]. Most of the lamellar structures in myocytes, however, were associated with mitochondria. They were located in the lumen of intracellular channels near mitochondria, in intracellular channels which were continuous with the outer mitochondrial chamber, in the outer mitochondrial chamber, and sometimes in the intermembrane space of cristae. Similar observations were made in incubated-fixed brown adipose tissue from young rats exposed to cold [27]. Although intracellular channels containing lamellar structures in myocytes were often located near mitochondria and lipid droplets, it was not possible to identify these channels either as part of the T-tubule system or as sarcoplasmic reticulum [3].

The relationship between intravascular lipolysis and development of lamellar structures was studied in unfixed heart tissue from fasted young rats injected with chylomicrons [3]. There were numerous chylomicrons in capillaries and virtually no lamellar structures in extracellular space or myocytes in heart tissue taken 1.5 min after injection and fixed immediately with OsO_4 (Fig. 3-C). Chylomicrons disappeared from the lumen of capillaries, and lamellar structures developed in the extracellular space of heart tissue which was incubated

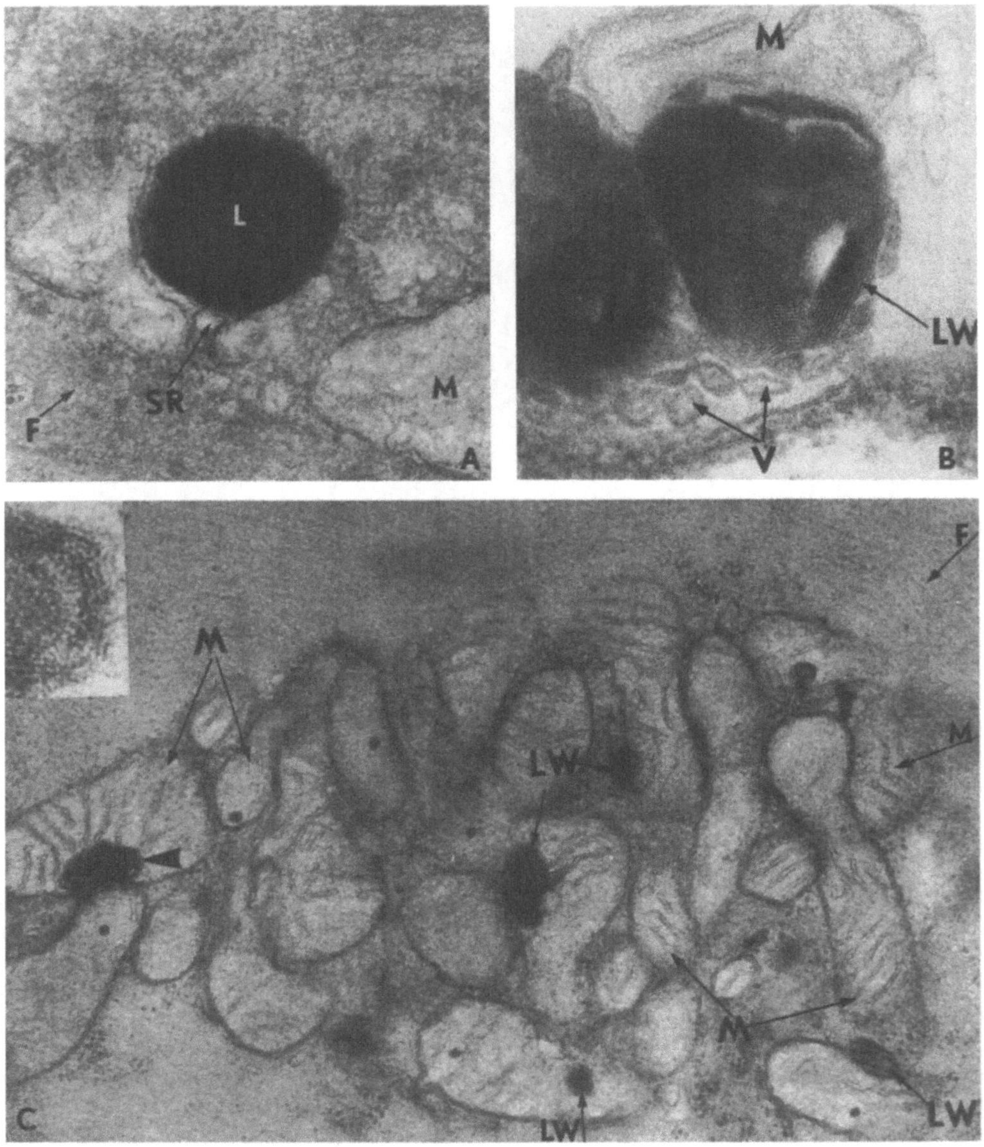

Fig. 4. Sections of a muscular (dorsalis pedis) artery taken from adult rats infused with a triacylglycerol emulsion (Intralipid) for 1.5–5 hr. The tissues were fixed with glutaraldehyde at 4° C, incubated 0 or 30 min at 25° C in sodium cacodylate solution containing tannic acid at pH 7.4, and postfixed with osmium tetroxide at 4° C. A. This section of unincubated tissue shows a lipid droplet (L) near an intracellular channel (IC) and a mitochondrion (M) in an arterial smooth muscle cell. F, myofibril. × 97,000. B. This section of tissue incubated 30 min after being fixed shows lamellar structures (LW) near a mitochondrion (M) in an arterial smooth muscle cell. V, vesicles. × 119,000. C. This section of tissue incubated 30 min after being fixed shows lamellar structures (LW) near and inside mitochondria (M) in a smooth muscle cell. Inset shows at a higher magnification the repeating periodicity of 40 Å in the lamellar structure marked by an arrowhead. F, myofibrils, × 47,000. Inset, × 250,000. (From Reinilä *et al.* [4].)

for 30 min in medium (without tannic acid) at pH 7.4 before being fixed with OsO_4 (Fig. 3-D). There was a small increase in volume of lamellar structures in capillaries but not in myocytes. It is not known whether fatty acids were oxidized or esterified in myocytes during incubation. Accumulation of fatty acids outside of the capillaries, not in the lumen where lipolysis occurred, indicates that transfer of fatty acids to the extracellular space in incubated tissues was a directed process. Such transfer was probably effected by lateral movement of fatty acids in a continuum of cell membranes, with formation of lamellar structures when fatty acids overcrowded the continuum near the extracellular space (Fig. 2).

In order to study the relationship between intracellular lipolysis and development of lamellar structures in myocytes, the triacylglycerol content of heart in adult rats was increased by infusing intravenously the animals with an emulsion of triacylglycerol [3]. The hearts were then fixed *in situ* by perfusion with ice-cold glutaraldehyde and blocks of tissue were incubated in tannic acid solution at pH 7.4 for 90 min at 25° C. During incubation, the amount of lamellar structures associated with mitochondria increased 60 fold and the total volume of lipid droplets in myocytes decreased 50%. Lamellar structures associated with mitochondria accounted for 70% of the lamellar structures in the tissue. Similar findings were made in heart of young rats injected with chylomicrons. These findings indicate that lamellar structures associated with mitochondria in incubated heart tissues were composed of fatty acids produced, at least in part, by lipolysis in myocytes.

Lamellar structures also developed in incubated muscular artery of rats infused intravenously with an emulsion of triacylglycerol before tissues were taken for study [4]. Arterial tissue from infused rats, but not control rats, contained lipid droplets at the luminal surface of arterial endothelial cells and lipid droplets in smooth muscle cells of the media (Fig. 4-A). Tissues incubated for 30 min developed lamellar structures at the luminal and abluminal surfaces of endothelial cells, within endothelial cells, in the extracellular space near plasma membrane of myocytes, and inside myocytes (Fig. 4-B), within intracellular channels, near intracellular lipid droplet and mitochondria, and inside mitochondria (Fig. 4-C). Lamellar structures were not found in incubated tissue from control (uninfused) animals. The distribution of lamellar structures in incubated tissue suggests that transport of fatty acids from the vascular lumen to the interior of medial smooth muscle cells in muscular arteries also occurs by lateral movement in a continuum of cell membranes (Fig. 1).

We have also studied the effect of pH on formation of lamellar structures from fatty acids produced by hormone-stimulated lipolysis in mouse adipose tissue [28]. Freeze-fracture electron microscopy showed lamellar structures in intracellular channels of adipocytes and endothelium, in extracellular space, and in capillary lumen of tissues processed at pH 8.1. Lamellar structures were not found, however, in tissues processed at pH 7.0. Instead, the fatty acids, which were protonated at this pH, formed smooth-surfaced lipid droplets in the extracellular space.

Fatty acid domains were demonstrated in the outer leaflet of cell membranes with freeze-fracture electron microscopy in mouse adipose tissue treated with isoproterenol and processed at pH 9.0 [29]. The particle-free areas were formed by fatty acid-soaps which entered the continuum of outer leaflets of membranes at sites of lipolysis and moved between and within cells in the plane of the continuum. When fatty acids overcrowded the continuum, they produced foldings in particle-free areas and later formed lamellar structures which projected into the lumen of intracellular channels, extracellular space and capillary lumen.

We conclude that the lamellar structures found in incubated heart and arterial tissue were composed of fatty acids which were produced by lipolysis in capillary lumen or myocytes, and transferred by lateral movement in an interfacial continuum to other sites where they formed bilayered extensions of the continuum. These findings demonstrate in fixed tissues transport of fatty acids from chylomicrons in the capillary lumen to the abluminal surface of endothelial cells and extracellular space, and from intracellular lipid droplets to the interior of mitochondria in myocytes. *In vivo,* overcrowding of membrane leaflets would be less, if not absent, because fatty acids would be removed from the membrane continuum by physiological processes such as esterfication and oxidation. Fatty acids would be present and flow in the continuum as 1:1 acid-soap complexes.

We propose that *in vivo* fatty acids and monoacylglycerols are transported in heart and muscular artery by lateral movement in cell membranes, with transport across endothelium in a continuum composed of the outer (luminal) leaflet of plasma and intracellular membranes of endothelial cells, and transport within myocytes in a continuum composed of the outer (luminal) leaflet of sarcoplasmic reticulum and the outer mitochondrial membrane (Fig. 1). At present, there is no evidence in heart for an interfacial continuum between endothelial cells and myocytes, or membrane continuity between the plasma membrane of myocytes and sarcoplasmic reticulum. We suggest that the route for transport of fatty acid/monoacylglycerol from capillaries to sarcoplasmic reticulum in heart and artery is probably similar to that demonstrated in adipose tissue.

Transport of lipoprotein lipase

Lipoprotein lipase is essential for uptake of fatty acids and monoacylglycerols from plasma triacylglycerol [1,

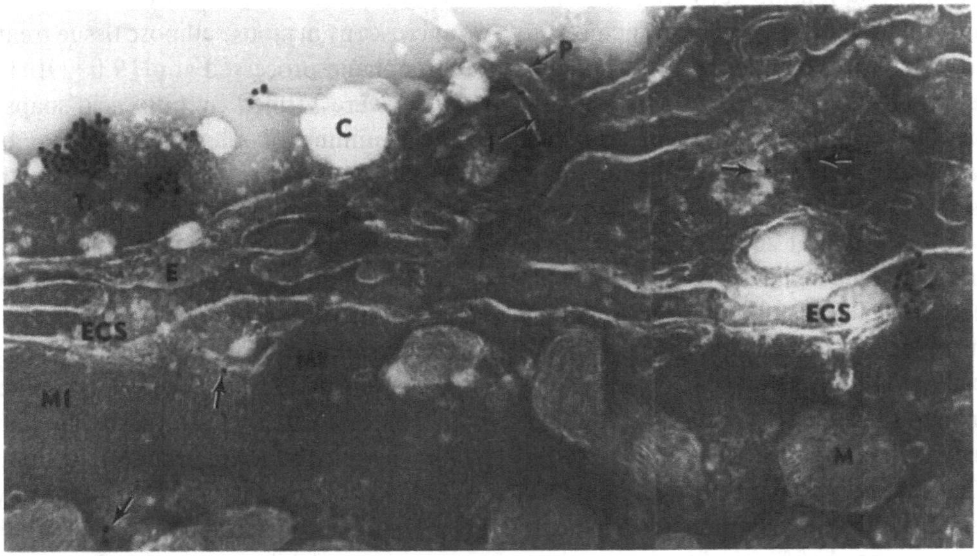

Fig. 5. Immunogold labeling of lipoprotein lipase in mouse heart. Note the string of gold particles (arrowheads) marking lipoprotein lipase molecules crossing the extracellular space (ECS), probably on the surface of a cellular projection, between a myocyte (My) and an endothelial cell. Also note the presence of immunolabeled lipase (arrows) in myocytes and endothelium, and the high concentration of lipase (T) at the luminal surface of the capillary endothelium. C, chylomicron; I, interendothelial junction; M, mitochondrion; Mf, myofibrils; P, endothelial process. X 69,000. (From Blanchette-Mackie *et al.* [18].)

2, 10, 17]. It is synthesized in the parenchymal cells of tissues and transported to the luminal surface of vascular endothelium where it acts [18]. Lipoprotein lipase is attached to the luminal surface of endothelial cells through an interaction with polysaccharide chains of proteoheparan sulfate [6, 39]. Lipoprotein lipase is readily released from endothelium into the blood stream by intravenous heparin.

Lipoprotein lipase in its normal active form is a noncovalent dimer of identical glycopeptides [39]. Intracellular events involved in the synthesis of lipoprotein lipase have been studied mostly in adipocytes [40]. The results show that lipoprotein lipase subunits are glycosylated in endoplasmic reticulum, the high mannose-type oligosaccharide chains are trimmed and processed to complex-type chains in Golgi, and the subunits are dimerized before the lipase is secreted. Heparin readily releases lipoprotein lipase from cultured cells, suggesting that the lipase is anchored to cell surfaces by proteoheparan sulfate [18, 39, 40]. Recent findings in cultured heart myocytes indicate that proteoglycans which bind lipoprotein lipase may be linked to cell surfaces by phosphatidylinositol-glycan anchors [41]. Accordingly, lipase was released from such cells by both heparin and phosphatidylinositol-specific phospholipase C. The extent to which double anchoring of lipoprotein lipase occurs is not known.

We studied the route of transport of lipoprotein lipase from myocytes to capillary lumen in heart of young mice [18]. The lipase was immunolocalized by electron microscopy, using a double antibody technique. About 78% of the lipase in heart was located in myocytes, 4% in extracellular space, and 18% in capillary endothelium. Lipoprotein lipase in myocytes was located primarily in sarcoplasmic reticulum, Golgi sacs, transport vesicles, and secretory vesicles at the cell periphery. Lipase in extracellular space was present near the orifice of secretory vesicles of myocytes and in narrow zones spanning the space between myocytes and capillary endothelium (Fig. 5).

About one-half of the lipase associated with capillary endothelium was inside cells, and the rest was mostly at the luminal surface of cells. Lipoprotein lipase inside endothelial cells was found in the lumen of vesicular structures, indicating that lipase was transported across endothelium via vesicles or transcellular channels. There was a concentration gradient of lipoprotein lipase between myocytes and the lumen of capillaries, with the lowest concentration in myocytes, in extracellular space and at the basal surface of endothelium, and the highest concentration at the luminal surface of endothelial projections (Fig. 5).

Our findings indicate that lipoprotein lipase in heart is synthesized and secreted by myocytes and transferred across extracellular space, probably along the surfaces of cellular projections, and across endothelial cells, via intracellular vesicles or transcellular channels, to the luminal surface of capillaries, where the lipase is concentrated at the surface of endothelial processes and acts on chylomicrons and VLDL [18]. Studies with bovine aortic endothelial cell monolayers showed that transport of active lipoprotein lipase from the basal to the apical (luminal) surface of endothelial cells required proteoheparan sulfate [42]. We propose that lipoprotein lipase molecules are transported by binding to successive molecules of proteoheparan sulfate located in or along plasma and intracellular membranes of myocytes and endothelial cells. Transport of lipase from myocytes to endothelium probably occurs at points of contact where lipase molecules could be transferred from proteoglycans of myocytes to proteoglycans of endothelial cells. These connections would involve processes of cells extending through the basement membrane to make contact with other cells. Both myocytes and endothelial cells synthesize proteoheparan sulfate needed for binding lipoprotein lipase.

Lipoprotein lipase molecules are suspended into the vascular lumen by proteoheparan sulfate molecules attached to endothelium [6, 39]. This positioning is thought to facilitate the interaction of chylomicrons and VLDL with lipoprotein lipase. Binding of these lipid particles to endothelium may also involve a direct interaction of apolipoproteins B and E with proteoheparan sulfate [39]. Fusion of the surface film of chylomicrons and VLDL with the outer leaflet of the luminal plasma membrane of endothelium (Fig. 1) is probably enhanced through the formation of lysophosphatidylcholine, 2-monoacylglycerol and fatty acid by the action of lipoprotein lipase on surface film and triacylglycerol core of these lipoproteins [30, 43, 44]. This fusion results in the direct flow of about one-half of the fatty acids and all of the monoacylglycerols produced, as well as some of the phospholipid and cholesterol of the surface film [19], into myocytes via an interfacial continuum of cell membranes. The rest of the fatty acids are transported to liver by albumin in the blood stream.

Transport of insulin

Synthesis of lipoprotein lipase and esterification of fatty acids in myocytes of heart require insulin [21, 22] Insulin is delivered to heart and arteries by the blood stream, and transported across the endothelium to myocytes by a receptor-mediated process [23, 24]. The receptors involved in transport of insulin across endothelial cells are synthesized by endothelial cells and are similar in structure to those found in other cells [23, 45–47]. The insulin receptor is a transmembrane integral protein, spanning both leaflets of the membrane. Insulin receptors are also present in heart myocytes [24, 48]. Studies in rat heart showed that I^{125}-insulin crossed from perfusing fluid to myocytes within 2 min [24].

Uptake and internalization of insulin by most cells involves receptor-mediated endocytosis [49, 50]. It is assumed that a similar process is involved in transport of insulin across endothelial cells [24, 46]. There is evidence that insulin receptors have considerable lateral mobility in cell membranes [49, 50]. Because insulin receptors span both leaflets of membranes, it is not feasible for receptors to flow in an interfacial continuum of membranes from endothelium to myocytes. However, points of contact between cells, resulting from cellular processes penetrating basement membrane, could facilitate transfer of insulin molecules from endothelial receptors to myocyte receptors.

Conclusions

Vascular endothelium is the dynamic interface in transport of lipid from blood to myocytes in heart and arteries. The luminal surface of endothelium is the site of action of lipoprotein lipase on plasma triacylglycerol, uptake of fatty acids from albumin, and capture of insulin from blood. Fatty acids and monoacylglycerols are transported to myocytes in an interfacial membrane continuum. Lipoprotein lipase is transferred from myocytes to the vascular lumen by proteoheparan sulfate in cell membranes. Insulin, needed for synthesis of lipoprotein lipase and esterfication of fatty acids, is captured from the blood and delivered to myocytes by endothelial insulin receptors. Fatty acids, monoacylglycerols, lipoprotein lipase and insulin are transported along the same route, but by different mechanisms. The route involves the plasma membrane of endothelium and myocytes, the membrane lining transendothelial channels, and intercellular contacts.

References

1. Scow RO, Blanchette-Mackie EJ, Smith LC: Role of capillary endothelium in the clearance of chylomicrons. A model for lipid

transport from blood by lateral diffusion in cell membranes. Circ Res 39: 149–161, 1976
2. Scow RO, Blanchette-Mackie EJ, Smith LC: Transport of lipid across capillary endothelium. Federation Proc 39: 2610–2617, 1980
3. Wetzel MG, Scow RO: Lipolysis and fatty acid transport in rat heart: electron microscopic study. Am J Physiol 246: C467–C485, 1984
4. Reinilä A, Blanchette-Mackie EJ, Scow RO: Uptake of plasma triacylglycerol by a muscular artery in the rat: an ultrastructural study. Virchows Arch [Cell Pathol] 47: 67–78, 1984
5. Scow RO, Blanchette-Mackie EJ: Why fatty acids flow in cell membranes. Prog Lipid Res 24: 197–241, 1985
6. Cryer A: The role of the endothelium in myocardial lipoprotein dynamics. Mol Cell Biochem 88: 7–15, 1989
7. Neeley JR, Morgan HE: Relationship between carbohydrate and lipid metabolism and the energy balance of heart muscle. Ann Rev Physiol 36: 413–459, 1974
8. Whitner JT, Idell-Wenger JA, Rovetto MJ, Neely JR: Control of fatty acid metabolism in ischemic and hypoxic hearts. J Biol Chem 253: 4305–4309, 1976
9. Zierler KL: Fatty acids as substrate for heart and skeletal muscle. Circ Res 38: 459–463, 1976
10. Robinson DS: Function of the plasma triglycerides in fatty acid transport. In: M Florkin and EH Stotz (eds) Comprehensive Biochemistry. Elsevier/North-Holland, Amsterdam, 1970, Vol 18, pp 51–116
11. Linder C, Chernick SS, Fleck TR, Scow RO: Lipoprotein lipase and uptake of chylomicron triglyceride by skeletal muscle of rats. Am J Physiol 231: 860–864, 1976
12. Idell-Wenger JA, Neely JR: Regulation of uptake and metabolism of fatty acids in muscle. In: JM Dietschy, AM Gotto Jr, and JA Ontko (eds) Disturbances in Lipid and Lipoprotein Metabolism. Am Physiol Soc, Bethesda MD, 1978, pp 269–284
13. Scow RO, Chernick SS: Mobilization, transport, and utilization of free fatty acids. In: M Florkin and EH Stotz (eds) Comprehensive Biochemistry. Elsevier/North-Holland, Amsterdam, 1970, Vol 18, pp 19–50
14. Spector AA, Fletcher JE: Transport of fatty acid in the circulation. In: JM Dietschy, AM Gotto Jr, and JA Ontko (eds) Disturbances in Lipid and Lipoprotein Metabolism. Am Physiol Soc, Bethesda MD, 1978, pp 229–249
15. Weisiger RA: Dissociation from albumin: a potentially rate-limiting step in the clearance of substances by the liver. Proc Natl Acad Sci USA 82: 1563–1567, 1985
16. Bassingthwaighte JB, Noodleman L, Van der Vusse G, Glatz JFC: Modeling of palmitate transport in the heart. Mol Cell Biochem 88: 51–58, 1989
17. Blanchette-Mackie EJ, Wetzel MG, Chernick SS, Paterniti JR Jr, Brown WV, Scow RO: Effect of the combined lipase deficiency mutation (*cld/cld*) on ultrastructure of tissues in mice: diaphragm, heart, brown adipose tissue, lung and liver. Lab Invest 55: 347–362, 1986
18. Blanchette-Mackie EJ, Masuno H, Dwyer NK, Olivecrona T, Scow RO: Lipoprotein lipase in myocytes and capillary endothelium of heart: immunocytochemical study. Am J Physiol 256: E818–E828, 1989
19. Scow RO, Chernick SS: Role of lipoprotein lipase during lactation. In: J Borensztajn (ed) Lipoprotein Lipase. Evener Publishers, Inc, Chicago, Ill, 1987, pp 149–185
20. Small CA, Garton AJ, Yeaman SJ: The presence and role of hormone-sensitive lipase in heart muscle. Biochem J 258: 67–72, 1989
21. O'Looney P, Maten MV, Vahouny GV: Insulin-mediated modifications of myocardial lipoprotein lipase and lipoprotein metabolism. J Biol Chem 258: 12994–13001, 1983
22. Schoonderwoerd K, van der Kraaij T, Hülsmann WC, Stam H: Hormones and triacylglycerol metabolism under normoxic and ischemic conditions. Mol Cell Biochem 88: 129–137, 1989
23. King GL, Johnson SM: Receptor-mediated transport of insulin across endothelial cells. Science 227: 1583–1586, 1984
24. Bar RS, Boes M, Sandra A: Vascular transport of insulin to rat cardiac muscle: central role of the capillary endothelium. J Clin Invest 81: 1225–1233, 1988
25. Cistola DP, Hamilton JA, Jackson D, Small DM: Ionization and phase behavior of fatty acids in water: application of the Gibbs phase rule. Biochemistry 27: 1881–1888, 1988
26. Blanchette-Mackie EJ, Scow RO: Lipolysis and lamellar structures in white adipose tissue of young rats: lipid movement in membranes. J Ultrastruct Res 77: 295–318, 1981
27. Blanchette-Mackie EJ, Scow RO: Movement of lipolytic products to mitochondria in brown adipose tissue of young rats: an electron microscopic study. J Lipid Res 24: 229–244, 1983
28. Amende LM, Blanchette-Mackie EJ, Chernick SS, Scow RO: Effect of pH on visualization of fatty acids as myelin figures in mouse adipose tissue by freeze-fracture electron microscopy. Biochim Biophys Acta 837: 94–102, 1985
29. Amende LM, Blanchette-Mackie EJ, Scow RO: Demonstration of fatty acid domains in membranes produced by lipolysis in mouse adipose tissue: a freeze-fracture study. Cell Tissue Res 246: 495–508, 1986
30. Scow RO, Olivecrona T: Effect of albumin on products formed from chylomicron triacylglycerol by lipoprotein lipase *in vitro*. Biochim Biophys Acta 487: 472–486, 1977
31. Sweetser DA, Heuckeroth RO, Gordon JI: The metabolic significance of mammalian fatty-acid binding proteins: abundant proteins in search of a function. Ann Rev Nutr 7: 337–359, 1987
32. Matarese V, Stone RL, Waggoner DW, Bernlohr DA: Intracellular fatty acid trafficking and the role of cytosolic lipid binding proteins. Prog Lipid Res 28: 245–272, 1989
33. Glatz JFC, Van der Vusse GJ: Intracellular transport of lipids. Mol Cell Biochem 88: 37–44, 1989
34. Scow RO, Desnuelle P, Verger R: Lipolysis and lipid movement in a membrane model: action of lipoprotein lipase. J Biol Chem 254: 6456–6463, 1979
35. Blanchette-Mackie EJ, Scow RO: Membrane continuities within cells and intercellular contacts in white adipose tissue of young rats. J Ultrastruct Res 77: 277–294, 1981
36. Blanchette-Mackie EJ, Scow RO: Continuity of intracellular channels with extracellular space in adipose tissue and liver: demonstrated with tannic acid and lanthanum. Anat Rec 203: 205–219, 1982
37. Blanchette-Mackie EJ, Scow RO: Sites of lipoprotein lipase activity in adipose tissue perfused with chylomicrons: electron microscope cytochemical study. J Cell Biol 51: 1–25, 1971
38. Belfrage P, Fredrickson G, Strålfors T, Tornquist H: Adipose tissue lipases. In: B Borgström and HL Brockman (eds) Lipases. Elsevier, Amsterdam, 1984, pp 365–416
39. Olivecrona T, Bengtsson-Olivecrona G: Lipoprotein lipase from milk – the model enzyme in lipoprotein lipase research. In:

Borensztajn (ed) Lipoprotein Lipase. Evener Publishers, Inc, Chicago, 1987, pp 15–58

40. Scow RO, Masuno H, Schultz CJ, Blanchette-Mackie EJ, Mateo C: Processing and transport of lipoprotein lipase in adipocytes. In: Y Oomura, S Tarui, S Inoue and T Shimuazu (eds) Progress in Obesity Research 1990. John Libbey, London, 1991, pp 229–232
41. Braun J, Severson DL: Diabetes reduces heparin- and phospholipase C-releasable lipoprotein lipase from cardiomyocytes. Am J Physiol 260: E477–E485, 1991
42. Saxena U, Klein MG, Goldberg IJ: Transport of lipoprotein lipase across endothelial cells. Proc Natl Acad Sci USA 88: 2254–2258, 1991
43. Scow RO, Egelrud T: Hydrolysis of chylomicron phosphatidylcholine *in vitro* by lipoprotein lipase, phospholipase A_2 and phospholipase C. Biochim Biophys Acta 431: 538–549, 1976
44. Smith LC, Scow RO: Chylomicrons: mechanism of transfer of lipolytic products to cells. Prog Biochem Pharmacol 15: 109–138, 1979
45. Jialal L, Crettaz M, Hachiya HL, Kahn CR, Moses AC, Buzney SM, King GL: Characterization of the receptors for insulin and the insulin-like growth factors on micro- and macrovascular tissues. Endocrinology 117: 1222–1229, 1985
46. Bottaro DP, Bonner-Weir S, King GL: Insulin receptor recycling in vascular endothelial cells: regulation by insulin and phorbol ester. J Biol Chem 264: 5916–5923, 1989
47. Bar RS, Dolash S, Dake BL, Boes M: Cultured capillary endothelial cells from bovine adipose tissue: a model for insulin binding and action in microvascular endothelium. Metab Clin Exp 35: 317–322, 1986
48. Forgue MD, Freychet P: Insulin receptors in the heart muscle: demonstration of specific binding sites and impairment of insulin binding in the plasma membrane of the obese hyperglycemic mouse. Diabetes 24: 715–723, 1975
49. Smith RM, Jarett L: Quantitative ultrastructural analysis of receptor-mediated insulin uptake into adipocytes. J Cell Physiol 115: 199–207, 1983
50. De Diego JG, Gorden P, Carpentier J-L: The relationship of ligand receptor mobility to internalization of polypeptide hormones and growth factors. Endocrinology 128: 2136–2140, 1991

Molecular and Cellular Biochemistry **116**: 193–196, 1992.

Malondialdehyde is a biochemical marker of peroxidative damage in the isolated reperfused rat heart

Donato Di Pierro, Barbara Tavazzi, Giuseppe Lazzarino and Bruno Giardina
Department of Experimental Medicine and Biochemical Sciences and Centro Interuniversitario per lo Studio del Danno da Ipossia e Iperossia, II University of Rome, Tor Vergata, Via O. Raimondo, 00173 Rome, Italy

Abstract

Concentration of MDA in isolated control, ischemic, and reperfused rat hearts was determined by using a new sensitive and reproducible HPLC method on the perchloric acid extract of the freeze-clamped tissues. By means of this HPLC assay for the direct measurements of MDA, concentrations of adenine nucleotide derivatives were also obtained in the same chromatographic run. Under the present experimental conditions, no detectable amount of MDA could be observed in control hearts while ischemic hearts showed 0.009 μmoles/g d.w. of MDA (s.d. = 0.001), this value representing the sensitivity limit of the method employed. On the contrary, reperfused hearts showed 0.118 μmoles/g d.w. of MDA (s.d. = 0.036), thereby indicating that this compound originates from an oxygen free radical-mediated breakdown of phospholipids and demonstrating the existence of quantifiable molecular damage occurring upon reperfusion. On the whole, our data demonstrate that MDA, if properly assayed, is a reliable index of peroxidative injury to biological systems. (Mol Cell Biochem **116:** 193–196, 1992)

Key words: malondialdehyde, peroxidative damage, reperfused rat heart

Abbreviations: Ado – Adenosine; d.w. – dry weight; HPLC – High-Performance Liquid Chromatography; Hyp – Hypoxanthine; Ino – Inosine; MDA – Malondialdehyde; Xan – Xanthine

Introduction

Production of oxygen-derived free radical species in the reperfused myocardium has been reported by several authors [1–3]. In spite of this finding the mechanism by which oxygen radicals are generated and, moreover, their eventual role in damaging reperfused heart tissue are still unclear. In fact, the transformation of xanthine dehydrogenase into xanthine oxidase, which has been addressed as the main source of superoxide anions also in the reperfused heart [4], has not been found to occur in some studies [5, 6] and, under similar experimental conditions, opposite results have been reported [7, 8]. Analogously, data reporting increase [7, 9–11] or no change [12, 13] of lipid peroxidation following myocardial reperfusion have been published in the last few years. Therefore, independently on the mechanism of their production, it appears that a close relationship

Address for offprints: D. Di Pierro, Department of Experimental Medicine and Biochemical Sciences, II University of Rome, Tor Vergata, Via O. Raimondo, 00173 Rome, Italy between oxygen radicals and biochemical damages of the reperfused myocardium is still lacking.

By using our new sensitive and reproducible HPLC assay for the simultaneous separation of MDA, and several adenine nucleotide derivatives [14] we investigated the eventual generation of MDA as a function of different perfusion conditions in isolated control, ischemic, and reperfused rat hearts. The results obtained in these experiments are the matter of the present paper.

Materials and methods

Male Wistar rats of 250–300 g b.w. were purchased from Morini (San Polo D'Enza, RE, Italy). Anesthesia was induced by i.p. injection of 200 mg/kg b.w. ketamine. After injection of 1000 IU heparin into the caudal vena cava, hearts were perfused by the non-circulating Langendorff technique, at a constant hydrostatic pressure of 7.85 kPa. The perfusion medium was a Krebs-Ringer solution composed by (mM): NaCl (119), $CaCl_2$ (2.5), KCl (4.8), $MgSO_4$ (1.18), $NaHCO_3$ (25), KH_2PO_4 (1.2), glucose (10) and 12 IU/l insulin, pH 7.3. The buffer was saturated with O_2/CO_2 (19:1; v:v) and kept at a constant temperature of 37° C.

Control hearts were perfused for 15 min. Ischemic hearts, after 15 min of equilibration, were subjected to 30 min of global normothermic ischemia, obtained by closing the flow rate above the aorta. Reperfused hearts, after 15 min of equilibration and 30 min of global normothermic ischemia, underwent to 30 min of reperfusion with the starting buffer. At the end of each period of perfusion, the hearts were freeze-clamped and then deproteinized by 0.6 M $HClO_4$ for the analysis of metabolites [15]. Samples (200 μl) were then directly analyzed for MDA, and adenine nucleotide derivatives by a HPLC apparatus Jasco 880-PU dual pump system (Tokyo, Japan), equipped with a LC 18-T 15 cm × 4.6 mm 3 μm particle size column (Supelco, Bellefonte, PA, USA), and connected to a Jasco 875-UV visible detector (Tokyo, Japan). Chromatographic conditions are essentially those reported in [14].

The between groups comparison of the data was performed by the 1-way analysis of variance.

Results

Table 1 reports values of MDA and of dephosphorylated compounds originating from ATP catabolism, i.e. Hyp, Xan, Ino, uric acid, and Ado. Referring to ATP catabolytes, a roughly 10 times increase of all the parameters considered occurred at the end of ischemia with respect to the corresponding values recorded in control hearts. Reperfusion induced a significant reduction of these compounds, although all of them resulted statistically different from the corresponding values determined in control hearts. Concerning MDA, a low but detectable amount could be recorded in ischemic hearts, albeit this value represents the minimal concentration falling in the range of linearity of a curve determined using standard MDA [14]. A totally different situation was obtained in reperfused hearts, where, with respect to ischemic hearts, 13 times higher amount of MDA was detected (0.118 μmoles/g d.w.; s.d. = 0.036).

The values of high energy phosphate compounds ATP, ADP, AMP, IMP, GTP, and GDP, as well as those of nicotinic coenzymes NAD and NADP, are reported in Table 2. A significant decrease of ATP, GTP, and GDP, and a concomitant significant increase

Table 1. Concentrations of MDA and dephosphorylated adenine nucleotides in isolated control, ischemic, and reperfused rat hearts as determined in perchloric acid extracts by HPLC

	MDA	Hyp	Xan	Ino	Uric acid	Ado
Control hearts	N.D.	0.038 (0.011)	0.021 (0.009)	0.049 (0.007)	0.140 (0.034)	0.072 (0.009)
Ischemic hearts	0.009[a] (0.001)	0.241[a] (0.052)	0.151[a] (0.039)	1.48[a] (0.290)	1.11[a] (0.400)	1.08[a] (0.400)
Reperfused hearts	0.118[ab] (0.036)	0.078[ab] (0.022)	0.054[ab] (0.017)	0.554[ab] (0.188)	0.584[ab] (0.154)	0.172[ab] (0.054)

Values are expressed as μmoles/g d.w. and represent the mean (s.d.) of 6 to 11 hearts. Perfusion conditions and HPLC assay are described in detail under Materials and methods. Abbreviations used: N.D. = not detectable. a) $p < 0.05$ with respect to control hearts. b) $p < 0.05$ with respect to ischemic hearts.

of ADP, AMP, and IMP was observed in ischemic hearts in comparison with the corresponding values recorded in control hearts. Reperfusion induced a normalization of ADP, AMP, and IMP, whilst ATP and GTP, although increased with respect to ischemic hearts, remained below the 50% of the pre-ischemic values. As previously reported, the concentrations of NAD and NADP (comprehensive of their respective reduced forms NADH and NADPH as a consequence of perchloric acid deproteinization of the tissue) were negatively influenced by reperfusion, thus probably contributing to the incomplete recovery of the activity of the reperfused myocardium.

Discussion

Molecular damages attributable to oxygen-derived free radicals, generated in the myocardium from the very early phase of reperfusion [1–3], have not yet been clearly demonstrated. Several studies, using different experimental models than the isolated organs, indicated a correlation between the production of oxygen radicals and irreversible modifications of membrane phospholipids. For instance, the increase of MDA [16, 17] or the appearance of an absorbing peak at 233 nm referred to conjugated dienes [18], have been reported as biochemical markers of phospholipid modification induced in the isolated microsomes by oxygen radicals. Conjugated dienes should derive as precocious species from the molecular rearrangement which follows the radical attack to a double bond of a fatty acid chain of phospholipids [17]. Differently, MDA would originate as a delayed product from the breakdown of the peroxidized phospholipids [17]. Therefore, it appeared reasonable to monitor the eventual modifications of these parameters, in order to ascertain the role of oxygen radicals as possible inducers of lipid peroxidation in the reperfused heart.

Unfortunately, data appeared in the literature gave conflicting results. Increase [7, 9–11] or no change [12, 13] of lipid peroxidation, as determined by either MDA or conjugated dienes, has been published. This confused situation rendered doubtful even the existence of molecular damages directly correlated to reperfusion [19].

Thanks to the new HPLC method that we recently set up [15], the results obtained in the present paper indicate two main interrelated conclusions:

i) in the isolated rat heart, lipid peroxidation is an actual phenomenon occurring to the membrane phospholipids, mainly as a consequence of reperfusion;

ii) MDA, if properly assayed, is a reliable biochemical marker of peroxidative injuries, also occurring in the reperfused myocardium.

The discrepancies reported on the occurrence of lipid peroxidation [7, 9–13] might be referred to the poor reliability and reproducibility of the MDA assays, i.e. both the colorimetric thiobarbituric acid reaction [20] and the most common HPLC methods at present published in literature [21–23], and to the low sensitivity of the conjugated dienes assay, whose method implies in addition a certain manipulation of the samples [18]. In this respect, our new HPLC method for the direct measurement of MDA and adenine nucleotide derivatives avoids any particular manipulation of the tissue out of the deproteinization by perchloric acid, and presents the characteristics of both high sensitivity and reproducibility [15].

The data obtained under the present experimental conditions strongly suggest that peroxidative phospholipid derangment mainly occurs upon readmission of

Table 2. Concentrations of high-energy phosphates and nicotinic coenzymes in isolated control, ischemic and reperfused rat hearts as determined in perchloric acid extracts by HPLC

	ATP	ADP	AMP	IMP	GTP	GDP	NAD	NADP
Control hearts	23.01	6.87	2.64	0.215	2.31	0.618	7.08	0.641
	(1.39)	(0.51)	(0.43)	(0.04)	(0.40)	(0.13)	(0.37)	(0.05)
Ischemic hearts	6.17[a]	8.69	5.60[a]	0.617[a]	0.87[a]	0.274[a]	6.12[a]	0.532
	(1.04)	(0.84)	(0.23)	(0.13)	(0.16)	(0.07)	(0.50)	(0.08)
Reperfused hearts	11.49[ab]	7.18	2.57[b]	0.292[b]	1.14[a]	0.295[a]	4.96[ab]	0.295[ab]
	(2.26)	(1.39)	(0.46)	(0.06)	(0.24)	(0.09)	(0.44)	(0.04)

Values are expressed as μmoles/g d.w. and represent the mean (s.d.) of 6 to 1 hearts. As a consequence of acid extraction of the tissue, NAD and NADP are comprehensive of the concentrations of their respective reduced forms. Perfusion conditions and HPLC assay are described in detail under Materials and methods. a) $p < 0.05$ with respect to control hearts. b) $p < 0.05$ with respect to ischemic hearts.

oxygen to post-ischemic heart, i.e. during reperfusion, thus indicating that oxygen-derived free radicals directly intervene in the onthogenesis of reperfusion-induced tissue injury. It is, however, worth noting that the low but detectable concentration of MDA, recorded after 30 min ischemia, again supports the theory of the involvement of oxygen radicals in myocardial damaging as the generation of significant amount of OH· radicals, already during ischemia, has recently been reported [13, 24].

In conclusion, although our data do not give an insight on the possible mechanism of oxygen radicals generation, they support, for the first time to best of our knowledge, both a clear and unquestionable evidence on the existence of reperfusion-related tissue damages and a strict correlation between reperfusion and MDA as an oxygen radical-induced peroxidative marker in a model of isolated ischemic and reperfused rat heart.

References

1. Garlick PB, Davies MJ, Hearse DJ, Slater TF: Direct detection of free-radicals in the reperfused rat heart using electron spin resonance spectroscopy. Circ Res 61: 757–760, 1987
2. Zweier JL: Measurement of superoxide-derived free radicals in the reperfused heart. Evidence for a free radical mechanism of reperfusion injury. J Biol Chem 263: 1353–1357, 1988
3. Baker JE, Felix CC, Olinger GN, Kalyanaraman B: Myocardial ischemia and reperfusion: Direct evidence for free radical generation by electron spin resonance spectroscopy. Proc Natl Acad Sci USA 85: 2786–2789, 1988
4. Chambers DE, Parks DA, Patterson G, Roy R, McCord JM, Yoshida S, Parmley LF, Downey JM: Xanthine oxidase as a source of free radical damage in myocardial ischemia. J Mol Cell Cardiol 17: 145–152, 1985
5. Keherer JP, Piper HM, Sies H: Xanthine oxidase is not responsible for reoxygenation injury in isolated-perfused rat heart. Free Rad Res Comms 3: 69–78, 1987
6. Bindoli A, Cavallini L, Rigobello MP, Coassin MG, Di Lisa F: Modification of the xanthine-converting enzyme of perfused rat heart during ischemia and oxidative stress. Free Rad Biol Med 4: 163–167, 1988
7. Tavazzi B, Cerroni L, Di Pierro D, Lazzarino G, Nuutinen EM, Starnes JW, Giardina B: Oxygen radical injury and loss of high-energy compounds in anoxic and reperfused rat heart: prevention by exogenous fructose-1,6-bisphosphate. Free Rad Res Comms 10: 167–176, 1990
8. Thompson-Gorman SL, Zweier JL: Evaluation of the role of xanthine oxidase in myocardial reperfusion injury. J Biol Chem 265: 6656–6663, 1990
9. Rao PS, Cohen MV, Mueller HS: Production of free radicals and lipid peroxides in early experimental myocardial ischemia. J Mol Cell Cardiol 15: 713–716, 1983
10. Roth E, Torok B, Zsoldos T, Matkovics B: Lipid peroxidation and scavenger mechanism in experimentally induced heart infarcts. Basic Res Cardiol 80: 530–536, 1985
11. Romaschin AD, Rebeyka I, Wilson GJ, Mickle DAG: Conjunated dienes in ischemic and reperfused myocardium: an *in vivo* chemical signature of oxygen free radical mediated injury. J Mol Cell Cardiol 19: 289–302, 1987
12. Brasch H, Schoenberg MH, Younes M: No evidence for an increased lipid peroxidation during reoxygenation in Langendorff hearts and isolated atria of rats. J Mol Cell Cardiol 21: 697–707, 1989
13. Maupoil V, Rochette L: Evaluation of free radical and peroxide formation during global ischemia and reperfusion in isolated perfused rat heart. Cardiovasc Drugs Ther 2: 615–621, 1988
14. Lazzarino G, Di Pierro D, Tavazzi B, Cerroni L, Giardina B: Simultaneous separation of malondialdehyde, ascorbic acid, and adenine nucleotide derivatives from biological samples by ion-pairing high-performance liquid chromatography. Anal Biochem 197: 191–196, 1991
15. Lazzarino G, Nuutinen M, Tavazzi B, Di Pierro D, Giardina B: A method for preparing freeze-clamped tissue samples for metabolite analyses. Anal Biochem 181: 239–243, 1989
16. Bidlack WR, Okita RT, Hochstein P: The role of NADPH-cytochrome b reductase in microsomal lipid peroxidation. Biochem Biophys Res Commun 53: 459–465, 1973
17. Buege JA, Aust SD: Microsomal lipid peroxidation. In: L Packer (ed.) Methods in Enzymology. Academic Press, New York, 1984, vol 105, pp 302–310
18. Recknagel RO, Glende EA Jr: Spectrophotometric detection of lipid conjugated dienes. In: L Packer (ed.) Method in Enzymology. Academic Press, New York, 1984, vol 105, pp 331–337
19. Kehrer JP: Concepts related to the study of reactive oxygen and cardiac reperfusion injury. Free Rad Res Comms 5: 305–314, 1989
20. Uchiyama M, Mihara M: Determination of malondialdehyde precursor in tissues by thiobarbituric acid test. Anal Biochem 86: 271–278, 1978
21. Esterbauer H, Lang J, Zadravec S, Slater TF: Detection of malonaldehyde by high-performance liquid chromatography. In: L Packer (ed) Methods in Enzymology. Academic Press, New York, 1984, vol 105, pp 319–331
22. Csallany SA, Der Guan M, Manwaring JD, Addis PB: Free malonaldehyde determination in tissues by high-performance liquid chromatography. Anal Biochem 142: 277–283, 1984
23. Bull AW, Marnett LJ: Determination of malondialdehyde by ion-pairing high-performance liquid chromatography. Anal Biochem 149: 284–290, 1985
24. Pietri S, Culcasi M, Cozzone PJ: Real-time continuous-flow spin trapping of hydroxyl free radical in the ischemic and post-ischemic myocardium. Eur J Biochem 186: 163–173, 1989

Molecular and Cellular Biochemistry **116**: 197–202, 1992.

Studies on the interaction of leucocytes and the myocardial vasculature

I. Effect of hypoxia on the adherence of blood granulocytes

Anneke Pietersma, Netty de Jong, Wim Sluiter and Johan F. Koster
Department of Biochemistry I, Faculty of Medicine and Health Sciences, Erasmus University Rotterdam, P.O. Box 1738, 3000 DR Rotterdam, The Netherlands

Abstract

Granulytes play an important role in increasing the infarct size after ischemia and reperfusion by the release of oxygen-derived free radicals (ODFR) and lysosomal enzymes. It has been shown that the number of granulocytes adhering to the vascular endothelium increases after occlusion of the coronary artery, and that the area of myocardial damage can be reduced by preventing granulocyte adherence with monoclonal antibodies directed against adhesion receptors. The underlying mechanism of granulocyte activation under these conditions is not yet known. We have investigated whether granulocytes can be activated directly by reduced oxygen tensions. Granulocytes were suspended in a hypoxic buffer and incubated on fibronectin and gelatin coated microtitre plates at 1–3% ambient oxygen to study their ability to adhere to these matrices. The results showed that the adherence of granulocytes to fibronectin was dependent on the duration of hypoxia. After 30 min of incubation under hypoxia granulocyte adherence increased 1.3 to 1.8 fold compared to the normoxic control. The adherence to fibronectin could be inhibited partially by anti-CD18 antibody, a monoclonal antibody to the common beta chain of a class of extracellular matrix receptors. This direct activation of granulocytes due to hypoxic conditions may have implications for the interaction of these cells with the vascular endothelium *in vivo,* (Mol Cell Biochem **116:** 197–202, 1992)

Key words: granulocytes, hypoxia, adherence, CD11/CD18

Introduction

Granulocytes are important mediators of ischemia/reperfusion injury. The invasion of granulocytes into the infarcted myocard has been known for many years and was first thought of as a beneficial part of the healing process [1, 2]. Recent studies, however, demonstrate that the ultimate size of the myocardial infarct after reperfusion can substantially be reduced by depletion of blood granulocytes beforehand or by preventing the attachment of granulocytes to the target tissue with monoclonal antibodies [2].

The adherence of granulocytes to the vascular endothelium is the first step in the migration of granulocytes.

Address for offprints: A. Pietersma, Department of Biochemistry I, Faculty of Medicine and Health Sciences, Erasmus University Rotterdam, P.O. Box 1738, 3000 DR Rotterdam, The Netherlands

In response to chemoattractants, e.g., complement fragments and bacterial products (formylated peptides), the number of granulocyte surface receptors increases which leads to increased adherence to the endothelium and eventually diapedesis and chemotaxis [3–5]. The mechanism of granulocyte activation in ischemia/reperfusion injury is not fully elucidated but activated complement, leukotrienes, prostacyclin (PGI_2) and/or cytokines are likely to be involved [4]. Ischemic cardiac muscle releases peptides that react with the first component of complement and increased complement binding by the ischemic myocard has been shown [2]. Furthermore, leukotriene B_4 synthesized by activated granulocytes stimulates the adherence of granulocytes to the endothelium [6]. The attachment of granulocytes to the endothelium is a prerequisite for mediating vascular damage by the production of potentially histotoxic agents like proteinases and ODFR [5]. Granulocyte adherence is mediated through a class of glycoprotein receptors referred to as the CD11/CD18 complex, which are members of the integrin family of adhesion molecules [2]. The granulocyte adhesion molecules are heterodimers consisting of a common beta subunit (CD18) with different alfa subunits (CD11a, b or c) that determine the molecular specificity [2]. The ligand on the endothelial cell membrane for the CD11a/CD18 complex (lymphocyte function-associated antigen, LFA-1) is the intercellular adhesion molecule-1 (ICAM-1) that is constitutively expressed at low levels but can be markedly upregulated by the cytokines interleukin-1 (IL-1) and tumour necrosis factor (TNF) [4].

Recent investigations have shown that adherence of granulocytes to the endothelium of the ischemic myocard may be an early event in myocardial ischemia [2, 6, 7]. This led us to investigate whether hypoxia directly stimulates the adherence of blood granulocytes.

Material and methods

Experimental conditions

All incubations under hypoxic conditions were performed in a cabinet that allowed handling of cells under controlled ambient conditions. This cabinet consisted of a conducting surface coupled to a thermostatted waterbath and a hood with two holes to fit a pair of latex gloves and a small lock chamber. During incubations a continuous flow of 99.9% N_2 (N_2-gas 3.0 technical, Hoekloos, Schiedam, The Netherlands), humidified and warmed to 37° C, was maintained. The ambient pO_2 was monitored with an oxygen analyzer (type OA 250, Servomex, Zoetermeer, The Netherlands) and kept at 1–3%. Buffers were rendered hypoxic by gassing with humidified N_2 to a final pO_2 of 25 mmHg as measured by acid-base laboratory analysis (ABL3, Radiometer, Copenhagen, Denmark). Next hypoxic buffers were equilibrated to ambient conditions of 1–3% O_2.

Experimental procedures

Isolation of cells

Granulocytes were isolated pyrogen free from citrate blood obtained from healthy human donors.

Blood was diluted 1:1 with phosphate buffered saline (PBS) pH 7.4 and layered onto a Ficoll hypaque gradient (Nycomed, Haarlem, The Netherlands). After centrifugation during 15 min at 800 × g, the pellet was washed with PBS and next the erythrocytes were lysed isotonically in ammoniumchloride solution (0.155 M NH_4Cl, 10 mM $NaHCO_3$, 0.1 mM EDTA, pH 7.4). Finally, granulocytes were washed and resuspended in Krebs Ringer phosphate buffer (KRP) pH 7.4 (120 mM NaCl, 4.8 mM KCl, 1.0 mM $CaCl_2$, 1.2 mM $MgSO_4$, 15 mM Na_2HPO_4 and 0.1% glucose). The granulocyte fraction obtained by this procedure in general consisted for 99% of granulocytes and contained >90% viable cells as determined by trypan blue exclusion.

Incubations of granulocytes

Granulocytes were brought into the cabinet, and diluted in hypoxic buffer (1 to 5 × 10^5 cells/ml) and next plated onto 96-well microtitre plates (Costar, Cambridge, UK) which were coated with 5 μg fibronectin/well, isolated as previously described [8] or 5 μg gelatin/well. In a different set of experiments granulocytes (10 × 10^6 or 20 × 10^6 cells/ml) were incubated in polypropylene (PP) tubes (Greiner, Alphen a/d Rijn, The Netherlands) in KRP without calcium. After 1 h of incubation in PP tubes the number of granulocytes in suspension was counted, using a Bürkner hemocytometer, and the release of lactate dehydrogenase (LDH) from the granulocytes was measured according to the method described by Metcalf [10]. LDH release was expressed as the percentage of the total amount of LDH of a sample. All control incubations were performed under normoxic ambient conditions in a thermostatted waterbath.

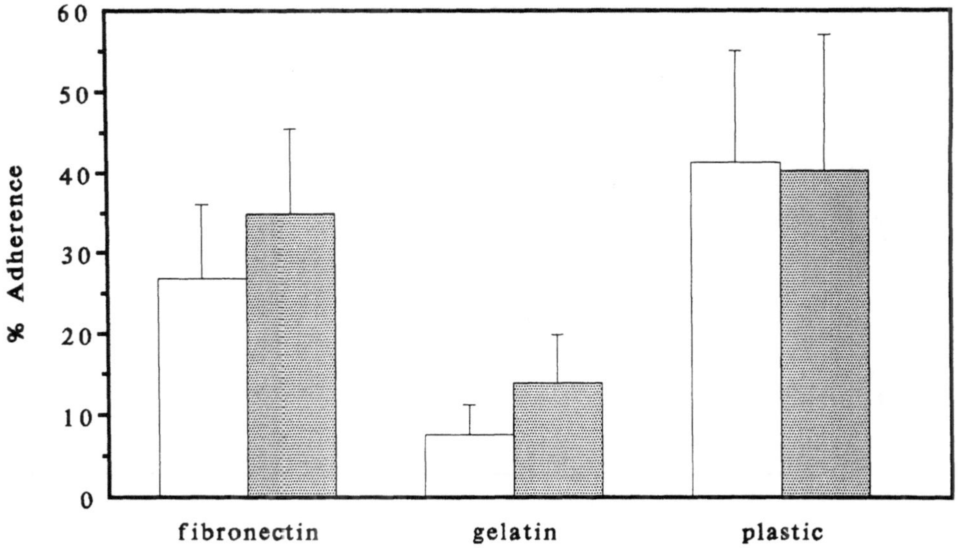

Fig. 1. Effect of hypoxia (1–3% ambient oxygen, open bars) and normoxia (21% ambient oxygen, dark bars) on the adherence of granulocytes to different substrates. Granulocytes were plated onto fibronectin and gelatin-coated or uncoated polystyrene microtitre plate wells and incubated for 30 min under hypoxic or normoxic conditions at 37° C. After extensive washing the number of adherent cells were quantitated.

Granulocyte adherence assay
After 30 min of incubation of granulocytes on fibronectin, gelatin or uncoated plastic (polystyrene) the nonadherent cells were removed by washing the microtitre plates with PBS pH 6.0. The adherent cells were quantitated according to the method described by Bath [9] using myeloperoxidase as an enzymatic marker. After lysis of the cells with 50 μl of 5% hexadecyltrimethylammoniumbromid (HTAB), 100 μl 0.2 mg/ml o-dianisidinehydrochloride), 0.4 mM hydrogen peroxide was added to each well to start the enzymatic reaction. The increase in absorption at 450 nm during 1.5 min was taken as a measure of myeloperoxidase activity. A standard curve of different concentrations of granulocytes in HTAB was used to estimate the percentage of adherent cells.

Inhibition of adhesion by monoclonal antibodies
Granulocytes were suspended at a concentration of 0.5×10^6 cells/ml in normoxic or hypoxic buffer together with various concentrations of anti-CD18 antibody (anti human LFA-1-beta1, Becton Dickinson, Etten-Leur, The Netherlands) and incubated for 30 min at 37° C in a shaking waterbath in polypropylene tubes wrapped in parafilm. Thereafter granulocytes were transferred to coated microtitre plates to study their capacity to adhere to fibronectin.

Microscopic evaluation
To allow light-microscopic examination of the adhering cells the adherence studies were repeated in fibronectin-coated (30 μg/well) 4-wells plates. After 5, 10, 20 and 30 min of incubation the wells were emptied and washed with normoxic or hypoxic KRP. Adherent cells were fixed with methanol, stained with May Grünwald-Giemsa and next counted in 5 fields of 100 mm² using an oculair grid.

Statistical analysis
Results are expressed as the mean and standard deviations (SD). Results from the adherence studies were analyzed with Wilcoxon's signed rank test for paired data. The results from the two other studies were analyzed using STATA (Computing Resource Centres, Los Angeles, USA) by linear regression analysis after rank transformation of the data. Results were considered statistically significant at $P < 0.05$.

Results

When granulocytes were incubated under hypoxic conditions for 1 h in PP tubes we observed a mean decrease in the number of cells of 17% (SD: 11) compared to 8% (SD: 8) for the normoxic cells. This decrease was not due to cell damage for reasons that the LDH release from cells incubated under hypoxic conditions did not differ from the normoxic control, i.e., 8.0% (SD: 3.5) and 8.5% (SD: 3.6) of total LDH, respectively. Therefore we decided to investigate whether the observed

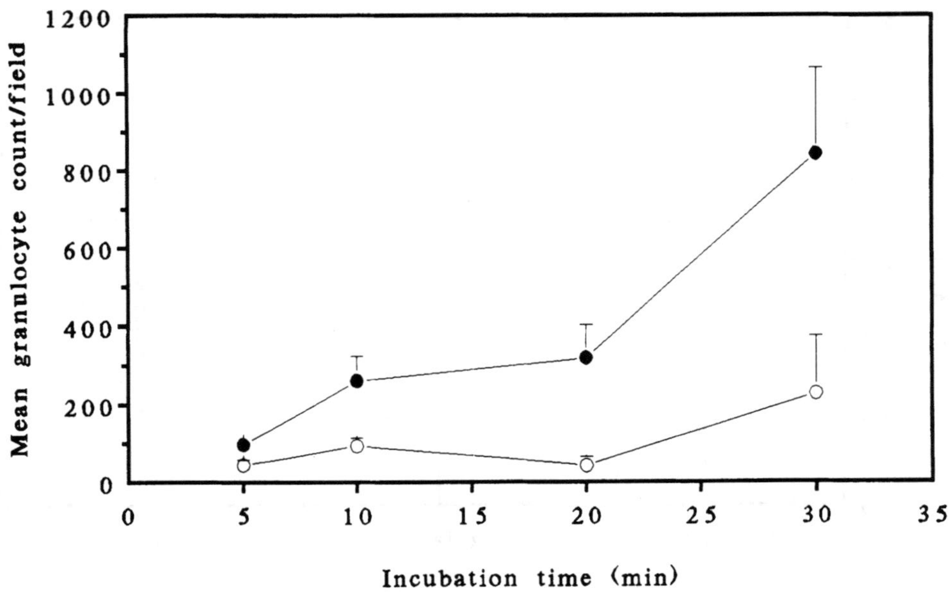

Fig. 2. The adherence of granulocytes in course of time under hypoxic (closed circles) and normoxic (open circles) conditions. Granulocytes were plated onto fibronectin coated 4-wells plates, after 5, 10, 20 and 30 min of incubation the wells were washed and adherent cells were fixed with methanol and microscopically quantitated.

decrease in the number of granulocytes after 1 h of incubation under hypoxic conditions, could be explained by an increased adherence.

Effect of hypoxia on the adherence of granulocytes to different matrices

After an incubation period of 30 min under hypoxia, granulocytes showed an increased adherence to the fibronectin matrix. The adherence of granulocytes significantly increased to 130% of their normoxic controls ($P < 0.05$; Fig. 1). Adherence of granulocytes to gelatin also significantly increased to 180% of the normoxic control ($P < 0.02$; Fig. 1). Fibronectin was found to be a better matrix for granulocytes, since adherence of granulocytes to this matrix was twice as high as to gelatin both under normoxic and hypoxic conditions. The adherence to polystyrene (uncoated microtitre plates) in normoxic and hypoxic incubations was 41.3% (SD: 13.7) and 40.3% (SD: 16.7), respectively, and exceeded binding to fibronectin, i.e., normoxia: 26.8% (SD: 9.3) and hypoxia: 34.9% (SD: 10.5), respectively. No difference due to hypoxia was observed in adherence of granulocytes to polystyrene.

To find out whether the increased adherence under hypoxic conditions is mediated through a rise in the intracellular calcium concentration, the granulocytes were incubated in calcium-free KRP with 1.5 mM ethylene glycol-bis-beta-amino ethyl ether (EGTA). The results showed that hypoxia stimulated the adherence of granulocytes to fibronectin to a similar extent in the presence or absence of calcium (130% and 150% of the corresponding normoxic control, resp.).

Effect of the duration of hypoxia on granulocyte adherence

To study granulocyte adherence in course of time, granulocytes were incubated for various time periods in 4-wells plates under normoxic and hypoxic conditions. The results presented in Fig. 2 show that the adherence of granulocytes to fibronectin under hypoxic conditions significantly increased with time ($P = 0.026$). After 30 min of hypoxic incubation the average number of granulocytes per field increased to 260% of the control value (range 160–370%). Granulocytes incubated under normoxic conditions did not exhibit a significant time-dependent increase in adherence ($P = 0.14$). In general, adherence of granulocytes was accompanied with changes in cell shape and formation of pseudopods (not shown).

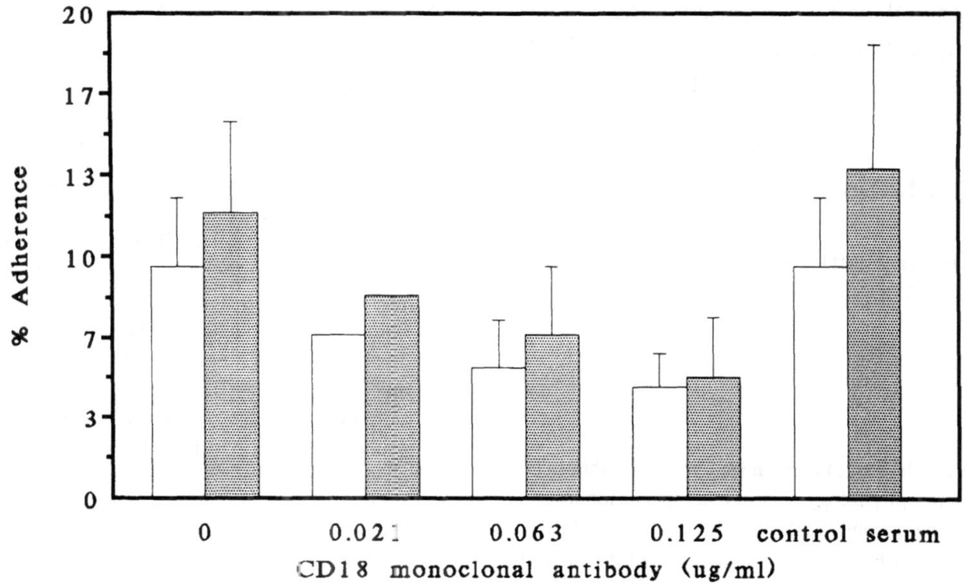

Fig. 3. Effect of blocking CD18 with anti CD18 monoclonal antibody (anti human LFA-1beta) on the adherence of granulocytes to fibronectin under normoxic (open bars) and hypoxic (closed bars) conditions. Granulocytes were preincubated with increasing concentrations of mAb for 30 min at 37° C and then plated onto fibronectin coated microtitre plate wells.

Effect of anti-CD18 antibody on granulocyte adherence to fibronectin

A significant decrease in adherence of granulocytes to fibronectin was found with increasing concentrations of anti-CD18 antibody ($P < 0.001$; Fig. 3). However, the efficacy of anti-CD18 antibody to inhibit the adherence of hypoxic granulocytes was less than that observed for normoxic granulocytes ($P < 0.01$). This indicates that another receptor that does not contain CD18 may be involved in the increased adherence of granulocytes to fibronectin under hypoxic conditions.

Discussion

The major finding of the present study was that, in comparison with normoxic granulocytes, granulocytes incubated under hypoxic conditions showed a 1.3 to 1.8 fold increase in the adherence to fibronectin and gelatin, respectively. Furthermore, the extent of adherence was dependent on the duration of the hypoxic incubation. We found that hypoxia probably directly affects granulocyte adherence.

Our results indicate that adherence in general is mediated through the CD11/CD18 class of cell surface receptors. Blocking the common beta chain of these integrin molecules with anti-CD18 antibody resulted in a significant decrease in granulocyte adherence to fibronectin for both control and hypoxic incubations. However, analysis of the decrease in granulocyte adherence by increasing amounts of anti-CD18 antibody showed that the difference between hypoxic and normoxic incubated cells remained. Conceivably the increased adherence due to hypoxia is independent of the CD18-containing class of adhesion receptors. Adherence of granulocytes to fibronectin can also be mediated by the fibronectin receptor [11]. In fact the fibronectin receptor and the CD11/CD18 complex may cooperate in the granulocyte adherence to this matrix. However, the fibronectin receptor can not be blocked by mAb to CD18 [11]. Therefore it is possible that the fibronectin receptor is responsible for the increased adherence of granulocytes due to hypoxia. Preliminary experiments indicated that depletion of intracellular calcium does not prevent the increase in adherence under hypoxic conditions. If granulocytes were depleted from intracellular calcium the difference between the adherence under hypoxic and normoxic conditions did not disappear. This indicates that the factor which causes the extra CD18-independent adherence is not mediated by the level of cytosolic calcium.

The present results are especially important in case the vessel wall is damaged during a period of ischemia/hypoxia. It is known that the subendothelial matrix contains fibronectin and collagen which is partly homologous to fibronectin [12, 13]. If during hypoxia the endothelium of the vessel wall is damaged in such a way that the subendothelial matrix is exposed to the circulation, this could implicate increased adherence and facilitated infiltration of granulocytes into the lesion.

Acknowledgements

We thank the Dutch Heart Foundation for financial support (grant no. 89.226).

References

1. Welbourn CRB, Goldman G, Paterson IS, Valeri CR, Shepro D, Hechtman HB: Pathophysiology of ischemia reperfusion injury: central role of the neutrophil. Br J Surg 78: 651–655, 1991
2. Werns SW, Lucchesi BR: The role of the polymorphonuclear leukocyte in mediating myocardial reperfusion injury. In: Singal PK (ed.) Oxygen radicals in pathophysiology of heart disease. Kluwer Academic Publishers, Boston, Dordrecht, Lancaster, 1988, pp 123–144
3. Springer TA: Adhesion receptors in the immune system. Nature 346: 425–434, 1990
4. Ward PA, Marks RM: The acute inflammatory reaction. Curr Op Immunol 2: 5–9, 1989
5. Haslett C, Savill JS, Meagher L: The neutrophil. Curr Op Immunol 2: 10–18, 1989
6. Mullane KM, Salmon JA, Kraemer R: Leukocyte-derived metabolites of arachidonic acid in ischemia-induced myocardial injury. Fed Proc 46: 2422–2433, 1987
7. Engler R: Consequences of activation and adenosine-mediated inhibition of granulocytes during myocardial ischemia. Fed Proc 46: 2407–2412, 1987
8. Engvall E, Ruoslahti E: Binding of soluble form of fibroblast surface protein, fibronectin, to collagen. Int J Cancer 20: 1–5, 1977
9. Bath PMW, Booth RFG, Hassall DG: Monocyte-lymphocyte discrimination in a new microtitre-based adhesion assay. J Immun Meth 118: 59–65, 1989
10. Metcalf JA, Gallin JI, Nauseef WM, Root RK: Laboratory manual of neutrophil function. Ravenpress, New York, 1986, pp 171–172
11. Nathan C, Srimal S, Farber C, Sanchez E, Asch A, Gailit J, Wright SD: Cytokine-induced respiratory burst of human neutrophils: dependence on extracellular matrix proteins and CD11/CD18 integrins. J Cell Biol 109: 1341–1349, 1989
12. Hynes RO: Integrins: A family of cell surface receptors. Cell 48: 549–554, 1987
13. Yamada KM: Adhesive recognition sequences. J Biol Chem 266: 12809–12812, 1991

Molecular and Cellular Biochemistry **116**: 203–209 1992.

Arachidonic acid incorporation in cardiomyocytes, endothelial cells and fibroblast-like cells isolated from adult rat heart

M.C.J.G. Linssen, P.H.M. Willemsen, V.V.Th. Heijnen and G.J. van der Vusse
Department of Physiology, Cardiovascular Research Institute Maastricht, University of Limburg, P.O. Box 616, 6200 MD Maastricht, The Netherlands

Abstract

The incorporation of radiolabeled arachidonic acid (^{3}[H]-AA) in normoxic cardiomyocytes (MC), cardiac endothelial cells (EC) and fibroblast-like cells (FL) isolated from adult rat heart was studied. Deposition of ^{3}[H]-AA in the cellular lipid pool was assessed with biochemical and autoradiographic techniques.

Extraction and subsequent analysis of lipids from the three different cell types revealed that MC contained significantly more triacylglycerols than EC and FL. The proportion of (unlabeled) AA was also higher in MC triacylglycerols than in EC and FL. The quantity of phospholipids did not differ among the three cell types studied. However, the content of (unlabeled) AA in the MC phospholipid pool was twice as high as in EC and FL.

The amount of ^{3}[H]-AA incorporated in the cellular lipid pool of MC, EC and FL depended on the concentration of AA in the incubation medium and the incubation time. In EC and FL incorporation of ^{3}[H]-AA was highest in the cellular phospholipid pool (0.01 μM AA, 30 min incubation). With increased concentration of AA and longer incubation times, the cellular triacylglycerol pool became more important as site of ^{3}[H]-AA incorporation. In MC, comparable amounts of ^{3}[H]-AA were incorporated in the cellular triacylglycerol and phospholipid pools (0.01 and 1 μM AA). At higher AA concentrations (10 μM) the triacylglycerol pool was the preferred site of ^{3}[H]-AA deposition.

Autoradiograhic analysis at the light microscopic level revealed that the extra-nuclear space was readily stained when the three cell types were incubated with ^{3}[H]-AA. These findings indicate that all cellular lipid pools and membranes are most likely site of deposition of radiolabeled arachidonic acid. (Mol Cell Biochem **116:** 203–209, 1992)

Key words: arachidonic acid, rat heart, endothelial cells, fibroblast-like cells, cardiomyocytes

Introduction

Phospholipids are the fundamental constituents of cardiac membranes and play an important role in the preservation of cellular integrity. Under pathophysiological conditions cellular integrity can become compromised and the membrane phospholipids can be chemically changed.

In the heart phospholipids are present in different subclasses, such as phosphatidylcholine, phosphatidy-

Address for offprints: M.C.J.G. Linssen, Department of Physiology, Cardiovascular Research Institute Maastricht, P.O. Box 616, 6200 MD Maastricht, The Netherlands lethanolamine and phosphatidylinositol, and have a different fatty acid composition. The fatty acid arachidonic acid (AA) comprises about 15% of the total fatty acid content in rat heart and the majority of AA is incorporated in the phospholipid fraction (96.4%), whereas 3.5% is found in the triacylglycerol pool and less than 0.1% appears as unesterified fatty acid [1].

Alterations in the myocardial membrane structure and function play most likely a crucial role in the development of myocardial cell injury during ischemia and reperfusion [2]. Lack of energy and intracellular accumulation of calcium have become increasingly evident in the development of irreversible tissue damage upon ischemia followed by reperfusion [3]. Several hypotheses have been proposed to explain the mechanism of cell death. The loss of control of the cytoplasmic calcium concentration could be attributed to the observed decline in energy status of the cell. This could lead to a reduction in the calcium-pumping activity of the sarcolemma, mitochondria and sarcoplasmic reticulum and possibly to partial opening of the slow calcium channels and other non-specific pores. Then the raise in calcium and other ions could cause a higher enzymatic activity of calcium-dependent enzymes such as phospholipases. As shown in previous studies, AA content rises after 30 min ischemia suggesting impaired phospholipid homeostasis during the ischemic insult [1, 4–6]. Changes in the content and composition of membrane phospholipids may play an important role in the transition from reversible to irreversible cell injury. As a result of stress on the membrane by osmotic pressure or contracture of the myofibrils, proper attachment of the membrane to the cytoskeleton can be further impaired causing disruption of the membrane [7]. Free radicals can cause additional damage of the membrane by attacking the unsaturated bindings in the fatty acyl chains of membrane phospholipids.

Information on the incorporation and release of AA is important to delineate the changes in phospholipids at the point of reversible to irreversible cell damage. Since the heart is composed of different cell types, i.e. endothelial cells, cardiomyocytes, fibroblasts, and so, inter-cellular differences in phospholipid and arachidonic acid homeostasis have to be anticipated. Knowledge about the AA incorporation under control conditions in the different cell types is still incomplete and needs further investigation. This knowledge is required to explain in a proper fashion changes in AA metabolism during early events of cell damage. For that reason the aim of the present study was to get more insight in the incorporation of AA in the cellular phospholipid pool under normoxic conditions. ^{3}[H]-AA has been incorporated in myocytes (MC), endothelial cells (EC) and fibroblast-like cells (FL) isolated from adult rat heart. Its cellular distribution has been analyzed by autoradiography on light microscopic level. The effect of the duration of the incubation and the concentration of AA on incorporation of labeled AA in the phospholipid and triacylglycerol pools has been monitored by biochemical measurements.

Materials and methods

Isolation and culturing of cells

The cells were isolated from 12 week old rats and cultivated on 3.5 cm diam. dishes or thermanox coverslips as previously described [8]. Briefly, hearts were perfused with a collagenase solution and endothelial cells (EC) were collected from the perfusion medium. The cardiomyocytes (MC) and fibroblast-like cells (FL) were harvested from the suspension, which was obtained by mincing the heart tissue. MC were kept only 20 hr in culture after isolation; then the attached MC were still rod shaped and excluded trypan blue. The non-attached cells were washed away. EC and FL were grown under tissue culture conditions. In the experiments described, EC and FL between passage 2 and 6 were used.

Labeling for autoradiography

Before labeling the cells were washed with Hanks' Balanced Salt Solution with 10 mM HEPES (HBSS). [5,6,8,9,11,12,14,15-^{3}H]-AA (209.6 Ci/mmol) from Amersham, England, was dissolved in ethanol in a concentration of 5 μM and was diluted in HBSS to 10 nM before the experiment.

For autoradiography 0.5 ml of HBSS, 1 μCi, 10 nM ^{3}[H]-AA was added to the cells on coverslips and incubations were performed for 30 min at 5% CO_2, 95% air, in a 37° C stove. Thereafter cells were fixed in 2% glutaraldehyde with 2% tannine and post fixed in 1% osmium tetroxide, stained en bloc with 2% uranyl acetate, dehydrated and embedded in epon. Ultrathin sections were covered with Ilford K_2 emulsion according to the dipping technique and exposed for 3–6 days. Then the emulsion was developed with D-76 (Kodak) 1:1 diluted, for 4 min and fixed in 1:8 diluted X55 profix (Amaloco) for 5 min. After washing with water for 20 min and drying, the coupes were stained with toluidine blue and examined under the light microscope.

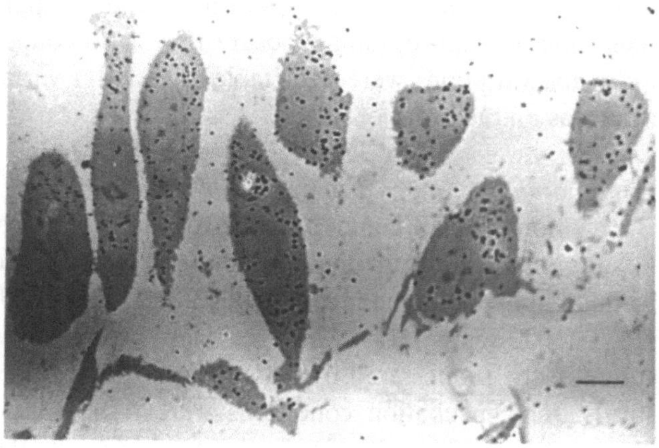

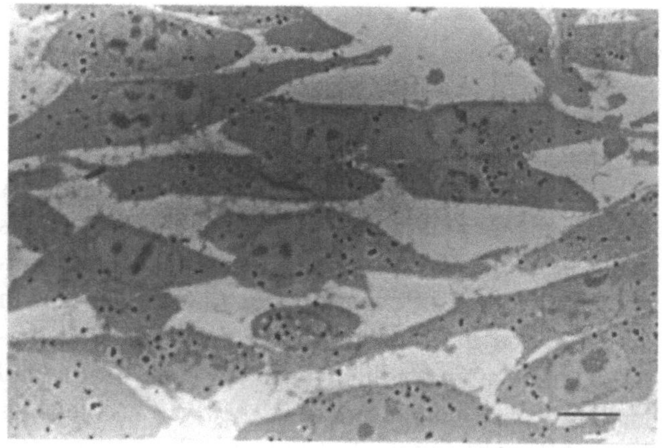

Fig. 1. Photographs of cells isolated from adult rat heart after incorporation of 0.01 μM ^{3}[H]-AA for 30 min and exposure to autoradiographic emulsions; a) endothelial cells; b) fibroblast-like cells; c) cardiomyocytes; bar represents 10 μm.

Labeling of endogenous lipid pool

For biochemical analysis of AA incorporation in the cellular lipid pool 0.5 ml HBSS + 1 μCi ^{3}[H]-AA, with or without additional unlabeled AA (1 μM or 10 μM) was put on 3.5 cm diam. dishes containing the cells under investigation. Incubations were performed for 30 or 90 min at 5% CO_2, 95% air, 37° C. After washing with PBS, cells were collected by scraping the cells in 0.25 ml 4° C PBS and additional washing of the dish was performed with 0.25 ml PBS.

The lipids were extracted from the 500 μl cell homogenate in 9.5 ml chloroform/methanol (2:1, by vol) + 0.01% butylated hydroxy toluene according to Folch [9]. The lipid classes were separated by thin layer chromatography (TLC) using plates coated with silicagel (Kieselgel 60 Merck). A mixture of fatty acids, triacylglycerols and phospholipids was used as external standard. The plates were predeveloped with chloroform/methanol/acetic acid/water (10:10:1:1, by vol) until the solvent front had moved 1 cm from its origin. After drying, the plates were further developed with petroleumether/diethylether/acetic acid/water (120:25:1:5, by vol). The lanes with the lipids standards were visualized with UV light after spraying with rhodamine G. Than comigrated spots on the parallel lanes were scraped from the plate and counted in a liquid scintillation counter (Beckman LS 3801). For these experiments protein was determined from parallel dishes according to Bradford [10].

Determination of the endogenous triacylglycerol and phospholipid pool

For analysis of the quantity of cellular triacylglycerols and phopholipids and the amount of AA present in these pools in the different cell types, in a subset of experiments lipids were extracted from homogenates of cells with hexane/isopropanol according to Rodriquez-Vico *et al.* [11]. The several lipid pools were separated by TLC as described above, scraped from the plates, methylated and quantified by gas chromatography as described previously [6]. The protein content in the pelleted residual fraction was dissolved and determined by the standard Lowry method [12].

Results

Autoradiographic analysis of ^{3}[H]-AA incorporation in isolated cells

When isolated cells were incubated with 1 μCi ^{3}[H]-AA for 30 min, in general a significant number of grains could be detected on the autoradiographic emulsions after 3 to 6 days of exposure (Figs 1A, B and C). In addition, the morphology of the cells was well preserved as seen under the light microscope. In EC and FL the nuclear compartment is very well visible (Figs 1A and B), whereas the lined up myofibrils, characteristic for striated muscle cells including cardiomyocytes, are clearly present (Fig. 1C).

When the number of grains were counted in the cells under investigation, the density of grains amounted to 0.21 ± 0.10, 0.30 ± 0.12 and 0.15 ± 0.06 grains per square μm cytoplasmic surface area, respectively. The difference did not reach the level of significance. The number of grains above the nuclear compartment (see Fig. 1A and B) was very low. The localization in the cytoplasmic extranuclear compartment (including intracellular membranes and plasmalemma) is very specific. On the average a grain density of 0.01 grains/μm^2 could be counted in the extracellular space (see also Figs 1A, B and C).

The values of the above mentioned cellular grain density were obtained by analyzing 15 to 45 individual cells. Considerable differences were observed between cell preparations incubated on different days. Figs 1A, B and C indicate that within one preparation also inter-individual differences in grain density are present. In some experiments sequential coupes were made of one cell. After covering these coupes with autoradiographic emulsion and appropriate exposure time, also coupe-to-coupe variations were detectable. The latter variation was on the order of 30%.

Incorporation of labeled AA in the cellular lipid pool

Incubation of MC, FL and EC with exogenous ^{3}[H]-AA results in a significant incorporation of label in the cellular lipid pool. On the average 600–8000, 200–2000, 1500–15000 dpm ^{3}[H]-AA/mg cellular protein was found to be sequestered by MC, FL and EC, respectively. The incubation conditions were 0.01, 10, or 10 μM AA for 30 or 90 min. In these experiments the amount of ^{3}[H]-AA incorporated in the total cellular lipid pool was only slightly influenced by the extracellular AA concentration and incubation time. The incorporated quantities correspond with 10–25% of the amount of ^{3}[H]-AA present in the incubation medium.

In general, ^{3}[H]-AA was mainly incorporated in the cellular phospholipid and triacylglycerol pool. At low AA concentrations (0.01 μM) ^{3}[H]-AA extracted by EC and FL was predominantly sequestered in the phospholipid pool. The proportion of labeled AA incorporated in triacylglycerols was small, resulting in a high phospholipid/triacylglycerol ratio (Table 2). When the incubation time was extended from 30 to 90 min a relative change of deposition of label could be observed in the direction of the triacylglycerols. When the extracellular AA concentration was increased from 0.01 μM to 1.0 μM, in EC and FL the preference of the phospholipid pool as site of AA incorporation was less marked. Longer incubation time did not influence the ratio phospholipid/triacylglycerol (^{3}[H]-AA) in FL, but caused a

Table 1. Light microscopic autoradiography in different cell types isolated from adult rat heart after 30 min, 1 μCi, 0.01 μM ^{3}[H]-AA incorporation or control incubation in the absence of labeled AA

Cell type	Condition	# cells	# grains/μm^2
EC	^{3}[H]-AA	45	0.303 ± 0.12
	control	22	0.005 ± 0.007
FL	^{3}[H]-AA	15	0.15 ± 0.06
	control	10	0.002 ± 0.003
MC	^{3}[H]-AA	24	0.2 ± 0.10
	control	20	0.006 ± 0.0013

EC, FL and MC refer to endothelial cells, fibroblast-like cells and cardiomyocytes, respectively. Data are expressed as number of grains per μm^2 cytoplasma, corrected for the number of grains counter per μm^2 in the extracellular space.

Table 2. The ^{3}[H]-AA phospholipid/triacylglycerol ratio in endothelial cells (EC), fibroblast-like cells (FL) and cardiomyocytes (MC) after 30 and 90 min of incubation. The concentration of added AA varied from 0.01 to 10 μM

Cell type	AA (μM)	n	Time (min)	
			30	90
EC	0.01	4	31.1 ± 28.1	15.1 ± 7.9
	1	6	4.9 ± 1.0	7.2 ± 0.9
	10	8	1.0 ± 0.4	1.4 ± 0.5
FL	0.01	4	158.7 ± 32.7	49.4 ± 30.8
	1	4	17.0 ± 4.8	18.0 ± 1.6
	10	4	2.3 ± 0.4	2.8 ± 0.5
MC	0.01	6	0.7 ± 0.4	0.9 ± 0.3
	1	6	0.8 ± 0.3	1.3 ± 0.7
	10	6	0.3 ± 0.1	0.4 ± 0.3

significant increase in EC from 4.9 ± 1.0 to 7.2 ± 0.9. In contrast, in MC no significant differences could be observed between the phospholipid and the triacylglycerol pool with respect to the amount of ^{3}[H]-AA incorporated. At 0.01 and 1.0 μM AA similar quantities of label were incorporated in the phospholipid and triacylglycerol pool during 30 and 90 min of incubation (Table 2). Raising the extracellular AA concentrations to 10 μM AA promoted the incorporation of ^{3}[H]-AA in cellular triacylglycerols. Under this condition the triacylglycerol pool was found to be the preferred site of deposition of label as indicated by ^{3}[H]-AA phospholipid/triacylglycerol ratio smaller than 1 (Table 2).

The content of phospholipids and triacylglycerol in isolated cells

Since differences in incorporation of labeled AA in the endogenous lipid pools in the various cell types can be caused by differences in pool size of triacylglycerol and phospholipids, the content of these esterified lipids was assessed (Table 3). In addition, the proportion of (unlabeled) AA in the triacylglycerol and phospholipid pool was monitored (Table 3).

The phospholipid content in the three cell types was on the order of 250–290 nmol fatty acyl chains per mg cellular protein, corresponding with about 125–145 nmol phospholipid per mg protein. No significant differences could be observed between the various cell types. In contrast, the size of the triacylglycerol pool was significantly higher in MC than in EC and FL. The difference between the latter two cell types did not reach the level of significance. Interestingly, the content of unlabeled AA in the cellular phospholipid pool was found to be twice as high in MC than in EC and FL (Table 3), despite similar total phospholipid levels in these cells. Likewise, the content of AA in the cellular triacylglycerol pool was relatively higher in MC than in the other two cell types as could be expected on the basis of the total triacylglycerol content. As a consequence, the ratio of AA present in the phospholipid over the triacylglycerol pool was significantly lower in MC (i.e. 20.5) than in EC and FL (i.e. 65.3 and 73.4, respectively).

Table 3. The cellular fatty acid content of triacylglycerols (TG) and phospholipids (PL) in isolated cardiac endothelial cells (EC), fibroblast-like cells (FL) and cardiomyocytes (MC) (panel A) and the content of arachidonic acid (AA) in these cell types (panel B). The ratios of the total content of PL and TG and the amount of AA in these lipid pools are given as well

A] Total content of fatty acyl moieties in TG and PL					
Cell type	TG	(n)	PL	(n)	PL/TG
EC	12.8 ± 6.2	(7)	297.7 ± 21.4	(6)	23.2
FL	10.6 ± 6.1	(6)	248.0 ± 46.8	(7)	23.5
MC	70.7 ± 24.7	(5)	275 ± 94.8	(3)	3.9
B] Total amount of AA in TG and PL					
Cell type	TG	(n)	PL	(n)	PL/TG
EC	0.32	(1)	20.90 ± 5.12	(6)	65.3
FL	0.30 ± 0.14	(4)	22.02 ± 8.44	(6)	73.4
MC	2.27 ± 0.46	(4)	46.63 ± 25.24	(4)	20.5

Data are expressed as nmol fatty acyl moieties per mg cellular protein; n refers to number of experiments.

Discussion

The present study shows that ^{3}[H]-AA added to cells isolated from adult rat hearts is readily incorporated in the cellular lipid pool.

Autoradiography at the light microscopic level revealed that label was mainly present in the extranuclear space of the cell. When cardiomyocytes (MC), endothelial cells (EC) and fibroblast-like cells (FL) were incubated under comparable conditions, the number of grains per square μm cytoplasmic surface were on the same order of magnitude in the three cell types. Since ^{3}[H]-AA is mainly incorporated in the cellular phospholipid and triacylglycerol pool (Table 2), the grains observed do most likely reflect deposition of labeled AA in the cellular membranes and lipid droplets being the storage sites of phospholipids and triacylglycerols, respectively. Although the light microscopic technique does not allow identification of the exact site of deposition of label in the cytoplasm the present findings strongly suggest that in addition to the sarcolemmal phospholipids also phospholipids in intracellular membranes have sequestered a substantial quantity of ^{3}[H]-AA. Autoradiography at the electron microscopic level is most likely required to discriminate between sarcolemmal deposition and sequestration in intracellular membranes in close vicinity of the plasmalemma (i.e. membranes surrounding the matrix of subsarcolemmal mitochondria). Discrimination between intracellular lipid droplets and neighbouring mitochondria also requires high resolution autoradiographic techniques. In this regard, Miyazaki and coworkers [13, 14] observed in cultured neonatal cardiomyocytes with electron microscopic autoradiography that deposition of ^{3}[H]-AA mainly occurred in mitochondrial and perinuclear membranes. Under normoxic conditions, the incorporation of label in the sarcolemma appeared to be minimal [13, 14].

Detailed chemical analysis indicated that the extracted ^{3}[H]-AA was mainly incorporated in the cellular phospholipid and triacylglycerol pool. The relative distribution of the labeled fatty acid over the two esterified lipid pools depended on a variety of factors such as type of cells studied, incubation interval and concentration of AA added to the incubation medium.

In general, phospholipids were the preferred pool of deposition of labeled AA in EC and FL when the extracellular concentration of AA was low. Obviously at higher concentrations the supply of AA to the cells exceeds the capacity to incorporate this fatty acid into phospholipids and esterification into the triacylglycerol pool is enhanced. It should be noted that the high phospholipid/triacylglycerol ratio of ^{3}[H]-AA incorporated corresponds with the relative abundance of unlabeled AA in the two lipid classes (Table 3). A marked difference was observed between MC on the one hand and EC and FL on the other. At AA concentration of 1.0 μM and lower, in MC the quantities of labeled AA sequestered in the phospholipid and triacylglycerol pools were comparable. At high AA concentration, the triacylglycerol pool was found to be the preferred site of incorporation. This is in agreement with the findings of Hohl and Rösen [15], who also demonstrated that in adult cardiomyocytes with increasing concentration of exogenous AA, a shift occurs in cellular storage from phospholipids to triacylglycerols. The relatively high rate of sequestration of ^{3}[H]-AA in the cardiomyocytal triacylglycerol pool can partly be explained by the fact that the total amount of triacylglycerol is about 7 times higher in MC than in EC or FL. Moreover, the percentage of (unlabeled) AA in cardiomyocytal triacylglycerols was approximately 3 times that in EC or FL. However, it is noteworth that the incorporation of labeled AA in the MC triacylglycerol pool is appreciably higher than could be expected on the basis of relative abundance of unlabeled AA in phospholipids and triacylglycerols, as the phospholipid/triacylglycerol ratio of ^{3}[H]-AA is on the order of 1 and the ratio for unlabeled AA is about 20. This difference most likely indicates that the cardiomyocytal triacylglycerol turnover is operating in a higher rate than the deacylation-reacylation cycle of membrane phospholipids. In contrast to the findings in adult cardiomyocytes, in cultured neonatal cardiomyocytes the deposition of labeled AA was most marked in the cellular phospholipid pool [13].

The relatively high rate of incorporation of labeled AA in the cellular triacylglycerol pool in adult cardiomyocytes should be taken into account when release of labeled AA from endogenous lipid pools are interpreted in terms of membrane damage evoked by insults like ischemia and reperfusion or energy depletion.

Acknowledgement

This study is supported by The Netherlands Heart Foundation, grant nr. 89213.

References

1. Van Bilsen M, van der Vusse GJ, Willemsen PHM, Coumans WA, Roemen THM, Reneman RS: Lipid alterations in isolated working rat hearts during ischemia and reperfusion: its relation to myocardial damage. Circ Res 64: 304–314, 1989
2. Van der Vusse GJ, Van Bilsen M, Reneman RS: Is phospholipid degradation a critical event in ischemia reperfusion induced damage? News in Physiol Sci 4: 49–53, 1989
3. Jennings RB, Steenbergen C, Kinney RB, Hill ML, Reimer KA: Comparison of the effect of ischaemia and anoxia on the sarcolemma of the dog heart. Eur Heart J 4 (suppl): 123–137, 1983
4. Verkleij AJ, Post JA: Physico-chemical properties and organization of lipids in membranes: their possible role in myocardial injury. Basic Res Cardiol 82 (suppl 1): 85–91, 1987
5. Chien KR, Han A, Sen A, Buja LM, Willerson JT: Accumulation of unesterified arachidonic acid in ischemic canine myocardium. Circ Res 54: 313–322, 1984
6. Van der Vusse GJ, Roemen THM, Prinzen FW, Coumans WA, Reneman RS: Uptake and tissue content of fatty acids in dog myocardium under normoxic and ischemic conditions. Circ Res 50: 538–546, 1982
7. Jennnings RB, Charles E, Murry MD, Steenbergen Jr., Reimer KA: Development of cell injury in sustained acute ischemia. Circ 82 (suppl II): II2–II12, 1990
8. Linssen MCJG, Vork MM, de Jong YF, Glatz JFC, Van der Vusse GJ: Fatty acid oxidation capacity and fatty acid binding protein content of different cell types isolated from rat heart. Mol Cell Biochem 98: 19–25, 1990
9. Folch J, Lees M, Sloane-Stanley GH: A simple method for the isolation and purification of total lipids from animal tissue. J Biol Chem 226: 497–509, 1957
10. Bradford MMA: A rapid and sensitive method for the quantitation of microgram quantities of protein utilizing the principle of protein-dye binding. Anal Biochem 72: 248–254, 1976
11. Rodriguez-Vico F, Martinez-Cayuela M, Zafra MF, Garcia-Peregrin E, Ramirez H: A procedure for the simultaneous determination of lipid and protein in biomembranes and other biological samples. Lipids 26: 77–80, 1991
12. Lowry O, Rosebrough NJ, Farr AL, Randall RJ: Protein mea-

surement with the folin phenol reagent. J Biol Chem 193: 265–275, 1951
13. Miyazaki Y, Gross RW, Sobel BE, Saffitz JE: Biochemical and subcellular distribution of arachidonic acid in rat myocardium. Am J Physiol 253: C846–C853, 1987
14. Miyazaki Y, Gross RW, Sobel BE, Saffitz JE: Selective turnover of sarcolemmal phospholipids with lethal cardiac myocyte injury. Am J Physiol 259: C325–C331, 1990
15. Hohl CM, Rösen P: The role of arachidonic acid in rat heart cell metabolism. Biochim Biophys Acta 921: 356–363, 1987

Molecular and Cellular Biochemistry **116**: 211–213, 1992.

Index

MIX
Papier aus verantwortungsvollen Quellen
Paper from responsible sources
FSC® C105338

If you have any concerns about our products,
you can contact us on
ProductSafety@springernature.com

In case Publisher is established outside the EU,
the EU authorized representative is:
Springer Nature Customer Service Center GmbH
Europaplatz 3, 69115 Heidelberg, Germany

Printed by Libri Plureos GmbH
in Hamburg, Germany